AF441837

Digital Broadcasting

WAVE SUMMIT COURSE

Satellite Communications
Takashi Iida ed.

Very Long Baseline Interferometer
Fujinobu Takahashi, Tetsuro Kondo,
Yukio Takahashi and Yasuhiro Koyama

Science of Space Environment
Tadanori Ondoh and Katsuhide Marubashi eds.

Modern Millimeter-Wave Technologies
Tasuku Teshirogi and Tsukasa Yoneyama eds.

Digital Broadcasting
Tadashi Shiomi and Mitsutoshi Hatori eds.

Mobile Communications
Hideichi Sasaoka ed.

Global Environment Remote Sensing
Ken'ichi Okamoto ed.

WAVE SUMMIT COURSE

Digital Broadcasting

Tadashi Shiomi and Mitsutoshi Hatori eds.

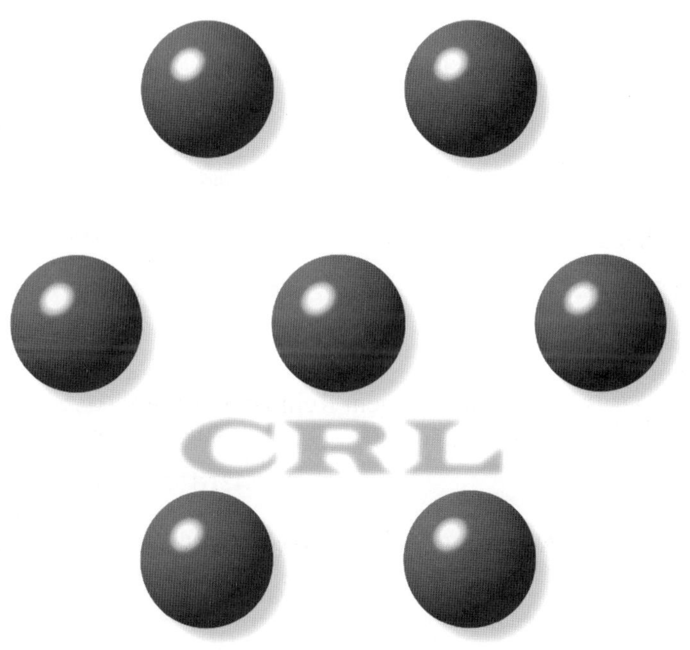

OHM
Ohmsha

IOS
Press

Wave Summit Course: Digital Broadcasting
© 2000 by Tadashi Shiomi and Mitsutoshi Hatori eds.

ISBN 4-274-90381-8 (Ohmsha)
ISBN 1-58603-099-X (IOS Press)
Library of Congress Catalogue Card Number 00-109994

Translated from the original Japanese edition:
"Wave Summit Kouza: Digital Broadcasting" published by Ohmsha, Ltd.
© 1998 by Tadashi Shiomi and Mitsutoshi Hatori eds.

Publisher
Ohmsha, Ltd.
3-1 Kanda Nishiki-cho
Chiyoda-ku, Tokyo 101-8460
Japan

Distributor

USA and Canada
  IOS Press, Inc.
  5795-G Burke Centre Parkway
  Burke, VA22015
  USA
  Fax: +1 703 323 3668

UK and Ireland
  IOS Press / Lavis Marketing
  73 Lime Walk
  Headington, Oxford OX3 7AD
  England
  Fax: +44 1865 75 0079

Germany
  IOS Press
  Spndauer Strasse 2
  D-10178 Berlin
  Germany
  Fax: +49 30 242 3113

Japan
  Ohmsha, Ltd.
  3-1 Kanda Nishiki-cho
  Chiyoda-ku, Tokyo 101-8460
  Japan
  Fax: +81 3 3293 6224

Netherlands and the rest of the World
  IOS Press
  Nieuwe Hemweg 6B
  1013 BG Amsterdam
  The Netherlands
  Fax: +31 20 620 3419

Far East jointly by
  Ohmsha, Ltd. and
  IOS Press

Printed in Japan

Foreword

In October 1896, one year after the success of Marconi's wireless telegraphy experiments, Japan started research into wireless telegraphy at the Electrical Technical Laboratory of the Ministry of Communications. The following year, Mr. Matsunosuke Matsushiro found success in his wireless telegraphy experiments. The Radio Research Laboratories, Ministry of Posts and Telecommunications, was established in 1952 after some reorganization of the laboratories researching radio communication technology. The Radio Research Laboratories was renamed the Communications Research Laboratory in 1988 to reflect the change in the environment for research following the privatization of the Nippon Telegraph and Telephone Corporation in Japan.

To celebrate 100 years of radio communications, we have spent several years at the Communications Research Laboratory planning and preparing to publish in book from the results of the research so far conducted. Entitled the "Wave Summit Course," it will contain a series of papers on various aspects of radio communications technology and is aimed at students of both the senior and graduate levels, that is, the generation of scientists who will lead the way in the 21st century.

Now the time for publication has arrived. This series consists of seven volumes: 'Satellite Communications,' 'Very Long Baseline Interferometer,' 'Science of Space Environment,' 'Modern Millimeter-Wave Technologies,' 'Digital Broadcasting,' 'Mobile Communications,' and 'Global Environment Remote Sensing.' We hope that these volumes will be widely read as useful guides to the different fields in which radio waves are used.

Finally, I would like to extend my deep gratitude to all those involved in enabling this project to come to fruition, including the authors, editors, and successive Director-Generals of the CRL and chairpersons of the publications committee, and of course to Ohmsha, Ltd. for their untiring assistance and encouragement.

Dr. Takashi Iida
Director-General
Communications Research Laboratory
Ministry of Posts and Telecommunications

Preface

The present world we are living in has gone through the material age and the energy age, and civilization is now said to be entering the information age. Throughout the course of mankind, information has generally been considered an essential part of the human life-support system. This is primarily due to the fact that information is required in order for us to interact with our external environment. Moreover, if we did not have various forms of internalized exchanges of information, a life-support system as we know it simply would not exist. Looking at information from a historical perspective, society appears to fully recognize the importance of the role that information plays in the development of civilization and human activities in the information age.

When we consider all forms of information available, it is difficult to measure the importance of the effects that broadcasting has had in the shaping of the world's societies and cultures. Certainly broadcast television, with its ability to present live images directly to a viewing audience has had a profound impact on every society. Not every effect has been positive, for we have witnessed the unfortunate events that have been attributed to children's cartoon programs, which demonstrates the negative effects television can have on society. We have also heard opinions on the effects of Western television programming on the societal turmoil in the former Soviet Union and the Eastern European nations. But these matters do not change the fact that television occupies a place in modern society, and broadcast television is now in transition from conventional analog systems to digital. Its integration with other communication systems and computers is one of the inevitable trends in the current age of information.

Broadcasting technologies cover a wide area, ranging from the processing of audio and video information to signal form, to transmitting and receiving those signals. This area has now expanded beyond the conventional techniques of utilizing ground-wave (terrestrial) propagation and cable transmission, to employing broadcast and communication satellites, and even broadcasting over computer networks. This text addresses the digital broadcasting techniques that have enabled these advances in broadcasting. Subjects ranging from the basic concepts to practical system technologies are explained in summary discourse form by authors working at the Communications Research Laboratory in the specialized fields on which they write. While we anticipate that the users of this text will be either university students or specialists and researchers working in the field of broadcasting, in the interest of advancing an appreciation for the dissemination of information, we also offer this text to anyone desiring to broaden his or her liberal education in the digital information age.

Chapter 1 presents an overview of broadcasting systems, starting with a brief history of the field. The differences between analog and digital systems are explained, as well as the advantages offered by digital systems.

Chapter 2 covers basic broadcasting techniques. This includes the processing of television signals, particularly with regards to utilizing MPEG-2 coding to compress the video and audio information content. Fundamentals of transmitting and receiving systems are also explained.

The modulation systems used to prepare the television signal for the transmission channel are described in Chapter 3. The discussion of this diverse subject ranges from basic concepts to practical systems, but the focus is primarily on digital modulation.

In Chapters 4, 5, and 6 a discussion of terrestrial, satellite, and cable network broadcasting systems is presented in chapter order. While most emphasis is placed on digital broadcasting systems, general techniques and technologies are also described. A good treatment of the OFDM system is given, and the latest technological developments in satellite systems are introduced, and working satellite systems are described. The cable television (CATV) network utilizes a different transmission medium than terrestrial or satellite broadcasts, and hence has its own peculiar problems with noise and distortion. The discussion of CATV covers some specific examples of these problems. The presentation of each of the above systems also includes an introduction to future trends in the applicable medium, both overseas and in Japan.

Chapter 7 describes how computers, digital technologies, and communications networks are being integrated with broadcasting to provide new services and information systems. The technological problems faced in the rapid development of these new systems are also addressed.

A broadcast television system includes many facets of operation, ranging from producing and transmitting information signals from images and sounds, to presenting the information to the viewer over a receiver. Complete coverage of all these subjects is not possible in this one text, but with regards to digital broadcasting technologies, we have attempted to provide the reader with a resource with which to obtain a general understanding of the subject at hand. The authors of this text are researchers and technicians working on the front lines of the broadcasting field. We are therefore confident that the information presented describes the latest technologies, and that its contents are accurate. However, the responsibility for any inconsistencies or areas that are difficult to understand must be borne by the editors.

Finally, the authors and editors present this book in the hope that it contributes to the knowledge people have of broadcasting technologies, and helps to advance the age of information.

October   2000   Tadashi Shiomi
                                                  Mitsutoshi Hatori

List of Authors

Chapter 1 : Aiichiro Tsuzuku

Chapter 2 : Shunichi Iisaku and Aiichiro Tsuzuku

Chapter 3 : Yukiyoshi Kamio

Chapter 4 : Aiichiro Tsuzuku

Chapter 5 : Hajime Fukuchi and Mitsugu Ohkawa

Chapter 6 : Isao Ishiguro and Masaaki Iguchi

Chapter 7 : Fumito Kubota

Contents

Chapter 1 Broadcast Systems Overview

1.1 Historical Background   1
    1.1.1 Radio Waves and Broadcasting   1
    1.1.2 The Transition from Radio to Television   3
    1.1.3 The Transition from Terrestrial to Satellite Broadcasting   4
    1.1.4 The Transition from Analog to Digital   5
1.2 The Digitalization of Broadcasting   6
    1.2.1 The Broadcasting System   6
    1.2.2 What is Digital Broadcasting?   7
    1.2.3 The Allocation of Broadcast Frequencies   9
1.3 Characteristics of Digital Broadcasting   11
    1.3.1 Multichannelization and the Effective Use of Frequency Resources 11
    1.3.2 Improving Picture and Sound Quality   13
    1.3.3 Improving System Performance   14
    1.3.4 Miscellaneous Problems in Digital Broadcasting   14
1.4 Conclusion   15
    • Chapter Questions   15
    • Chapter References   15

Chapter 2 The Fundamentals of Broadcasting Techniques

2.1 The Television Video and Audio Signals   17
    2.1.1 The Standard Television Video Signal   17
    2.1.2 The HDTV Video Signal   24
    2.1.3 The Audio Signal   25
2.2 High-Efficiency Coding   27
    2.2.1 High-Efficiency Coding of the Video Signal   27
    2.2.2 High-Efficiency Coding of the Sound and Audio Signal 34
    2.2.3 MPEG   37
2.3 Transmission and Reception Techniques   44
    2.3.1 Transmitting Equipment   44
    2.3.2 Receiving Equipment   47
    2.3.3 Transmission Media and Modulation Systems   51
2.4 Conclusion   53
    • Chapter Questions   53
    • Chapter References   53

Chapter 3 Digital Modulation and Error Correcting Codes

3.1 Digital Modulation Fundamentals   55
    3.1.1 What is Digital Modulation?   55
    3.1.2 Basic Modulation Systems   57
    3.1.3 Spectrum Distribution and Bandlimiting   61
    3.1.4 Demodulation Systems   64
    3.1.5 Characteristics Evaluation   66
3.2 The Basic Principles and Features of *M*-ary Modulation Systems   69
    3.2.1 Quadrature Amplitude Modulation   69
    3.2.2 M-ary Vestigial Sideband (VSB) Transmission   70
3.3 Forward Error Correction   72
    3.3.1 Error Detection, Error Correction, and Hamming Distance   73
    3.3.2 Block Codes   74
    3.3.3 Difference Set Cyclic Codes   81
    3.3.4 Convolutional Codes   83
    3.3.5 Techniques Related to Error Correction   89
    3.3.6 Combinations of Codes   90
3.4 Trellis-Coded Modulation   92
    3.4.1 Basic Principles of Trellis-Coded Modulation   93
    3.4.2 Trellis-Coded Modulation   93
    3.4.3 Trellis-Coded Modulation Performance   96
    3.4.4 Decoding of Trellis-Coded Modulation   97
3.5 Conclusion   98
    • Chapter Questions   98
    • Chapter References   98

Chapter 4 Terrestrial Broadcasts

4.1 Terrestrial Digital Television   101
    4.1.1 Broadcasts using Ground-Wave Signals   101
    4.1.2 Problems with Ground-Wave Broadcasts   102
4.2 OFDM Modulation System   105
    4.2.1 The History of OFDM   106
    4.2.2 The Principles of OFDM   106
    4.2.3 OFDM Features   112
    4.2.4 Techniques Used to Improve OFDM Performance   113
    4.2.5 OFDM Problem Area   119
    4.2.6 OFDM-related Questions & Answers   123
4.3 Propagation of Ground-Waves and Delayed Waves   126
    4.3.1 Propagation of Ground-Waves   126
    4.3.2 What are Delayed Waves (Ghosts)?   129
    4.3.3 The Effects of Delayed Waves on Bit Error Probability   130
    4.3.4 The Link Budget   136
4.4 Digital Broadcast Television Utilizing Mobile Receivers   139
    4.4.1 Radio Wave Propagation   139

4.4.2 Problems in Mobile Receivers — 140
4.4.3 Theoretical Analysis of Bit Error Rate Characteristics — 145
4.4.4 Link Budget (Mobile Receiver) — 147
4.5 Developments in Digital Television — 149
4.5.1 The International Scene — 149
4.5.2 The Domestic (Japan) Scene — 151
4.6 Conclusion — 154
• Chapter Questions — 154
• Chapter References — 154

Chapter 5 Satellite Broadcast Television

5.1 An Overview of Digital Satellite Broadcasting — 157
5.1.1 Introduction — 157
5.1.2 Analog Satellite Broadcasting Systems — 161
5.1.3 Digital Satellite Broadcasting Systems — 163
5.2 Wave Propagation in the Satellite Link — 166
5.2.1 Rain Attenuation Characteristics — 167
5.2.2 Rain Attenuation Prediction — 169
5.2.3 Rain Attentuation Countermeasures — 173
5.3 Satellite Digital Broadcasting Systems (DBS) — 175
5.3.1 Satellite Features — 177
5.3.2 Earth-Station Description — 180
5.4 Satellite Digital Broadcast System Link Budget — 184
5.4.1 Degradation Factors and Quality Standards — 184
5.4.2 Link Budget Example — 186
5.5 Future Digital Satellite Broadcasting Systems — 190
5.5.1 Digital Broadcasts over Broadcast Satellites (BS) — 190
5.5.2 21-GHz Band Satellite Broadcasting — 193
5.5.3 Mobile Digital Satellite Sound Broadcasting — 199
5.6 Appendix: Frequency-Dependence of Rain Attenuation Coefficients — 201
5.7 Conclusion — 202
• Chapter Questions — 202
• Chapter References — 202

Chapter 6 CATV (Cable Television)

6.1 CATV – History and Background — 205
6.1.1 A Brief History — 205
6.1.2 CATV Systems — 209
6.1.3 CATV Transmission Network — 214
6.2 Digital CATV — 219
6.2.1 Information Source Coding — 219
6.2.2 Multiplexing System — 220
6.2.3 Channel Coding — 224
6.2.4 The Set Top Box (STB) — 226

6.3 CATV Transmission Quality ... 227
  6.3.1 The Effects of Thermal Noise ... 227
  6.3.2 The Effects of Nonlinear Distortion ... 228
  6.3.3 The Effects of Phase Noise ... 230
  6.3.4 Interference Caused by Undesired Signal (s) ... 233
6.4 CATV Link Budget ... 239
  6.4.1 64-QAM Transmission Characteristics ... 240
  6.4.2 The Objectives of Transmission Quality ... 241
  6.4.3 Formulating the Link Budget ... 243
  6.4.4 Cautionary Notes Regarding Budgets ... 246
6.5 Developing Digital CATV and Establishing Standards ... 247
  6.5.1 Transmitting Digitally over CATV ... 247
  6.5.2 International Standards for Digital CATV ... 247
  6.5.3 Domestic (Japanese) Standards for Digital CATV ... 247
6.6 Conclusion ... 248
  • Chapter Questions ... 248
  • Chapter References ... 249

Chapter 7 Future Trends in Broadcasting Services

7.1 The Integration of Communications and Broadcasting ... 251
  7.1.1 Converting the Communications Network to Digital ... 253
  7.1.2 The Influence of Computer Technologies ... 255
7.2 Interactive TV Services ... 256
  7.2.1 Mass-Calling Service ... 256
  7.2.2 Intertext Broadcasting ... 258
7.3 Video on Demand Service ... 259
  7.3.1 System Structure ... 260
  7.3.2 Network Technology Developments and Related Problems ... 264
7.4 Internet Broadcasting ... 265
7.5 ISDB ... 266
7.6 Conclusion ... 266
  • Chapter Questions ... 266
  • Chapter References ... 267

Answers to Chapter Questions

Answers to Chapter Questions ... 269

Authors' Profile

Authors' Profile ... 273

Index

Index ... 277

Interludes

Japan Receives its First Broadcast – "Yukata Style" ... 4
The First TV Image was the Katakana Character "イ" ... 5
The Price of a TV Set ... 6

Selling Colorful Antennas for Color Television   51
Is the Analog Generation in the Immediate Future?   97
Simulcast Frequencies   106
Can OFDM also be Employed in Communications Applications?
Shaking Buildings, Dancing Ghosts   114
Clark's Early Predictions of a Global Communications Network using
  Satellites – The Father of Communications, John Pierce
  [Clark, 1993]   158
HD-SAT Project   195

Chapter 1
Broadcast Systems Overview

A brief overview of the history of broadcasting will be presented in this chapter and the differences between analog and digital broadcast systems will be explained. We shall also discuss some of the unique features of digital systems, along with the problems inherent in digital broadcasting.

1.1 Historical Background

It has been almost 50 years since the first television (TV) program was broadcast in Japan. In the intervening years, television has played an increasingly prominent role as a source of information, as well as providing entertainment, education, and cultural programming for the general populace. Recent news events such as the political upheaval in Eastern Europe and the Persian Gulf War have brought the international stage directly into the home's of viewers everywhere. The technological developments making these satellite broadcasts possible bring a new dimension to the traditional role of television, providing a medium which has the power to influence international politics, social norms, and cultures around the world.

As we enter the 21st century, the field of broadcasting is going through many changes. The traditional services of television, wireless (radio) communications, and computers are being replaced by services popularly referred to as multimedia, interactive media, and advanced TV. What has made all this possible has been the realization of genuine digital technology, introduced right at the point when "conventional" broadcast technologies had reached their zenith. In the discussion that follows, we shall trace the history of broadcasting from the earliest radio to the latest digital broadcasting techniques.

1.1.1 Radio Waves and Broadcasting

In 1888, Heinrich Hertz, a professor at Bonn University verified experimentally the existence of the electromagnetic waves predicted earlier by the British physicist James Maxwell (Fig. 1.1). Seven years later, the Italian inventor Guglielmo Marconi used a coherer (an electromagnetic wave detector originally invented in 1890 by the French physicist Edouard Branly (Fig. 1.2)) to successfully conduct the first wireless transmission. Marconi later made a number of improvements to his transmission and detection techniques, and by 1899 he was able to send signals 50 kilometers across the English Channel, and in 1901, conducted the first radio transmission across the Atlantic.

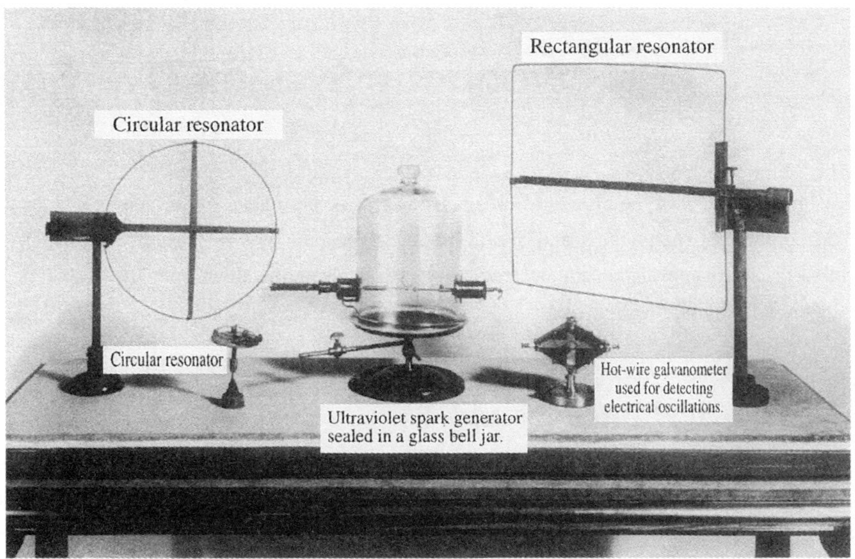

Fig. 1.1   Hertz' Experimental Device (Courtesy of the Munich, Germany Museum)

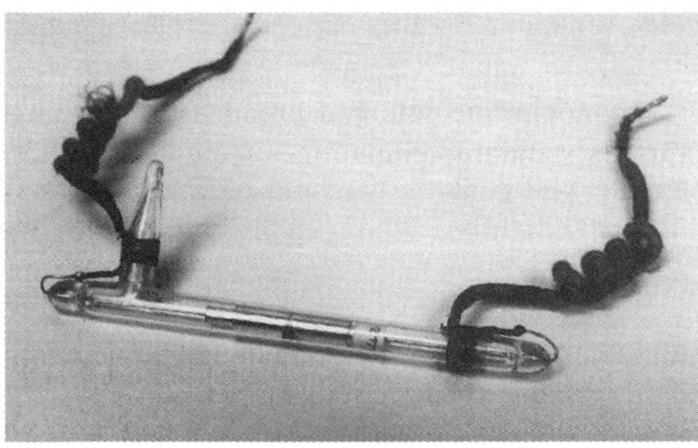

Fig. 1.2   Coherer (Courtesy of the Communications Research Laboratory)

The ability to send messages over long distances was greatly enhanced by the invention of the vacuum tube, or valve (the British electrical engineer John Fleming invented the "diode" in 1904; the American inventor Lee De Forest invented the "triode" in 1906), and the first radio station (KDKA, a Westinghouse station in Pittsburgh, PA, USA) started regular operations in 1920. The first radio broadcast in Japan was conducted in 1925 on March 22, the date that Japan marks as "Broadcast Commemoration Day."

1.1.2 The Transition from Radio to Television

The desire to view from one's own home events that occur somewhere else appears to stem from people's natural curiosity. Thus, when science gave us the ability to transmit sounds over electromagnetic waves (i.e., radio broadcasts), the next step was to naturally investigate the possibility of transmitting images to accompany those sounds, and research toward that end was conducted by scientists of many countries. In Japan, research in image transmission began in about 1925, primarily at the Hamamatsu Technical High School (currently the Shizuoka University Faculty of Engineering), Waseda University, and the predecessor to today's Communications Research Laboratory (CRL). In 1927, Professor Kenjiro Takayanagi of the Hamamatsu Technical High School used a crude camera tube containing a Nipkow disk (a mechanically scanned rotating disk) as a transmitter and a cathode-ray tube (CRT) as a receiver screen to electrically produce an image. In 1933, the Russian-born American inventor Vladimir Zworykin invented the iconoscope (an electrically scanned camera tube), an event which hastened considerably the realization of practical television. The first experiments in the transmission of television signals were conducted by NHK (Nippon Hoso Kyokai) in 1939. The signals transmitted in that experiment were picked up at a receiver located 13 kilometers from the transmitter. World War II temporarily interrupted work in the area of television, but research resumed shortly after the end of the war. By 1953, Japan had commenced regular TV broadcasts.

Research in color television broadcast was being conducted in many countries even before regular monochrome (black-and-white) service began, but since service was initially in monochrome, color receiver systems were designed for compatibility with monochrome sets and broadcasts. The three color systems adopted by the various standards committees were NTSC (National Television System Committee) in the United States, PAL (Phase Alternation by Line) in Europe, and SECAM (Sequentiel Couleur a Memoire) in France. Since Japan chose to use the same system as the US for black-and-white transmission (525 scanning lines, 30 frames-per-second), it also adopted the NTSC system for later color broadcasts. Trial broadcasts in color began in Japan in 1957, with regular commercial programming commencing in 1960 and continuing to the present.

In localities where signal reception (from terrestrial (non-satellite)) broadcasts was poor, common-use (master) antennas were set up to serve multiple households, thus introducing CATV (community antenna TV). CATV has evolved into what we now refer to as cable TV, but the original intention was to provide the signal reception facilities for viewers in areas that did not otherwise have television available. In the 1960s, high-quality co-axial cables and transistor amplifiers increased the popularity of CATV.

Japan Receives its First Radio Broadcast – "Yukata Style"

By 1924, radio stations operating in the middle-frequency bands were gradually increasing in number in Europe and the United States, but Japan had yet to establish a broadcasting program. In holding off on inaugurating radio services, however, developments in receiver designs made very rapid progress over the following year. During that time of receiver development, rumors began to spread that technicians working with radio had picked up what they thought was a broadcast signal from the US. The Ministry of Communications subsequently set up a detailed experiment in an attempt to capture this broadcast. The experimental receiver was located at the Hiraiso Branch of the Communication Ministry's Electrical Experiment Laboratory (presently the CRL's Hiraiso Solar Terrestrial Research Center), and the station from which the signal was received was KGO located in Oakland, California, USA. This special broadcast was conducted from 1:00 AM to 3:00 AM, Pacific Standard Time, and 6:00 PM to 8:00 PM, Japan Standard Time. This time-slot was selected because it was night throughout the signal propagation path over the Pacific Ocean, and absorption of radio waves is reduced at night. The photograph below shows Mr. Marumo, head of the Hiraiso Branch Lab at the receiver, and technician Takagishi observing. The date was August 30, 1924 [Wakai, 1987].

Receiving Japan's First Radio Broadcast [Communications Research Laboratory, 1987]

1.1.3 The Transition from Terrestrial to Satellite Broadcasting

With the 1962 launching of the Telestar satellite (AT & T, USA), it was finally possible to watch news in "real-time" from anywhere in the world right in the comfort of one's own home. Satellite link capability brought a new dimension of excitement to the world of broadcast television. The first joint experiment between the US and Japan was conducted in 1963 using the Relay-1 satellite (Fig. 1.3), and the first direct news which arrived from America carried the shocking events of the President Kennedy assassination.

Fig. 1.3   The First Japanese Television Broadcast over a Satellite Link (Courtesy of KDD)

In 1977, the World Administrative Radio Conference (WARC) met to settle the issues surrounding the use of broadcasting satellites operating in the 12 GHz band, establishing technical standards and allocating frequencies and satellite orbits to the participating nations. Using the broadcasting satellite (BS) Yuri-1 launched in 1978, Japan conducted the first of numerous experiments, and utilized the results of those experiments to start regular broadcast service from Yuri-2 commencing in 1984. Broadcasting satellites are lifted into geostationary orbits 36,000 km above the equator, and operating in the SHF band (12 GHz frequency band), provide coverage for all of Japan with just 120 W of transmission power.

1.1.4 The Transition from Analog to Digital

Even though it has been over 70 years since radio broadcasting first began, the most popular modulation systems remain analog types, very similar to methods used in the earliest practical radios. More recently, however, a portion of the signal is

The First TV Image was the Katakana Character "イ"

In Japan, research on television systems was actually an on-going effort even before the first radio broadcast took place. Professor Kenjiro Takayanagi of the Hamamatsu Technical High School (presently the Shizuoka University Faculty of Engineering) used a Nipkow disk and a cathode-ray tube (CRT) receiver to produce an image of the katakana character [イ]. The Nipkow disk (invented in 1884 by the German university student Paul Nipkow) is a disk having perforations in its perimeter. Light passing through the holes as the disk is mechanically scanned is converted into electrical signals, which produce line images of the picture being viewed. Professor Takayanagi later wrote, "After I had completed the experiment and went outdoors, newspaper extras announcing the passing of Emperor Taisho were everywhere." [Takayanagi, 1971]. The date was December 25, 1926 (15th year of Taisho).

digitally modulated and carried between the frequencies of TV and FM broadcasts. This limited level of digital broadcasting has been going on since 1985 in the form of television character multiplexing (teletext). The teletext signal is an amplitude modulated digital signal transmitted during the vertical retrace interval of the TV signal.

With regards to audio programs, research on fully digital broadcasting started in the mid-80s with the Eureka-147 Project in Europe. In the field of broadcast television, work began with DigiCipher (the motion picture coding system developed in the US) and MPEG (moving picture image coding experts group, details provided in Ch. 2) in the latter part of the 1980s. Research in digital broadcasting in the US and Europe brought about the establishment of a working group under the auspices of ITU-R, the Radio Communications Sector of the International Telecommunication Union (ITU). This body functions to set technical standards and to make recommendations to the communications industry.

Digital broadcasting actually commenced in 1994 when the United States began service over a communications satellite (CS). Japan initiated its digital broadcasting satellite services in October, 1996. The transition from conventional to fully digital broadcasting was made possible by the development of digital signal processing techniques which permit the compression of picture data, forward error correction, and the rapid advances made in digital modulation and demodulation techniques. These advances were coupled with the tremendous progress made in microcircuits and semiconductors, which gave us the optimum hardware packaging for the ideas underpinning the digital broadcasting movement.

Terrestrial applications of digital broadcast systems will be discussed in Chapter 4, satellite operations in Chapter 5, and CATV is covered in Chapter 6.

The Price of a TV Set

Almost anyone can afford a television set today, but that has not always been the case. For example, when television service first began in 1953, a 17-inch black-and-white set (all sets were black-and-white then; color TV was not introduced until 1960) cost 240,000 yen and a 14-inch set cost 175,000 yen. The average salary of a college-educated company employee at that time was 15,000 yen per month. One can therefore understand why a TV set was then considered such a status symbol!

1.2 The Digitalization of Broadcasting

1.2.1 The Broadcasting System

As illustrated in Fig. 1.4, the broadcasting system consists essentially of three major elements: production equipment, transmission equipment, and reception equipment. Excepting that the modulators and demodulators operate on different principles, this basic structure applies to both analog and digital broadcasting systems.

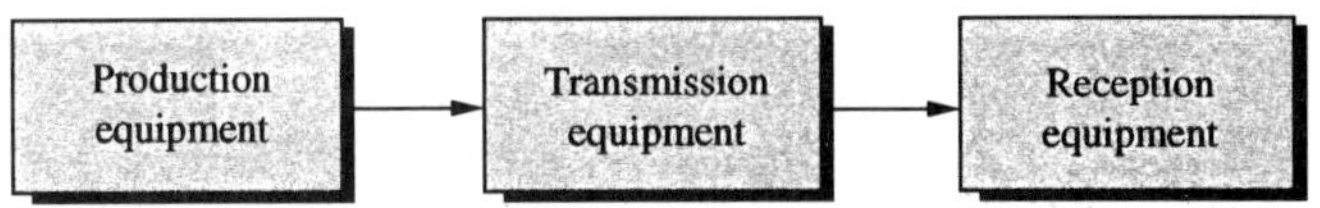

Fig. 1.4   The Broadcasting System

Production Equipment
Production equipment consists of the devices and facilities required to produce the programming, and include the studio, video tape recorders (VTR), the films and tapes required for recording, storage, and playback, and the portable production equipment used for on-the-scene reporting.

Transmission Equipment
The transmission equipment functions to modulate the carrier with signals produced by the production equipment, and then transmit the resulting signal from the facility's antenna. In a terrestrial broadcast, the transmitter antenna's height, transmission power and broadcast frequency are determined by regulation, and to most effectively utilize the radio spectrum. On isolated islands and in mountainous areas where normal signal reception is poor, television broadcast is provided through remote satellite stations (repeaters or translators). The signal from the main station is fed to the satellite station either through a microwave link, communications satellite link, or a direct signal may be utilized.

Communications satellite signals broadcast to individual households are transmitted over frequencies established by international law, and transmission power depends on the receiving equipment (120 W with BS-3). For CATV broadcasts, a cable network from the station to subscriber households replaces the antenna system.

Reception Equipment
The reception equipment includes the antenna which picks up the broadcast signal, the demodulator which restores the modulated signal, and a monitor (including speakers for audio) to view the picture.

1.2.2 What is Digital Broadcasting?
A Comparison of Digital Signals and Analog Signals
The term "digital" stems from its dictionary definition of "fingers," or "to count with one's fingers." An analog signal is a signal that has a specific value at any instant in time, which makes it a continuous signal. The simplest way to compare the concept of "digital" and "analog" is to compare how digital and analog clocks display time (Fig. 1.5). Let us use, for example, the two clocks displaying the time of 10:20. When a clock with hands (analog) shows a time of 10:20, the hour hand is between 10 and 11, slightly closer to the 10. Without have to look too closely at the minute hand, one can readily estimate that the time is about 10:20. The digital clock, however, indicates the hour precisely as 10. Since the hour indication will remain the same until 11 o'clock, one must actually look at the minute display in order to

Fig. 1.5   Comparing an Analog Clock with a Digital Clock

know what time it is. We can see by this example that the analog clock indicates time as a continuous event, but the digital clock indicates time in discrete units, making the jump from integer to integer. This does not imply that the digital clock is less precise than the analog clock, as the indicated values can be divided (quantized) into units that are infinitesimally small. For example, it is common for "professional" stop watches to indicate time in units as small as 1/1000th of a second.

When information is thus indicated using discrete values, if we are talking about a binary numbering system, the only possible values are 0 and 1. In digital systems, data is stored, transmitted, and reproduced using only a number set consisting of 0 and 1. Data processing tasks are therefore simple, and the chances of errors are greatly reduced. Digital data systems can also easily be made compatible with digital computers, which also operate on the same numerical principles. The use of computers greatly simplifies editing and data manipulation tasks.

The Signal Flow in a Digital Broadcasting System

The signal flow in a digital broadcasting system is illustrated in Fig. 1.6. In a digital system, the analog information contained in the picture and accompanying sound is converted to digital. High-efficiency coding procedures are then used to digitally compress certain portions of the signal, and multiplexing is used to combine several regions of the compressed data. Forward error correction codes are next added to correct any errors incurred during transmission, and the signal is digitally

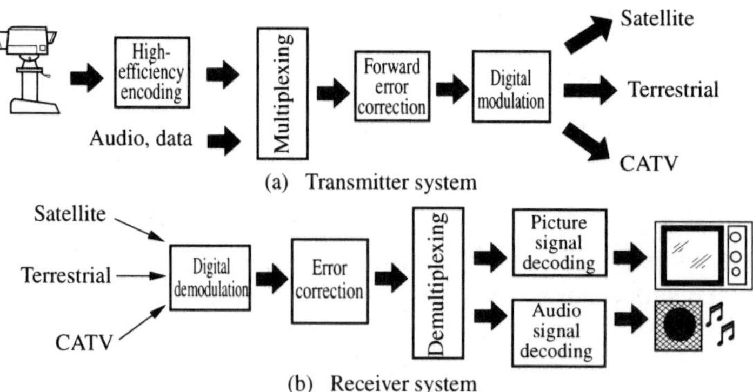

(a)   Transmitter system

(b)   Receiver system

Fig. 1.6   Signal Flow in a Digital Broadcasting System

modulated so as to efficiently utilize the frequency spectrum over which the signal is transmitted. At the receiver end, this operation is reversed, with the signal going through digital demodulation, error correction, demultiplexing, and decoding in order to restore the original video and audio signals. Encoding and multiplexing procedures are the same for all transmission modes (ground-wave terrestrial, satellite, and cable TV), and standards are being developed to cover these common areas. FEC and digital modulation systems, on the other hand, are designed specifically for a particular mode of transmission, and systems which best suit the application in which they are used are under development.

Digital Modulation Systems
Current television broadcasts are generally categorized as terrestrial (ground-wave) broadcasts, satellite broadcasts, or CATV, depending on the particular mode of transmission. Digital modulation systems are specifically designed for optimum operation in a given mode of transmission. The typical modulation systems and the modes of transmission for which they are best suited are listed in Table 1.1.

Table 1.1   Transmission Modes – Modulation Systems

Mode	Advantage	Disadvantage	Modulation system
Satellite	Wide service area. No ghost images. Wideband possible.	High rain attenuation. Not suited for mobiles. Limited by satellite power output, nonlinear at high output.	QPSK TC8PSK
Terrestrial	Area-specific service. Suited for mobiles. No output restrictions.	Ghost image problem. Complex channel plan.	OFDM 8-VSB
Cable	High-quality network. Bidirectional transmission capability. Wideband possible.	No mobile service. Cable installation expensive.	64QAM 16-VSB

In terrestrial systems, other than vestigial side band (VSB) system, orthogonal frequency-division multiplexing (OFDM) is used because of its resistance to ghosts caused by delayed waves and other factors that produce interference in mobile applications. In satellite broadcasts, quadrature phase shift keying (QPSK) or trellis-coded 8-phase shift keying (TC8PSK) is used because of power-limiting at the satellite and the nonlinear characteristics of the traveling wave tube amplifier (TWTA). Since CATV generally features a high-quality transmission network, *M-ary* quadrature amplitude modulation (QAM) can be used. QPSK and QAM signaling is covered in Chapter 3, and OFDM in Chapter 4.

1.2.3 The Allocation of Broadcast Frequencies
Of the three broadcasting modes listed above, the signals from satellites and

terrestrial broadcasts are carried by radio waves, which requires the allocation of broadcast frequency bands. Frequencies are allocated in accordance with international agreements, with each country assigning domestic frequencies within the framework of those agreements. Such agreements are necessary in order to provide some measure of order in the various areas of communications, and to prevent interference between signals, which do not recognize national boundaries. It is the job of the ITU, which is one of the functions of the United Nations, to make these international agreements. More specifically, frequency allocation falls under the responsibility of the Radio Communications Sector of the ITU (ITU-R) and the International Frequency Registration Board (IFRB), whose tasks are also to set technical standards and to solve problems of radio interference that have international implications.

In communications, since the nature of radio waves vary greatly depending on frequency, frequencies must be selected based on how they behave in a given

Table 1.2 Frequency Band Allocations and their Intended Uses

Frequency range	Frequency band	Wavelength	Uses
3 ~ 30 kHz	VLF (Very Low Freq)	10 ~ 100 km	Maritime (mobile) comm.
30 ~ 300 kHz	LF (Low Freq)	1 ~ 10 km	Aviation (mobile) comm.
300 ~ 3000 kHz	MF (Medium Freq)	100 ~ 1000 m	AM radio
3 ~ 30 MHz	HF (High Freq)	10 ~ 100 m	Shortwave radio
30 ~ 300 MHz	VHF (Very High Freq)	1 ~ 10 m	FM-TV
300 ~ 3000 MHz	UHF (Ultrahigh Freq)	10 ~ 100 cm	TV, mobile comm.
3 ~ 30 GHz	SHF (Superhigh Freq)	1 ~ 10 cm	Satellite comm.-broadcasting, radar
30 ~ 300 GHz	EHF (Extremely High Freq)	1 ~ 10 mm	Radio astronomy, radar

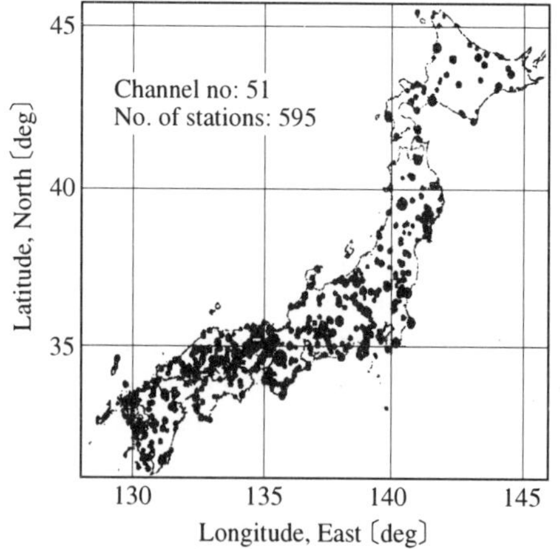

Fig. 1.7   TV Broadcasting Station Distribution

application (i.e., by how different signals are propagated at particular frequencies). As noted in Table 1.2, the communications frequencies in Japan are allocated as follows: AM radio, 531 ~ 1,602 kHz; FM radio, 76 ~ 90 MHz; TV broadcast, 90 ~ 108 MHz and 170 ~ 222 MHz and 470 ~ 770 MHz; satellite broadcasting, 11.7 ~ 12.2 GHz.

With present broadcasting and communications traffic loads, the existing spectrum is extremely crowded, and analog and digital broadcasting was initially forced to go through a simultaneous broadcasting period at the time digital was introduced. There simply was not room in the broadcasting spectrum for terrestrial broadcasts to be assigned new frequency bands (Fig. 1.7) [NHK, 1997]. Under such circumstances, digital broadcasting services had to innovate and use portions of the frequency spectrum allocated for analog broadcasts.

1.3 Characteristics of Digital Broadcasting

The primary reason that the CD (compact disk) so quickly replaced the older analog record was the superior aural fidelity offered by the CD player itself, which features digital circuit processing capability. (The CD is also much easier to handle.) We see many of the same features in digital broadcast television that have made CD players popular. These features will be briefly explained in the discussion that follows.

1.3.1 Multichannelization and the Effective Use of Frequency Resources

High-Efficiency Coding

There are certain mechanisms in the human senses of perception (both sight and hearing) that allow much of the normal information contained in the video and audio of a television picture to be omitted without the picture (or sound) suffering a perceptible loss of quality. The ability to omit this redundant information greatly reduces the amount of information that we normally consider as "required" in the transmitted signal. Digital signal processing techniques make it possible to reduce the amount of transmitted information even further. The high-efficiency coding techniques used to compress these broadcast signals are generally referred to as *MPEG*. As an example of what can be achieved using MPEG, a television picture at normal resolution is transmitted at a rate of 270 Mbit/s (at 10-bits, the luminance and chrominance signals digitized to 13.5 megasamples at 8-bit resolution converts to a rate of 216 Mbit/s). Using MPEG, this rate can be reduced to between 4 and 15 Mbit/s. Thus, by using the MPEG compression coding techniques, it is possible to carry numerous channels over a frequency band of limited width. (More in Sec. 2.1.)

Modulation Systems and Spectrum Efficiency

In conventionally modulated analog broadcasts, one channel occupies a bandwidth of 6 MHz. The bandwidth of the video signal occupies 4.2 MHz of the 6 MHz bandwidth. In digital broadcasts, the modulation system is designed so that digital information can be transmitted at a rate of 10 ~ 30 Mbit/s. Note that the units used

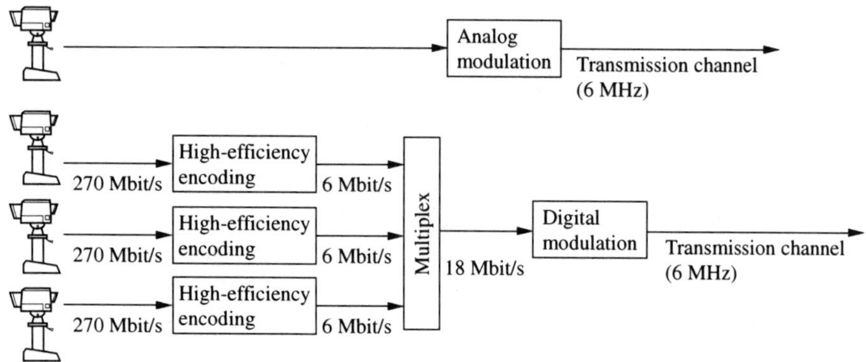

Fig. 1.8   High Spectrum Efficiency Signal Transmission

to measure information differs between analog and digital systems. The volume of digital information that can be transmitted over a channel of a given bandwidth of course depends on the quality of the circuit, but with digital signals, this volume is measured as the amount of information (bits) transmitted per-second, per-hertz of the bandwidth (bit/s/Hz). This measure defines the system's spectrum efficiency, which indicates how effectively the frequency resource is being used.

Spectrum efficiency in digital broadcasting can range from 2 to 5 bit/s/Hz, depending on the type of modulation system being used. For example, let us assume that the system has a spectrum efficiency of 3. In this case, the transmittable bit-rate becomes 18 Mbit/s. The subsequent use of the MPEG compression standard can compress one program to 6 Mbit/s, thus making it possible to broadcast three programs over one channel of normal bandwidth (Refer to Fig. 1.8).

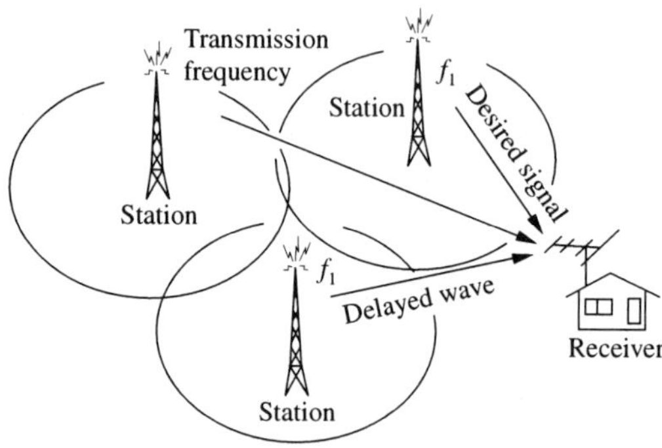

Fig.1.9   Single Frequency Network (SFN)

Interference Countermeasures
Digital modulation systems offer superior immunity to signals produced by interference. In conventional terrestrial analog broadcasts, considerable effort is expended by station personnel in assigning channel frequencies so as to avoid

cochannel and adjacent channel interference. In digital broadcasting, as is illustrated by Fig. 1.9, the single frequency network (SFN) permits several neighboring to broadcast over the same frequency band, which makes very effective use of frequency resources.

1.3.2 Improving Picture and Sound Quality
Encoding for High Definition Television
High Definition Television (HDTV) is encoded using the MPEG video compression standards. By employing MPEG, the ~1.2 Gbit/s of information required for HDTV operation (at 8-bit resolution, the luminance and chrominance signals are digitized to 74.5 megasamples) can be reduced to a rate of between 20 and 120 Mbit/s. This high-efficiency coding technique makes the broadcasting of HDTV possible.

Anti-Noise Measures
Digital modulation techniques are inherently superior to conventional analog modulation with respect to system noise. This is because the digital values contained in the modulated signal can be stripped of noise in the demodulation process. Robustness against noise depends on the type of digital modulation used, but as a comparison, 16-QAM is about 100-times better than analog modulation, and QPSK about 1000-times better. The result of this is that in transmitting a picture of equal quality, a digital transmission requires several hundred times less power than its equivalent analog transmission.

Forward Error Correction
In analog broadcasts, reception of a good picture drops gradually as the strength of the signal decreases on the fringes of the service area. However, the use of forward error correction (FEC) in digital broadcasts results in a uniformly good picture quality over the station's complete service area. At the point where signal strength drops below a given level, error correction suddenly fails to function and reception stops completely (Fig. 1.10). This phenomenon is called the cliff effect because of its cliff-like graph.

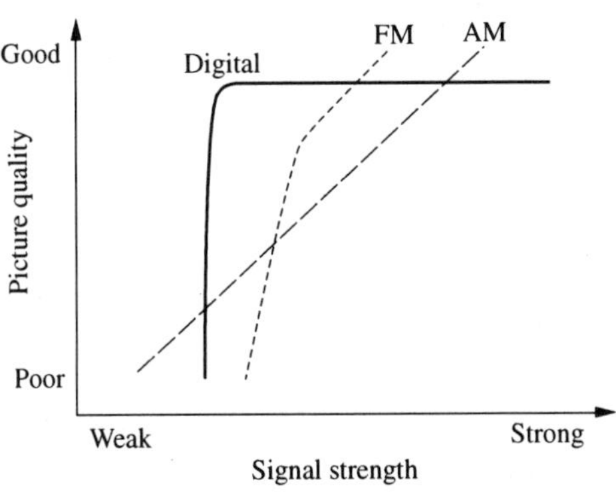

Fig.1.10　The Cliff Effect

1.3.3 Improving System Performance

The Combining of Video, Audio, and Data

The use of encoders to convert signals into a digital format permits video and audio signals to be processed using the same digital techniques. It also allows HDTV programming to be multiplexed and carried on the same channel as conventional broadcasts. With the flexibility offered by the latest receiver designs, programming identification information added at the transmitter allows the viewer to receive a program of choice without even knowing which channel is broadcasting the program.

Mobile Receivers

As was noted previously, given the robustness offered by digital modulation with respect to noise and interference, the interference caused by multipath reception and the rapid variation in signal strength – a common problem in mobile receivers – is virtually eliminated. Good reception is assured on receivers installed in automobiles, buses, trains, and other mobile applications.

Conditional Access and Signal Scrambling

The scrambling of the transmitted signal (by changing the order of the data stream sequence and mathematical processing) so that only specific subscribers can receive the picture is referred to as conditional access. Paid subscribers are provided with the means to unscramble the received signal. Local (area-specific) broadcasts are intended for subscribers residing within a given area.

1.3.4 Miscellaneous Problems in Digital Broadcasting

Defining the Service Area

In digital broadcasting, since reception of a given channel is either good or impossible (due to the cliff effect, Sec. 1.3.2), the standards used to define the service area are different from those used to define the analog television service area. The service area for analog television is defined as that area in which a channel can be received at an acceptable quality level at least 50% of the time by at least 50% of all receiver locations. Of course, this definition is insufficient for digital television, as it implies the possibility of only receiving 30 minutes of a one-hour program. The matter of service area standards for digital television (time and spatial) is still under consideration.

Mobile Receivers

As noted earlier, with the ability to correct errors through digital signal processing, we can expect a higher level of performance from digital modulators used in mobile receivers than from conventional analog devices. However, mobile receivers do have the inherent disadvantages of short antennas, and consequently, smaller antenna gains. The fading resulting from delayed wave reception also causes signal strength to vary considerably. The effects of these factors mean that, compared to their fixed-location counterparts, transmission power to mobiles must be 100- to

1000-times greater (for the same area of coverage).

Copyright Matters
Repeated copying of a digital program causes no degradation of later generations of the recorded material. A program broadcast by a station can therefore be recorded on a household VTR with no loss in original quality. Illegal copying is a violation of copyright laws. Consequently, including a copy inhibit flag as part of the control information is being considered as a means to prevent illegal copying.

The Financing of Digital Broadcasting Facilities
The process of converting broadcasting facilities to digital is steadily progressing. About 30% of the image switchers and VTRs, and about 70% of the CM banks (commercial message banks, a device in which commercials are stored for ready playback) are now digital (according to a 1996 survey by the Association of Radio Industries and Businesses). However, problems in acquiring financing for expensive items such as digital modulators and repeaters means that full digitalization will require additional time.

Broadcast Frequency Problems
Japan has been allocated 62 channels for terrestrial television broadcasts and eight channels for satellite broadcasting. Considering the state of television set ownership, it is not realistic to think that one day analog broadcasts will suddenly disappear and digital broadcasts will replace them. A simultaneous broadcast period will be required for some time in the future. Given the lack of frequency resources, it is also inconceivable that the introduction of digital broadcasting will be accompanied by its own band of allocated frequencies; in which case digital broadcasts must share the existing frequencies with conventional analog television.

1.4 Conclusion

This chapter has provided an overview of the history of television and its associated systems. Some of the differences in analog and digital systems were also presented, as were the advantages offered by digital television with respect to the effective use of frequency resources, quality of service, and performance characteristics.

Chapter Questions
Q1: Explain how digital broadcasting makes better use of frequency resources.
Q2: List the advantages offered by digital broadcasting.
Q3: List the basic technologies exploited by digital television.

Chapter References
[NHK, 1997] Nippon Housou Kyoukai: "Zenkoku Terebijyon – FM – Rajio Housoukyoku Ichiran", NHK Integrated Technology, Inc. (1997) (in Japanese)
[Takayanagi, 1971] K. Takayanagi, et al.: Terebijyon Gijutsushi Henshuu Iinkai: "Terebijyon

Gijutsushi", The Institute of Image Information and Television Engineers (1971)

[Wakai, 1987] N. Wakai: "Denpa-tte Naani", Denki Tsuushin Shinkoukai (1987) (in Japanese)

Chapter 2

The Fundamentals of Broadcasting Techniques

An overview of the basic technologies supporting digital broadcasting will be presented in this chapter. The video and audio signals of standard and enhanced-performance television will also be discussed briefly, along with the high-efficiency coding techniques used to enable the transmission of television signals over narrow frequency bands.

2.1 The Television Video and Audio Signals

2.1.1 The Standard Television Video Signal

Basic Principles

The television video signal is constructed from a three-dimensional picture signal, which conveys information describing the original scene in the horizontal (x), vertical (y), and time (t) directions. Special techniques are used to convert the three-dimensional picture signal into the television video signal, which is unidimensional. As the scene is sampled in the time direction, the three-dimensional picture signal may be thought of as a two-dimensional time sequence of information, describing the horizontal and vertical elements of the scene in detail. As illustrated in Fig. 2.1(a), the two dimensions of the scene are converted to a video signal through the process of scanning. Each horizontal scan across the picture is referred to as a scanning line (Fig. 2.1(b)). As the scanning lines progress sequentially down the two-dimensional screen (sampling the scene in the vertical direction), a time-sequential signal of the complete scene is produced (Fig. 2.1(c)). In the scanning lines, the picture is resolved horizontally into a set of points called picture elements, which are commonly abbreviated as pixels (Fig. 2.1(d)). Thus, by using the scanning method of sampling, the scene appears to the viewer as having two spatial dimensions and a time dimension, which accounts for motion in the scene. To transmit the video signal through a circuit or over radio waves, each pixel in a scanning line is converted to a voltage proportional to the image intensity at that spot, and the television video signal is then tranmitted as a sequential signal replicating the scanning operations.

The visual quality (definition) of a television picture depends on the number of pixels, which is a function of the number of scanning lines in one picture frame, and the number of pixels per line. The number of frames transmitted per second also determines how uniform the motion in a scene appears to the viewer (the higher the frame rate, the better), and a higher frame rate reduces the line flicker of the picture. In order to satisfy viewing quality requirements, the Japanese standard for "normal" television uses 525 scanning line per frame, a frame rate of 30 frames per second, and an aspect ratio (horizontal-to-vertical dimension ratio) of 4:3. The European standard for television specifies 625 lines per frame and a frame rate of 25 frames per second.

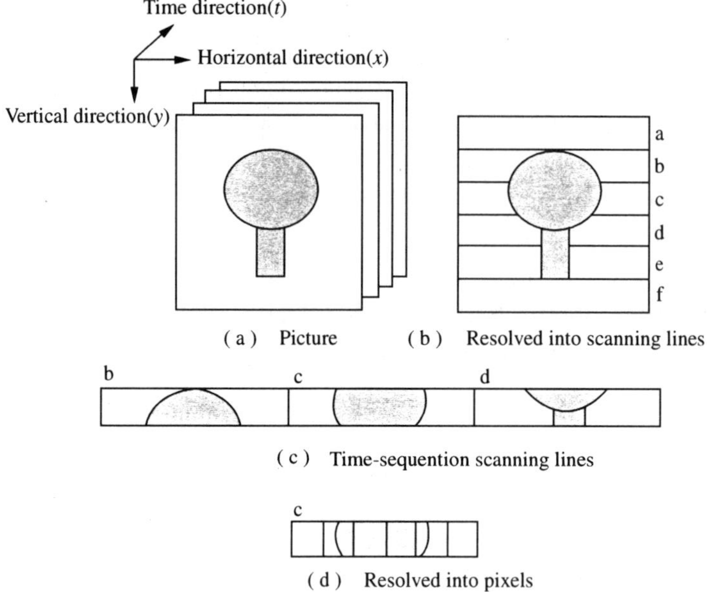

Fig. 2.1   Television Video Signal Information

Scanning

Scanning of the television scene proceeds sequentially from left to right in creating the horizontal scanning lines, and then top to bottom of the screen to complete the full set of scanning lines for one picture. As shown in Fig. 2.2, with progressive scanning (also called sequential scanning), the full frame of horizontal lines are scanned top to bottom in one operation. The dashed lines (e.g., a' to b, and b' to c) represent the brief time interval in which the scanning beam returns from the right-hand edge to the left-hand edge of the screen after completion of a scanning line, and is referred to as the horizontal retrace, or the blanking retrace. The interval in which the scanning beam returns from the lower right corner of the screen to the upper left corner (i.e., e' to a) is referred to as the vertical retrace, or the vertical blanking retrace.

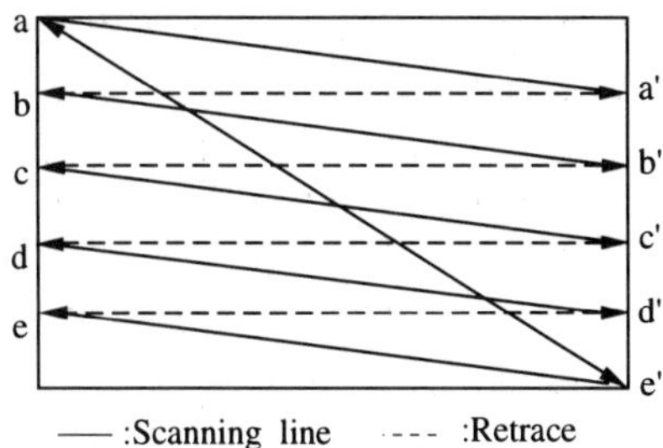

Fig. 2.2   Progressive (Sequential) Scanning

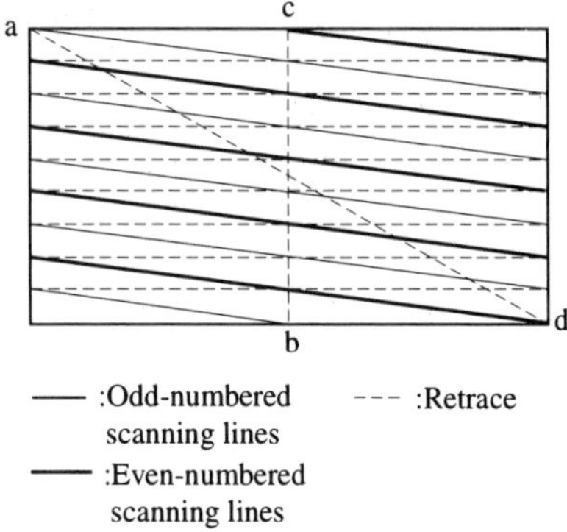

Fig. 2.3   Interlaced Scanning

Motion in the scene is rendered smoother, and picture flicker is less evident at higher frame rates. However, the problem is that higher frame rates require a wider bandwidth for signal transmission. Transmission bandwidth can be reduced to approximately one-half of the progressive scanning requirement with little degradation in picture quality by using the interlaced scanning technique (Fig. 2.3). In interlaced scanning, the first set of scanning lines begins at point a and ends at point b (represented by the fine lines in the diagram). During the vertical retrace interval, the scanning beam returns to point c in the center of the top of the screen. From point c, the heavy lines are scanned, ending at point d, to create the second set of scanning lines. The vertical retrace then returns the scanning beam to point a in preparation for the next scanning frame. Interlaced scanning thus creates two half-frames, the first consisting of the odd-numbered lines, and the second consisting of the even-numbered lines.

The coarse pictures made up of the odd-numbered and even-numbered scanning lines are called the odd field and the even field in interlaced scanning. A frame consists of two fields. The television signal transmits the picture at a frame rate of 30 frames per second, which means that the vertical scanning rate (or field rate) is 60 fields per second. At this rate, picture flicker is not noticeable to the viewer. As illustrated in Fig. 2.3, the scanning process used to create one frame is referred to as 2:1 interlaced scanning, which is the current world-wide television standard.

Standard Television Video Signal

The television video signal is partitioned by scanning line, and by field. The horizontal scanning line repeat rate is referred to as the line frequency (or horizontal frequency), and under the Japanese television standard, is specified as 15.734264 kHz. The field repeat rate is referred to as the field frequency (or vertical frequency), and is specified as 59.94 Hz. Note that the frame frequency which we refer to commonly as 30 Hz is actually 29.97002996 Hz.

In the video signal, individual pixels in a scanning line are represented by a voltage level. The higher the intensity (brightness) of the pixel, the higher the voltage level. Conversely, the lower the intensity, the lower the voltage. The sampling of a scanning line thus creates a video signal similar to the one illustrated in Fig. 2.4(a). A blanking pulse is inserted during the retrace interval in order to suppress the sampling signal, which prevents any disruptions in the picture at the receiver end (Fig. 2.4(b)). The interval over which the blanking pulse is effective also carries the synchronizing (sync) pulses used to synchronize playback of the picture at the receiver with the transmission of the scan signals (Fig. 2.4(c)). The horizontal sync pulse is inserted in the blanking interval between the scanning lines, and the vertical sync pulse is inserted in the blanking interval between fields. In order to separate the horizontal and vertical sync pulses, the pulse width of the vertical sync pulse is much wider than the width of the horizontal sync pulse. The

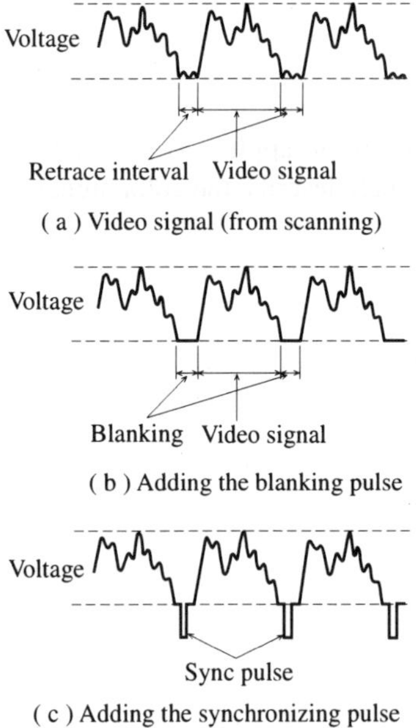

(a) Video signal (from scanning)

(b) Adding the blanking pulse

(c) Adding the synchronizing pulse

Fig. 2.4   Video Signal Waveforms

waveform produced by these operations comprises the television video signal.

Maximum frequency of the standard television video signal (f_{max}) is given as follows:

$$f_{max} = 1/2 \; KN^2 f_p \, (b/h) \, (k_v/k_n), \tag{2.1}$$

where K is a correction coefficient, N represents the number of scanning lines, f_p is the frame rate (fps), b/h is the aspect ratio, k_v is the effective scanning rate in the vertical direction, and k_h is the effective scanning rate in the horizontal direction. The coefficient of 1/2 stems from the fact that two pixels form a black and white pair (one cycle), which means that the total number of pixels generated at maximum frequency is $1/2$ x . $(N^2 f_p \, (b/h))$. K is a coefficient assigned as a result of conducting viewer evaluations, and reflects the upper limit of resolution of the scanning lines created by vertical sampling. K is normally assigned a value in the neighborhood of 0.7. k_v and k_h are correction factors that account for the subtraction of the vertical and horizontal blanking (retrace) interval, plus the portion of the scanning lines that are not usable, and take the value of $k_v/k_h = 0.92/0.835$. Substituting these values, the 525 scanning lines, frame rate = 30, and an aspect ratio of $b/h = 4/3$ into eq. (2.1), maximum frequency of the standard television video signal occupies a channel width of approximately

$$f_{max} = 4.2 \text{ MHz}. \tag{2.2}$$

Since the sync pulse is inserted in the video signal, minimum frequency must be considered as extending to the dc level. In terrestrial television broadcast systems, the video signal is prepared for transmission by modulating the video carrier with the video signal (AM), which is followed by the partial suppression of one side of the carrier which results in vestigial sideband (VSB) modulation. The vestigial side of the signal is occupied by the audio carrier. Together, these signals form the television signal, and occupy a channel bandwidth of 6 MHz. (Refer to Fig. 4.16 in Ch. 4.)

The preceding description applies to the monochrome television signal. In the following section, we shall describe the color signal.

Color Signal

The color signal is used to convey color information in the televised scene by using the three primary colors of red (R), green (G), and blue (B). Note however, that a signal for each specific primary color is not necessary in order to obtain a color picture. The most common method of creating the color signal is by using the same luminance signal (Y) as used in monochrome pictures to express brightness. Y is then also used as a component of two color difference signals ($R - Y$ and $B - Y$), and together, these three signals (Y, $R - Y$, $B - Y$) comprise the three tristimulus values needed by a color picture tube. Because of the way human visual perception works, the luminance signal can be transmitted at a bandwidth of 4.2 MHz and the two color difference signals can be transmitted at a narrower bandwidth (each at 2 MHz), and picture quality will still remain at an acceptable level.

In the current television standard, the luminance signal and the two color difference signals are transmitted as a frequency-division multiplexed signal. This is accomplished by establishing one color subcarrier (f_{sc} = 3.58 MHz), then by shifting phase and amplitude modulating the signal with the two color difference signals. A quadrature phase modulation system is thereby realized. (This will be decsribed mathematically in eq. (2.8).) The resultant signal is transmitted over one channel by frequency-division multiplexing with the luminance signal, which together is referred to as the composite signal. All three television system standards (NTSC, PAL, SECAM) are based on the composite signal concept, and feature the capability of displaying color programs in shades of gray (black and white) on monochrome receivers, and black and white programming on color receivers. Japan has adopted the same NTSC standard as used in North America.

The luminance signal (Y) is the same signal as used in the NTSC monochrome standard, and is derived as follows:

$$Y = 0.30R + 0.59G + 0.11B. \tag{2.3}$$

The two color difference signals, $R - Y$ and $B - Y$, are then

$$R - Y = 0.70R - 0.59G - 0.11B, \tag{2.4}$$

and $\qquad\quad B - Y = -0.30R - 0.59G + 0.89B. \tag{2.5}$

Note however, that the three primary color signals, R, G, and B, are corrected for gamma characteristics (gamma-corrected), which adjusts for proper luminescence output in response to the electrical signal input to the CRT display. In the NTSC system, the $R - Y$ and $B - Y$ color difference signals are further converted as shown in eqs. (2.6) and (2.7), which results in the two color (or chrominance) signals, I and Q.

$$I = 0.74(R - Y) - 0.27(B - Y) = 0.60R - 0.28G - 0.32B. \tag{2.6}$$
$$Q = 0.48(R - Y) + 0.41(B - Y) = 0.21R - 0.52G + 0.31B. \tag{2.7}$$

This color conversion process produces the I signal (in-phase signal) and the Q signal (quadrature phase signal). Bandwidth of the I signal is 1.5 MHz, a frequency to which visual perception is sensitive, and the Q signal has a bandwith of 0.5 MHz, placing it at a frequency where visual perception is less sensitive. Limiting of the bandwidths of these signal is a measure to conserve space in the channel band. The resulting NTSC television video signal, $g(t)$, is thus expressed as

$$g(t) = Y + I \cos (2\pi f_{sc}t + \theta) + Q\sin (2\pi f_{sc}t + \theta), \tag{2.8}$$

where $\theta = 33 \times (2\pi/360)$.

As is evident from eq. (2.8), the two chrominance signals (I and Q) are used to amplitude modulate the two subcarriers ($\cos (2\pi f_{sc}t + \theta)$ and $\sin (2\pi f_{sc}t + \theta)$), which operate at the same frequency, shifted in phase by 90-degrees.

The various specification parameters for the three standard television systems are listed in Table 2.1. The broadcast spectrum for the Japanese standard is shown in Fig. 4.16 (Ch. 4). The PAL and SECAM standards specifiy 625 scanning lines, a field frequency of 50 Hz, and a line frequency of 15.625 kHz. Unlike NTSC, the PAL system does not convert the color difference signals to an I and Q signal, but uses them in unmodified form ($R - Y$, $B - Y$), and a quadrature phase modulation system similar to that used by NTSC is employed to multiplex the color difference signal with the luminance signal. In SECAM, the two color difference signals ($R - Y$, $B - Y$) are assigned different subcarriers (4.40625 MHz, 4.25 MHz),

Table 2.1   Standard Television Systems

System	Scanning lines	Field freq.	Line freq.	Interlace ratio	Aspect ratio	Color signal transmission	Audio transmission
NTSC	525	59.94 Hz	15.73426 kHz	2:1	4:3	Quadrature phase modulation (I, Q)	FM
PAL	625	50 Hz	15.625 kHz	2:1	4:3	Quadrature phase modulation $(R - Y, B - Y)$	FM
SECAM	625	50 Hz	15.625 kHz	2:1	4:3	Line sequential FM	AM

and are frequency modulated and multiplexed with the luminance signal.

Digitally Coding the Standard Television Video Signal
The television signals used for standard broadcasting in the NTSC, PAL, and SECAM systems each consists of the luminance signal and the two color difference signals, frequency-division multiplexed as the color composite signal. We additionally have the three signals that convey the color in the picture, namely the luminance signal (Y), and the two color difference signals $(R - Y, B - Y)$, and these signals are referred to as the component signals. There are two methods used for digitally coding the television video signal: one is called component coding, and involves coding each of the component signals separately, while the second method is composite coding, in which the composite signal (a combination of the three component signals) is coded as a single entity. From TV-telephones to Hi-Vision (the Japanese version of HDTV), the current method of digital coding is based on component coding.

The recommended procedures for component coding are provided by ITU-R, recommendation 601. Among other things, recommendation 601 specifies such matters as the sampling frequency at which the luminance and two color difference signals are converted from analog to digital signals, generally in a ratio of 4: 2: 2 for $(Y, R - Y, B - Y)$. In this case, Y is sampled at the basic sampling frequency, and the two color difference signals are sampled at one-half the basic sampling frequency. Sampling ratios of 4: 4: 4 are also specified.

The major parameters for 4: 2: 2 sampling are listed in Table 2.2. At the basic sampling frequency, given the number of effective pixels (number of samples) per line, the rate for sampling the luminance signal occupying a 4 MHz bandwidth is 13.5 MHz, and 720 pixels per line. For the color difference signals, which occupy a bandwidth of 2 MHz, the sampling rate is 6.75 MHz, and 360 pixels per line. This standard is applicable for both the 525 line/60 Hz system (Japan, North America), and the 625 line/50 Hz system (Europe). In other words, the basic sampling frequency for both systems is derived as an integer-multiple of 3.375 MHz, which is the least-common-denominator of the horizontal scanning frequency of both systems. In the 4:2:2 sampling ratio, the sampling frequency becomes 4-times, 2-times, and 2-times 3.375 MHz, respectively. The number of quantizing bits is 8 bits per pixel (10-bit quantization is also possible) for each of the component signals. Consequently, the transmission bit-rate of the source signal (source rate) using 8-bit quantization is $(13.5 + 6.75 \times 2) \times 8 = 216$ Mbit/s.

Table 2.2   4:2:2 Ratio System Coding Parameters (ITU-R recommendation 601)

Parameter		525 lines/60 Hz system	625 lines/50 Hz system
Component signals		Luminance signal (Y), color difference signals ($R - Y, B - Y$)	
Pixels per scanning line	Y	858	864
	$R - Y, B - Y$	429	432
Effective pixels per scanning line	Y	720	
	$R - Y, B - Y$	360	
Sampling frequency	Y	13.5 MHz	
	$R - Y, B - Y$	6.75 MHz	
Coding method		Linearly quantized PCM	
Quantization bits		8 bits/pixel (10-bit possible), all signals	
Relationship between video signal level and quantizing level (8 bit/pixel)	Quantized signal level	1 ~ 254 (0, 255 used for sync)	
	Y	220 levels (16: black, 235: white	
	$R - Y, B - Y$	225 levels (128: achromatic color)	
Transmission bit-rate		216 Mbit/s (8-bit quantization) 270 Mbit/s (10-bit quantization)	

2.1.2 The HDTV Video Signal

HDTV Video Signal

The world-wide standards for High Definition Television (HDTV) are specified in ITU-R recommendation 709, and there are currently two standards: the Japanese Hi-Vision system and the European system. The Japanese HDTV video signal calls for 1,125 scanning lines and an aspect ratio of 16:9, which provides a high-quality image on a wide viewing screen. Like conventional television, interlaced scanning at a field frequency of 60 Hz is used. This results in a line frequency of 33.75 kHz ($= 1,125 \times 30$ Hz). The maximum transmission frequency (f_{max}) for the HDTV video signal, including all correction functions, is 30 MHz. The signaling system employs the component signals of luminance (Y) and color difference signals ($R - Y, B - Y$), but not the composite signal used in the NTSC standard.

The coefficients used for the three signals differ from the NTSC standard, and the signals are described as follows:

$$Y = 0.2125R + 0.7154G + 0.0721B. \tag{2.9}$$

$$R - Y = 0.7875R - 0.7154G - 0.0721B. \tag{2.10}$$

$$B - Y = -0.2125R - 0.7154G + 0.9279B. \tag{2.11}$$

The three primary color signals (R, G, B) are gamma-corrected as in the NTSC system.

In the European HDTV system, the video signal specifications are 1,250 scanning lines (625×2), 50 Hz field frequency, and a line frequency of 31.25 kHz. A summary of the scanning parameters for the two systems is provided in Table 2.3. Note also that the signal levels and synchronization procedures of the video signal and sync pulse also differs between the two signals, as does the handling of analog information. It is thus evident that we do not have a universal standard for HDTV either.

Table 2.3   HDTV Scanning Signal Parameters

Parameter	1,125 lines/60 Hz	1,250 lines/50 Hz
Aspect ratio	16 : 9	
Interlace ratio	2 : 1	
Scanning lines (total)	1,125	1,250
Scanning lines (effective)	1,035	1,152
Field freq.	60 Hz	50 Hz
Line freq.	33.75 kHz	31.25 kHz

Digitally coding the HDTV Video Signal

Component coding, in which the luminance signal and two color difference signals are separately encoded, is used to digitally encode the HDTV video signal. The procedures for component coding are provided in ITU-R recommendation 709, and the principle parameters are listed in Table 2.4.

The sampling frequency and the effective number of pixels per scanning line for the luminance signal are 74.25 MHz and 1,920 pixels/line, respectively. The same parameters for the color difference signals are 37.125 MHz and 960 pixels/line, respectively. These parameters are applicable to both the 1,125 line/60 Hz and 1,250 line/50 Hz systems. The number of quantizing bits for all three signals is 8 bits per pixel, but 10-bit quantizing is also possible. The transmission bit-rate for the primary signals using 8-bit quantization is therefore calculated as $(74.25 + 37.125 \times 2) \times 8 = 1.188$ Gbit/s.

2.1.3 The Audio Signal

Television Audio Signal Broadcasting

The audio accompanying terrestrial television broadcast is a frequency modulated (FM) signal located 4.5 MHz above the video carrier. (The broadcasting frequency spectrum is shown in Fig. 4.16.) The fully modulated audio carrier has a maximum frequency deviation from center of ±25 kHz, and the video carrier-to-audio carrier power ratio is 6 dB.

Digitally Coding the Audio Signal

The factors affecting television sound quality are the frequency bandwidth of the

Table 2.4   Coding Parameters for the HDTV Signal Using 4:2:2 Format Sampling (ITU-R recommendation 709)

Parameter		1,125 lines/60 Hz system	1,250 lines/50 Hz system
Component signals		Luminance signal (Y), color difference signals ($R - Y, B - Y$)	
Pixels per scanning line	Y	2,200	2,376
	$R - Y, B - Y$	1,100	1,180
Effective pixels per scanning line	Y	1,920	
	$R - Y, B - Y$	960	
Sampling frequency	Y	74.25 MHz	
	$R - Y, B - Y$	37.125 MHz	
Coding method		Linearly quantized PCM	
Quantization bits		8 bits/pixel (10-bit possible), all signals	
Relationship between video signal level and quantizing level (8 bit/pixel)	Quantized signal level	1 ~ 254	
	Y	220 levels (16: black, 235: white)	
	$R - Y, B - Y$	225 levels (128: achromatic color)	
Transmission bit-rate		1.188 Gbit/s (8-bit quantization) 1.486 Gbit/s (10-bit quantization)	

audio signal and the dynamic range of the sound reproduction. A person with normal hearing ability is capable of distinguishing audio frequencies over a range of 15 Hz to 20 kHz, and covering a dynamic range of 80 to 120 dB. Sampling frequency and the number of quantizing bits must account for this range. In AM broadcasts, the audio signal bandwidth is 9 kHz, but for high-quality sound systems, a bandwidth of 15 ~ 20 kHz is required. With regards to digitalization of the audio signal, the process was initially started using broadcast-grade VTRs, so the sampling frequency is derived as a simple integer fraction of the line frequency of the video signal. Three different sampling frequencies are provided for: 48 kHz, 44.1 kHz, and 32 kHz. Using the 48 kHz or the 44.1 kHz sampling frequency, an audio signal bandwidth of 20 kHz is realizable, but with the 32 kHz rate, this drops to 15 kHz. Based on the encoding devices presently available in broadcasting studios, ITU-R recommendation 646 specifies 48 kHz as the sampling frequency for audio signal encoding.

Note that the dynamic range of FM broadcasts is about 73 dB. However, in high-quality digital sound the typical range is 86 dB (14-bit quantization) or 98 dB (16-bit quantization). ITU-R recommendation 646 specifies 16-bit linear quantization for this standard. The audio signal coding parameters are listed in Table 2.5.

Table 2.5   Video Signal Coding Parameters

Parameter	Value	Notes
Standard coding frequencies	48 kHz	$(1,144/375) \times$ line freq. (525/60 system) $(384/125) \times$ line freq. (625/50 system)
	44.1 kHz	$(147/160) \times 48$ kHz
	32 kHz	$(2/3) \times 48$ kHz
Coding system	Linearly quantized PCM	
Quantization bits	16-bit	
Transmission bit-rate (2 channels of audio)	1.536 Mbit/s 1.4112 Mbit/s 1.024 Mbit/s	48 kHz $\times$ 16-bit $\times$ 2Ch (48 kHz standard) 44.1 kHz $\times$ 16-bit $\times$ 2Ch (44.1 kHz standard) 32 kHz $\times$ 16-bit $\times$ 2Ch (32 kHz standard)

2.2 High-Efficiency Coding

2.2.1 High-Efficiency Coding of the Video Signal

As noted in the previous section, using the component signals, the transmission bit-rate for the standard television video signal is 216 Mbit/s, and for the HDTV signal is 1.188 Gbit/s. In order to transmit these signals over frequency bands of acceptable widths, it is essential to reduce the transmission bit-rate of the video signal. This is accomplished by high efficiency coding. With regards to the actual practice of high efficiency coding, we have drawn on the experience gained in low bit-rate coding of TV-telephone applications and video conferencing systems. These techniques have now been expanded to the high bit-rate encoding methods employed by broadcast television and HDTV. The international standards organizations have also been active in this area, and various recommendations have been made in order to accommodate the amount of information involved in digitally coding the television video signal.

The basic parameters for the high efficiency coding standard applicable to the digital television signal are provided in ITU-R recommendation 601 (Table 2.2), and in ITU-R recommendation 709 for HDTV (Table 2.4). The background for establishing digital standards can be found in the digitalization of studio imaging equipment, and in the digitalization of the transmission channels of the integrated services digital networks (ISDN), which covers a relatively narrow field in communications. The experience gained in these early efforts has been supplemented by the digitalization of broadcast signals, and storage media such as the compact disk (CD). A significant event in the advancement of digital coding techniques was the establishment of MPEG (Motion Pictures Experts Group), a working group set up through the combined cooperation of the ISO (International Standards Organization) and the IEC (International Electrotechnical Commission).

Basic Structure

Encoding of the picture information (video) centers on two major principles: reducing the redundancy of elements of a statistical nature in the picture, and reducing the redundancy of elements permitted by human psycho-visual perception characteristics. The redundancy in statistical elements of the picture include factors that affect the picture at the pixel level, such as the systematic error (bias) in the distribution of the luminance (brightness) information. The spatial redundancy in pixel correlation, and the temporal redundancy in frame correlation also fall into this category. Note in particular that the correlation between adjacent pixels and between successive frames (or fields) is especially strong, and eliminating redundancies, both spatial and temporal, is a practical means of accomplishing this. It is therefore possible to reduce the number of bits (which are redundant) required for coding the television signal without any degradation in picture quality.

A block diagram illustrating the sequence of high efficiency coding is given in Fig. 2.5. The transform process in the first stage and encoding of the video signal in the third stage function to reduce the statistical redundancies in the picture, and the quantization process in the second stage reduces the psycho-visual redundancies. The only process that subjects the image to distortion is the quantization process in stage two. The other operations only involve a reduction in the volume of picture data.

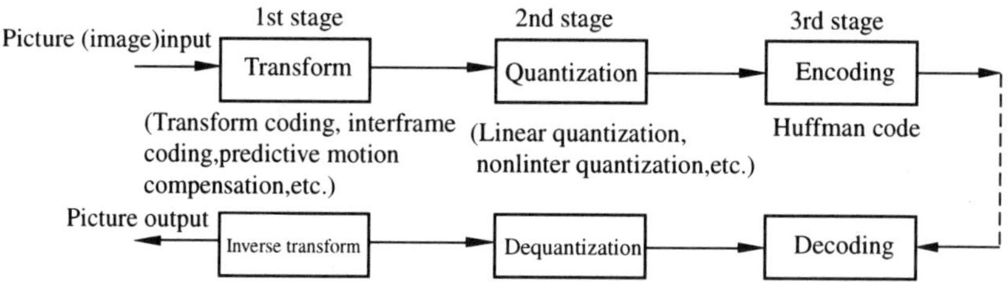

Fig. 2.5   High Efficiency Coding Process

Differences in Coding Techniques for the Various Imaging Media

Since the pictures associated with TV-telephone and video conferencing systems generally contain a limited amount of motion, coding techniques that exploit finite frame differentials in the time domain have proved effective. Advantage can also be made of the fact that the camera is stationary, and background image information is usually superfluous, and can thus be largely omitted from the transmission. On the other hand, animation of the picture in standard television and HDTV involves reproducing the complete picture, and since this information must be coded for high image quality, coding efficiency is substantially lower than in applications requiring a lower level of quality. Television pictures also require encoding in a real time environment.

The coding system used for information storage media must be compatible with a real time decoder, but real time operations are not necessarily required of the encoders storing this information. Thus, selecting the right coding/decoding techniques for the job at hand requires some thought.

Transform Coding

With regards to video signal correlation characteristics, most of the signal power tends to be concentrated in the lower spatial frequency components in the frequency domain of the pixel block matrix. Code developers use this fact, assigning fewer coding bits to areas of low frequency components. The advantage gained here is that human vision is also relatively insensitive to components in the higher frequencies, and fewer bits are therefore required to encode the picture information of those frequencies, so there is an overall gain (on both ends of the signal frequency domain) in the reduction of the number of bits needed in the compressed signal. The method used to accomplish this is what is referred to as transform coding (TC).

In transform coding, the picture is divided into several tens of pixel blocks, ranging in size from 4×4 pixels to 16×16 pixels. The pixel value of each block is subjected to an orthogonal transform, and the transform coefficients are quantized accordingly, and coded for transmission. The axis of the transformed block is rotated in relation to the old block axis, which is translated to a specific transform coefficient corresponding to the signal power at that point. The higher power levels represented by this rotation are coded by more bits, and lower signal power levels by a smaller number of bits. The result is a lower overall average number of coding bits required. A block diagram of the orthogonal transform coding process is shown in Fig. 2.6.

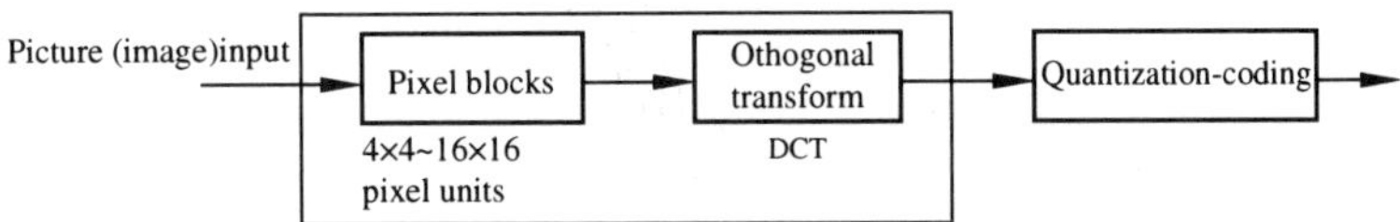

Fig. 2.6   Orthogonal Transform Coding

The various types of orthogonal transforms include the discrete cosine transform (DCT), the Hadamard, Fourier, and wavelet transforms. The Karhunen-Loeve transform [Netravali, 1994] is considered the ideal orthogonal transform, but its required processing time has prevented the realization of a practical system. The DCT approaches the Karhunen-Loeve transform in performance, and offers the best coding efficiency among the other transforms. In general, the difference in transform encoder performance lies in their ability to concentrate signal power into a small number of transform coefficients. In other words, the better an encoder is at directing signal power into a small number of transform coefficients, the more efficient it will be in encoding the signal. DCT is used in the MPEG standard for reducing redundancies in the spatial frame. Reducing redundancies in the time frame is also accomplished by using the DCT in predictive coding for motion compensation, which we will discuss next.

Predictive Coding

Predictive coding consists of predicting the values of the next pixel to be encoded, based on the value of the corresponding pixel in the previous frame (or field). The

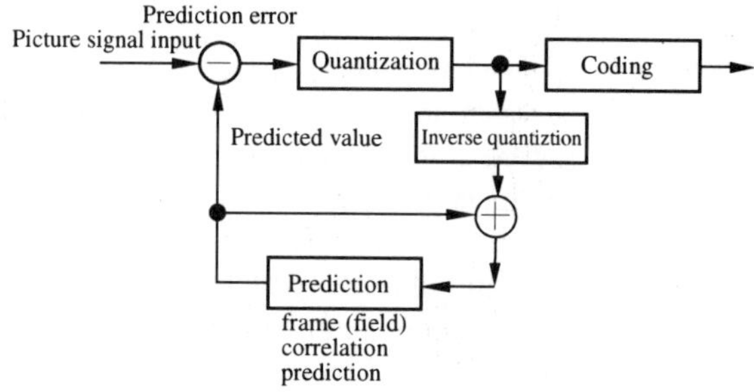

Fig. 2.7   Predictive Coding

prediction error, which is the difference between the predicted value and the actual
value is then encoded for transmission. As used here, "predictive" simply implies
estimating the differences in pixel data between frames (fields). A block diagram of
the predictive coding process is shown in Fig. 2.7. Here, the difference is taken
between the picture signal input and the predicted value. This value is encoded as
the prediction error, and sent to the transmission channel. This is specifically
accomplished by taking the sum of the inverse quantization of the quantizer output
and the predicted value, yielding a regenerated signal value consisting only of the
prediction error. This signal is then applied to input for use in predicting the value
of the next pixel.

If the pixel correlation between consecutive frames (fields) is high, it means
that the prediction was close, in which case the signal power of the prediction error
is smaller than that of the actual pixel in the picture. Fewer bits can therefore be
used to transmit the corresponding information, resulting in a reduction in the
temporal redundancies. Compared with the actual pixel, the power of the prediction
error is concentrated in the regions near the zero level. The prediciton process using
pixels in the previous field is referred to as interfield prediciton, while using pixels
from the previous frame is called interframe prediction.

Predictive Motion Compensation
Predictive motion compensation is a widely used technique for coding motion in the
picture, which was originally one of the weak points of interframe coding. In
predicting the pixel value for coding, the motion compensation process searches the
prior frame or field for the pixel having a value most closely resembling the present
pixel, and translates the spatial offset to a two-dimensional motion vector (MV) that
represents the motion information at that point. The two problems involved are 1)
determining the scope of the pixel search, which is a function of the degree of
precision of the reproduced scene motion, and 2) determining the pixel group unit
used in transmitting the MV (which relates the number of the transmitted bits to the
amount and rate of motion). With regards to the search area, a search of ±15 pixels
(horizontal/vertical) in the actual television signal should be sufficient to cover all
levels of picture motion. With regards to the number of bits needed to represent the

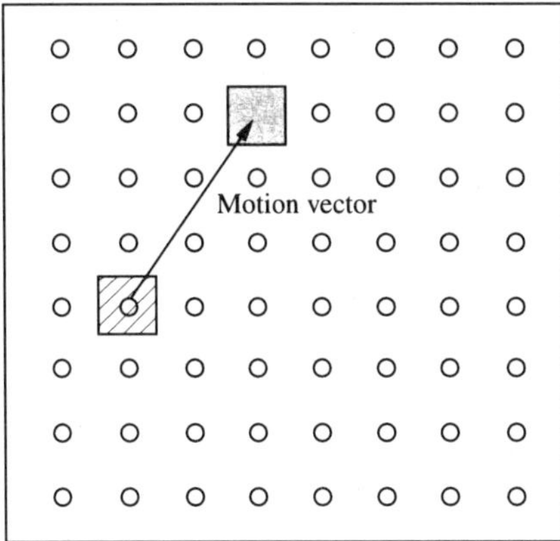

□ : Present pixel location (due to picture motion)
▨ : Location of most similar pixel in previous frame
○ : Pixel　　Sesrch area:8×8 pixels

Fig. 2.8　Predictive Motion by Pixel Units

amount of motion, using a DCT to code the motion vector in block units (8×8 or 16×16 pixels) is much more efficient than coding the MV in single pixel units. Figure 2.8 illustrates the procedure of encoding the MV in pixel units. The small circles represent pixels, and in this case, the search area is 8×8 pixels.

Bidirectional Predictive Motion
Simple predictive coding is based on information contained only in the previous frames or fields of the picture. Bidirectional predictive coding uses an expanded version of the predictive coding algorithm. Using this algorithm, the process skips from the previous pixel to the corresponding pixel in the second future frame to predict motion. It then backs up and predicts the values of the pixels that were originally skipped. Since the prediction process works in both directions, it is called bidirectional predictive coding. Seen from the immediate pixel, the coding of that pixel will reflect both past and future information. This concept, as applied to interframe coding is employed in the MPEG standards for coding, and the principle is illustrated in Fig. 2.9. Note that coding of the signal starts from the I picture. The process then jumps to the P picture, then goes back to the B pictures, whose predictions are based on the I and P information. (B pictures are not used for subsequent coding, so do not propagate coding errors.)

Applying Human Visual Perception Characteristics to the Coding Process
Using the high efficiency coding techniques discussed up to this point, the source data rate can be reduced to a rate of 1/10th to 1/20th of its unmodified value. In order to further cut redundancies, however, we must take advantage of the psycho-visual characteristics listed below:
1: The human eye is less sensitive to structural information occurring at high spatial

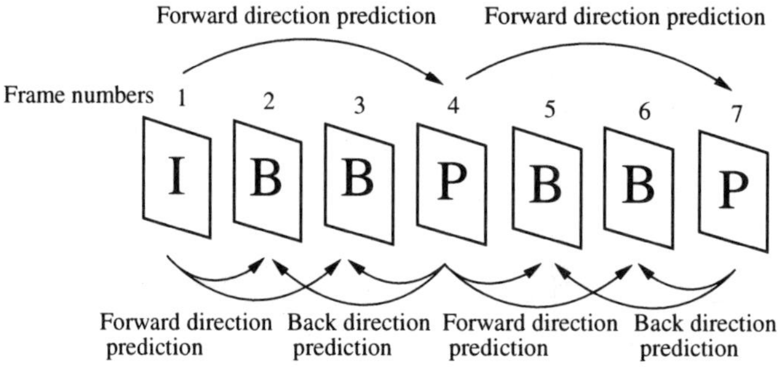

Fig. 2.9   Bidirectional Predictive Coding

frequencies than to information at low spatial frequencies.

2: Tonal distortion is more noticeable in the gray areas than in the black and white areas.

3: A sudden change in luminance levels is more difficult to visually perceive than is a constant or gradually changing level.

When applying the visual characteristics of the human eye to the coding process, the following facts need to be considered:

The coding process is based on the the degree of correlation of the subject picture. Where low spatial frequency components are present, correlation of the picture is strong, and hence, signal power is concentrated in these areas. Consequently, coding efficiency is diminished in areas where the picture pattern has fine detail, and correlation is low. On the other hand, psycho-visual characteristics reduce the perception of coding noise in areas where fine detail is present more than in the areas where highly correlated, flat patterns predominate. We can therefore use coarser quantization in these areas, which reduces the number of bits required to transmit the picture.

Quantization

After redundancies in the picture have been eliminated, the process of quantization is used to reduce the amount of information transmitted in the television signal. Generally speaking, the process of quantization uses sampling to break the continuous analog signal information into an n number of levels. At each sampling point, the analog signal is converted to a discrete value that closely approximates the analog value at that instant (called the quantized value). The quantization processor was shown as the second stage in the block diagram of Fig. 2.5. Figure 2.10 illustrates the two types of quantization. In linear quantization, the quantizing process is conducted in steps having equal intervals. In nonlinear quantization, the quantizing step intervals vary to reflect changes in the input signal.

Quantization processes also take into consideration the psycho-visual characteristics of humans. In predictive coding, for example, the distribution of the prediction error is heavily concentrated on values near zero, which corresponds to the area that contains the most visually important components. The quantization

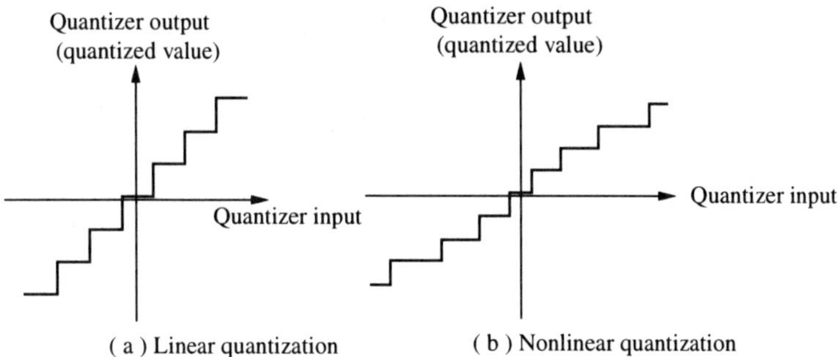

Fig. 2.10   Quantization

process responds by increasing the number of quantizing bits in this region. In areas where prediction errors are large, the quantization process is more coarse. Also, in the orthogonal transform, signal power is concentrated in the dc and low spatial frequency components of the transform coefficients, which are important to the visual senses. Consequently, the quantization process is finer in the dc and low frequency component regions, and coarser in the high spatial frequency component regions. Quantization values in the areas of high spatial frequency components are therefore, for the most part, zero.

Binary Coding

In order to transmit a digital signal, the quantized values must be converted to binary code. This binary code can be represented in two different forms. In the first method, called fixed-length code, all signal values (quantizing bits) are assigned to a code having a fixed-length bit sequence. In the second method, the length of the sequence varies in response to the probability of occurrence of the input signal values. This is called variable-length coding. In variable-length coding, where there is a bias toward the probability of occurrence of certain quantizing values or motion vectors, short codes are used to represent components with a high frequency of occurrence, and long codes to represent components with a low frequency of occurrence. These measures achieve an overall reduction in the transmission bit-rate. Variable-length coding is most often used in conjunction with with linear quantization, and compared with fixed-length/nonlinear quantization, achieves approximately a 1-bit reduction effect.

One of the variable-length codes capable of reducing to a minimum the average code bit length is the Huffman code, a code which is widely used in MPEG and other coding applications. Given the data and the probability of occurrence of the group of symbols shown in Table 2.6, the Huffman code is generated by the following procedure:

Step1: Tabulate the given symbols in descending order of their probabilities of occurrence. Join (by adding) the two lowest probabilities and affix a binary 0 and 1 to the nodes of the probability values in descending order (the higher value takes a 0, the lower a 1).

Step 2: Take the sum of the two probability values added, and rearrange the list again according to descending order of probability of occurrence. (If the added value is equal to another probability value, move the added value above the other.)

Step 3: Repeat Step 2 until only two probabilities of occurrence remain. Follow the branches back to the origin, forming the code from the sequence of binary digits.

The Huffman code can also be generated from a coding tree diagram as shown in Fig. 2.11, advancing from left to right. Read the code (consisting of a sequence of 0s and 1s at the nodes where probabilities of occurrence were added) from right to left to find the code symbol and the bit length of the code. Average code length is found by multiplying the number of bits (bit length) by the original probability of occurrence values and adding the results. Average bit length is 1.8 in this example. Compared with the bit length of a fixed-length code (2 bits), a substantial reduction in the number of transmitted bits can be achieved using the Huffman code.

Table 2.6   Huffman Code

Symbol	Probability of occurrence	Fixed-length code	Huffman code
a	0.45	00	1
b	0.3	01	00
c	0.15	10	010
d	0.1	11	011
Ave. bit length		2	1.8

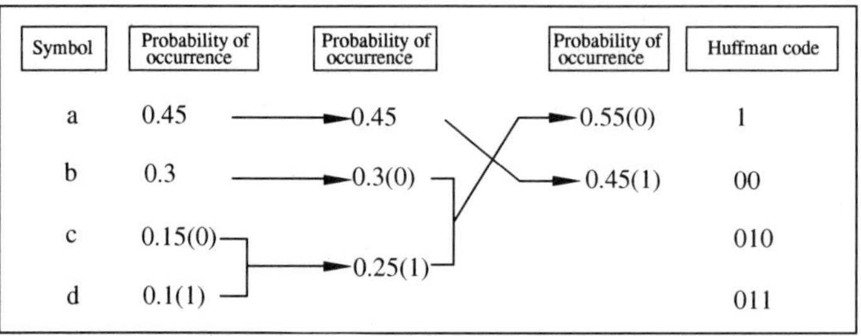

Fig. 2.11   Huffman Code Generation

2.2.2 *High-Efficiency Coding of the Sound and Audio Signal*

The audio signal is digitally transmitted in a standard frequency band of 48 kHz. At 16-bit quantization, one channel of audio is transmitted at a bit-rate of 768 kbit/s. Because of these specifications, highly efficient coding systems are essential for transmitting high-quality sound/audio signals over an acceptable bandwidth, and for recording into limited capacity media. Encoders used for this purpose utilize the statistical nature of the audio signal, and function by either reducing statistical (and other) redundancies, or by exploiting the auditory senses of humans, and eliminating the inaudible information from the signal. (This latter approach is called perception or sensory coding.) These techniques are also used in combination, taking advantage of the effects offered by both approaches to signal coding.

The Sense of Hearing and the Masking Effect
High efficiency coding of sound and audio signals is accomplished by subband coding or transform coding, techniques which exploit the following human auditory perception characteristics:

1: In quiet surroundings, the minimum threshold of audibility for humans is a given value that depends on the frequency of the tone. Since this minimum level of audibility differs depending on the frequency band of the signal, it is possible to vary coding bit distribution specific to the frequency band of sound signals being coded.

2: Where a loud tone and a quiet tone are simultaneously present, depending on specific conditions, the human auditory senses render the quiet tone difficult to perceive. This is referred to as the masking effect.

For example, as illustrated in Fig. 2.12, the high level sound (A) near the 0.4 kHz frequency masks sound (B), which has its frequency component marked by the dashed line. Note however, that sound (C) (frequency marked by the solid line) is strong enough to overcome the masking effects, and thus remains audible even though it is at a lower sound level than is sound (B).

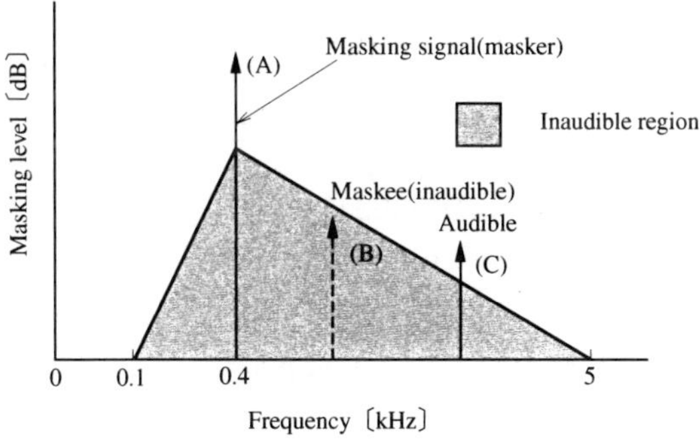

Fig. 2.12   Masking Effects

Therefore, the sounds which exceed the masking level depend on frequency, but only the portion of the sound which exceeds the minimum threshold of audibility and exceeds the masking level requires encoding. Data within certain frequency bands need not be transmitted at all. This eliminates a substantial number of bits that would otherwise be included in the signal.

Subband Coding

In subband coding (SBC), the frequency spectrum of the input signal is divided into a number of subbands, which are independently encoded with the signal energy biased toward the more acoustically important low frequencies. Reductions in the coding bits are possible using this technique, because the bit distribution in the various subbands is dependent on the signal energy present in a given band.

One of the subband coding techniques is based on the MPEG-1 Layer 2 algorithm. In this system, the frequency spectrum of the audio signal is divided into 32 subbands. Taking into consideration the masking effect, the quantizing bits are dynamically allocated to cover the range between the minimum audible level and the maximum level of the signal.

The input signal in Fig. 2.13 is a linearly quantized 16-bit PCM digital audio signal (source coding rate: 768 kbit/s). Passing the input signal through a filter bank (32 filters) divides the spectrum into 32 subbands of equal 750 Hz bandwidths. Of the 36 samples per subband (corresponding to one frame), one sample per block (= 12 samples) is extracted as a scale factor, which is used to sample the maximum signal level. Three samples are taken for each subband contained in one frame. Also, as determined by scale factor selection information, two or three successive scale factors from each subband of a data frame are checked to determine the maximum signal level, and the resulting information is output for transmission. Masking information is also applied to the 1024-point FFT (fast Fourier transform), which operates parallel to the filter bank. Each masking level for the frequency components of the input signal is thus computed and combined with the minimum audible level information, which yields the overall masking curve for the input signal. Parts of the sound signal, and the noise that falls under the overall masking curve are inaudible, masked by the sound signal itself. The number of quantizing bits for each subband are dynamically allocated, selected so that quantizing noise remains under the masking threshold. The sound signal is requantized using the remaining quantizing bits, and then multiplexed with the bit allocation and scale factor data to become the output signal.

The use of subband coding, at a standard frequency of 48 kHz, 16-bit

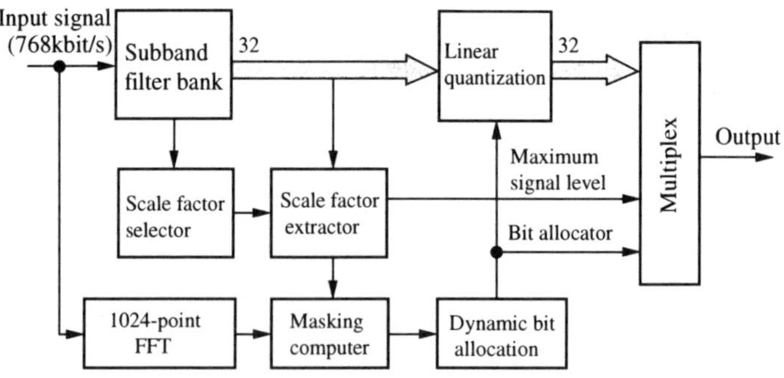

Fig. 2.13   MPEG-1 Layer 2 Encoder

linearly quantized PCM, and at a source coding rate of 768 kbit/s reduces the digital sound signal to about 1/6th the original information volume, and a 128 kbit/s coding bit-rate per audio channel is realizable.

Transform Coding

The United States standard for advanced television (ATV) employs the AC-3 (Audio Coding-3) transform code as a high efficiency audio signal encoder. With this system, it is possible to transmit a 6-channel audio signal at a bit-rate of 320 ~ 384 kbit/s. Figure 2.14 gives a block diagram of the AC-3 encoder. The encoder operates using a type of discrete cosine transform called a TDAC (time-domain aliasing cancellation) transform.

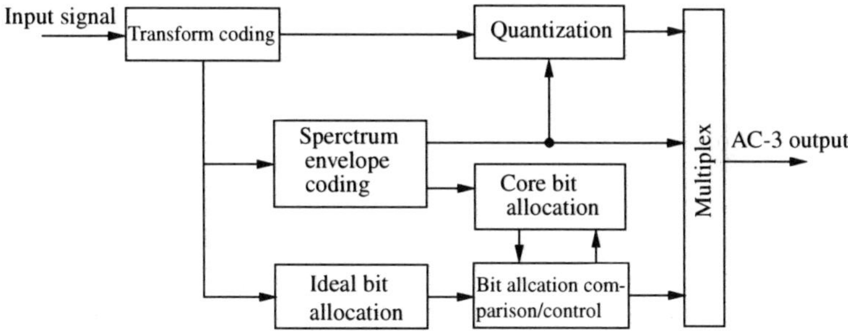

Fig. 2.14   AC-3 Encoder

The TDAC transform divides the input signal into blocks of 512 samples (at a standard 48 kHz, approx. 10 ms intervals). A DCT is carried out every 512 samples to yield a frequency-domain signal with 256 DCT coefficients. The transform uses a window function for 50% overlap of adjacent blocks, which serves to prevent block distortion at decoding, and provides a smoother transition between blocks. The coefficients resulting from the DCT are encoded as an exponent part and a mantissa part. The exponent part (i.e., the logarithm of the absolute value) represents the spectrum envelope of the audio signal. Compared with subband coding in which the frequency spectrum is divided into 32 uniform channels (750 Hz each), AC-3 uses the exponentially varying sensitivity characteristics of human hearing, separating the spectrum envelope into 40 divisions, with detailed divisions in the lower frequencies and coarse divisions in the frequencies. Since the amount of variation in the envelope between adjacent frequency points is small, compression of the code is derived from differentially encoding adjacent coefficient exponents. Quantizing bits are allocated to the coefficient mantissas after roughly computing the masking level from the 40-division spectrum envelope.

2.2.3 MPEG

MPEG is the ISO/IEC group whose purpose is to develop the international standards for high efficiency coding techniques for video and audio signals. The standards released started with MPEG-1, the 1.5 Mbit/s video-audio coding standard used primarily in CD-ROMs and other digital storage media. This was followed by the MPEG-2 (Motion Pictures Experts Group-2) standard, which specifies the general coding techniques used in high bit-rate applications such as digital broadcasting, data storage, and communications. Television signals are generally coded using the MPEG-2 compression standard, so it will be that standard on which the following description is based.

Video Format

ITU-R recommendation 601 specifies the MPEG-2 video format as follows: the 4: 2: 0 format (as converted from the 4: 2: 2 format) luminance signal consists of 720 pixels × 480 lines, and the two color difference signals consist of 360 pixels × 240 lines. These specifications provide the spatial resolution and other image quality characteristics which make the standard suitable for coding HDTV-grade broadcast signals.

(a) Progressive/interlaced scan: MPEG-2 specifications provide for both interlaced and progressive scanning.

(b) Luminance-color difference signal formats and sampling points: The basic MPEG-2 format for the luminance and color difference signals is 4: 2: 0, but specifications for the 4: 2: 2 and 4: 4: 4 formats are also provided. Human visual characteristics are more sensitive to the luminance than to the color information, so it is therefore possible to sample the signal at sub-Nyquist rates (subsampling), thereby reducing the volume of coded information. As compared to 4: 2: 2, the color resolution in the 4: 2: 0 format is slightly degraded, but the difference is barely perceptible to the human eye. In the 4: 2: 0 format, subsampling of the color difference signals, both in the horizontal and vertical directions, reduces the sample size to one-half that of the luminance signal. Figure 2.15(a) shows the spatial orientation of the luminance and color difference signal sampling points for the 4: 2: 0 format. In the 4: 2: 2 format, only the horizontal direction of the color difference signal is subsampled, and as shown in Fig. 2.15(b), for every two luminance signals sampled in one scanning line, the color difference signal is sampled just once. In the 4: 4: 4 format, the sampling points for the luminance and color difference signals are equal (Fig. 2.15(c)).

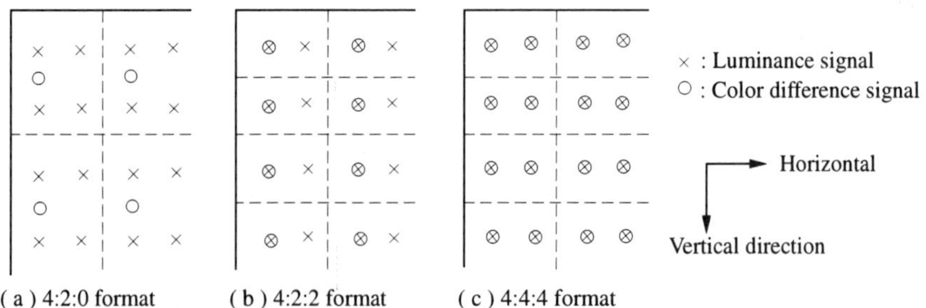

Fig. 2.15   Luminance Signal and Color Difference Signal Formats and Sampling Points

Picture Data Structure

The structural arrangement of the picture data is shown in Fig. 2.16. Picture processing under the MPEG standard is carried out using the following structural units:

(a) Picture: The picture consists of the luminance signal and two color difference signals for one frame or field. There are three types of pictures: the I (intra-coded) picture, the P (predictive-coded) picture, and the B (bidirectionally predictive-coded) picture.

(b) GOP: The GOP (Group of Pictures) refers to a set (group) of pictures. The first picture independently encoded in the GOP becomes the I picture. When the picture is edited, the I picture serves as the access point. The GOP structure is defined by the parameter pair, (M, N). M represents the I/P picture interval (i.e., the number of pictures between the I picture and the P picture), and N represents the number of picture in the bidirectional GOP sequence (i.e., the length of the GOP sequence). In Fig. 2.16, $M = 5$, $N = 15$.

(c) Sequence: The sequence begins with the sequence header, signifying the start of a GOP, and the sequence end code is the terminator. Generally, a sequence represents one complete video program. The sequence header contains the information defining the encoded picture size, aspect ratio, frame frequency, and bit-rate.

(d) Block: A block consists of 8×8 pixels, on which the luminance signal (Y) and one of the two color difference signals ($R - Y$ or $B - Y$) are operative. A block is the unit on which the DCT is carried out.

(e) Macroblock: In the 4: 2: 0 format, a macroblock is a large (16×16 pixel) block consisting of four 8×8 luminance signal blocks and two 8×8 color difference signal blocks (one $R - Y$, one $B - Y$), for a total of six blocks. Using the 4: 2: 2 format, the macroblock consists of four luminance blocks and four color difference signal blocks (two $R - Y$, two $B - Y$). In the 4: 4: 4 format, there are four luminance signal blocks and four blocks of each color difference signal, for a total of 12 blocks. The macroblock is the unit used for predictive motion compensation and adaptive quantization.

(f) Slice: A slice consists of a set of macroblocks. When the picture is divided into a number of slices, an error occurring in a given slice will not be propagated, as error recovery is effected with the sync at the beginning of the next slice.

DCT Coding

The DCT-based code as specified by MPEG is derived from the mathematical operations performed by the two-dimensional discrete cosine transform. The transform and inverse transform performed by the two-dimensional DCT is given in the following equations:

$$F(n,m) = 1/4\, C(n)C(m)\sum_{i=0}^{7}\sum_{j=0}^{7} x(i,j)\cos\{(2i+1)n\pi/16\}$$

$$\cdot \cos\{(2j+1)m\pi/16\}, \tag{2.12}$$

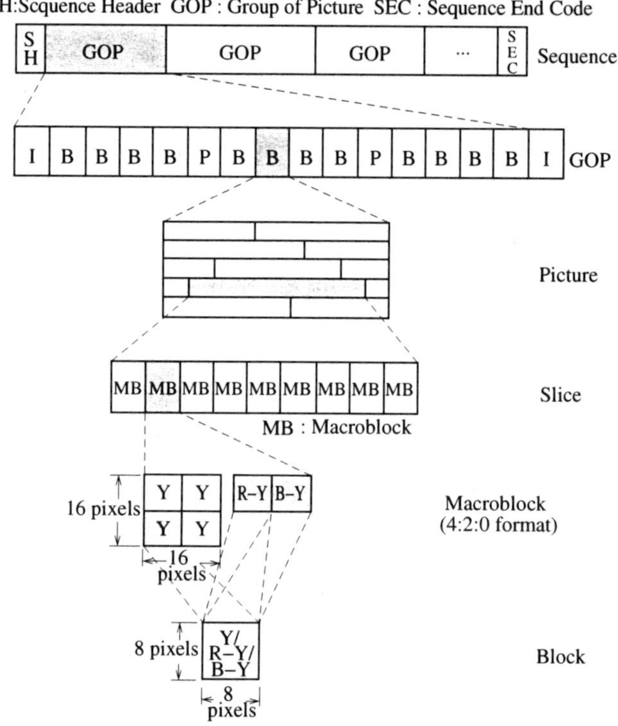

Fig. 2.16   Picture Data Structure

where $C(n)$ and $C(m) = 1/\sqrt{2}$ $(n, m = 0)$

$\qquad\qquad\qquad\quad = 1$ $(n, m \neq 0)$,

and

$$x(i,j) = 1/4 \sum_{n=0}^{7} \sum_{m=0}^{7} C(n)C(m)F(n,m)\cos\{(2i+1)n\pi/16\}$$

$$\cdot \cos\{(2j+1)m\pi/16\}. \tag{2.13}$$

In the above equations, $x(i, j)$ represent the pixel values for the i rows and the j columns of the 8×8 pixel block. F(n, m) represent the DCT coefficients of the transform, where F(0, 0) is the dc component of the block. Increasing numerical values of both n and m indicate higher frequencies in both horizontal and vertical directions, and blocks other than F(0, 0) represent non-dc components.

The DCT is relatively compatible with the statistical nature of natural images, and accomplishes its intended task with fewer computations than other transform methods. Consequently, hardware implementation of the high-speed arithmetic processes are possible, and a number of custom LSIs are already available. DCT is thus currently the preferred transform coding method. However, DCT produces a non-reversible code, and the original information cannot be completely decoded at the receiver end. Also, interblock dc levels and possible differences in the low frequency components (unwanted biproducts of block processing) can result in block distortion, which appears as a mosaic (blocking) pattern. The bit distribution in the high frequency components is also generally "thin," which can result in quantizing errors. The effects of this is called "mosquito noise," as it has the appearance of a mosquito flying around the periphery of the block. Generally, such image degradation problems become more severe as the compression ratio of the coded signal increases.

Quantizing the DCT Coefficients
The 64 DCT coefficients can be quantized at a different level (degree of fineness) for each coefficient by using the intraframe or interframe modes. At default, the intraframe mode specifies coarse quantizing of the DCT coefficients in the high frequency components, and the interframe mode calls for quantizing all coefficients at the same level of fineness. In the MPEG-2 standard, the characteristics of the picture are analyzed block-by-block and quantized accordingly, which means that quantization is step-switched in macroblock units. In addition to linear quantization, nonlinear quantizing methods can also be used.

Predictive Motion Compensation and Detection of Motion
Predictive motion compensation uses the correlative characteristics between frames (or fields), and operates on the luminance component contained in the 16×16 pixel macroblocks to reduce the level of redundancies. Detection of the motion vector (MV) is possible at the half-pixel (or half-pel, pel is a further abbreviation of pixel) unit, and the range over which motion detection operations are carried out is established beforehand. The standard used for macroblock matching is by determining the mean absolute error (MAE) of each pixel within the block, and minimizing the MAE. This is referred to as the block matching method.

Variable Length Coding
Following the two-dimensional DCT, the resulting coefficients are quantized, and as illustrated in Fig. 2.17(a), starting from the low frequency components, a progressive zig-zag scan is used to convert the two-dimensional data to a unidirectional bit stream. By making a few modifications, we can code the low frequency components more efficiently without adversely affecting picture quality. Using the zig-zag scan to quantify the high frequency components of the DCT coefficients, the most commonly occurring level is zero (typically there are long strings of zeroes). However, instead of adding each zero in a sequence to the bit stream as it occurs, the code is made more efficient by simply transmitting the number of continuous zeroes contained in a sequence. Transmitting the number of continuous zeroes in a sequence is referred to as zero run length coding. In zero run length coding, a coefficient value that is not a zero level and the number of zeroes occurring up to that point (the run) is combined as a set, and the Huffman code is used in a variable length encoder to construct the code, based on frequency of occurrence principles.

The field DCT results in a stronger correlation in the vertical direction than does a frame DCT. Because of this, the DCT coefficients are concentrated in the low vertical frequencies, and a more efficient code is created by modifying the order of progression of the zig-zag scan to account for the order in which the DCT coefficients appear. The MPEG-2 standard for field DCT is as illustrated in Fig. 2.17(b), and is referred to as vertical direction priority scanning. The two types of scan progressions are selectable by individual pictures.

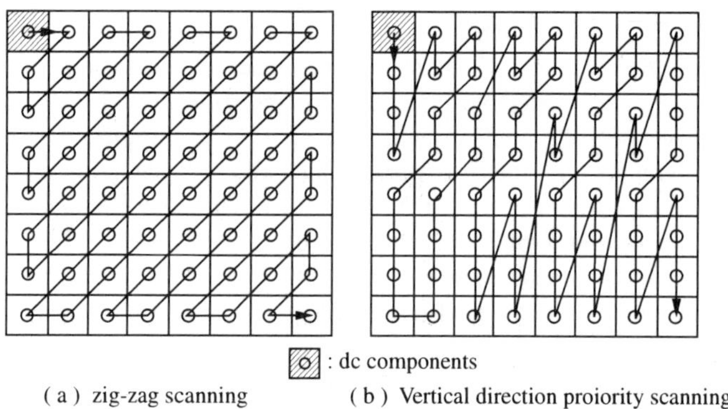

(a) zig-zag scanning   (b) Vertical direction proiority scanning

Fig. 2.17   Scanning Progressions for DCT Coefficients

The MPEG-2 Encoder and Scalability

The two-dimensional DCT, predictive motion compensation, and variable length coding functions that have been discussed so far are included in the MPEG-2 encoder, whose block diagram is shown in fig. 2.18. However, the MPEG-2 standard also provides for scalability of the system, as outlined in Table 2.7. The scalable functions of the MPEG-2 signal provide for partial decoding of the video bit stream, which results in a step-wise variation in SNR (signal-to-noise ratio), and spatial and temporal resolution. The scalable profiles are realized at the receiver end by selecting only a specific bitstream for decoding. The following is a description of SNR and spatial scalabilities, as specified in the MPEG scalable profile.

(a) SNR scalability: The SNR scalable profile is realized by step-grading the quantization precision factor. For example, the base layer calls for a coarser level of DCT coefficient quantizing, while the enhanced layer provides for a finer level of quantization. Decoding only the base layer produces a picture having a lower SNR, but decoding both the base and the enhanced layers results in a picture with a high SNR.

(b) Spatial scalability: The spatial scalability specified by MPEG-2 provides for different levels of spatial resolution from the decoded bit stream. The base layer allows transmission of a basic quality picture (in spatial resolution). Addition of supplementary information in the enhances layer provides an HDTV-grade picture.

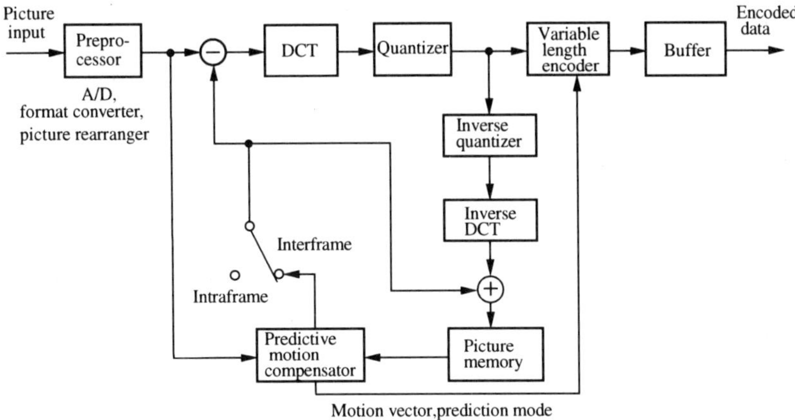

Fig. 2.18   MPEG-2 Encoder

Table 2.7   MPEG-2 Scalable Functions

	Scalable function	Base layer	Base + Enhanced layer
Spatial scalability	Spatial resolution	Low resolution picture	High resolution picture
SNR scalability	Precision of quantization of DCT coefficients	Higher level of coding distortion	Lower level of coding distortion
Temporal scalability	Temporal resolution (frame freq.)	Less refined picture motion	Smooth picture motion

MPEG Levels and Profiles

MPEG-2 is a broad-based standard that specifies numerous functions and equipment compatibility by defining levels and profiles (Table 2.8). Five profiles specify the degrees of code compression, ranging from the simple profile to the high profile. Any profile that is more advanced (toward the right side in the table) is upward compatible with a lower function profile. For example, the spatial scalable profile (in the 4th column) provides all the functions included in the simple/main/SNR scalable profiles. The simple profile differs from the other profiles in that it only uses I/P pictures, and consequently does not have bidirectional predictive capabilities. All profiles other than high specify only the 4: 2: 0 format.

There are four different levels for controlling resolution specified in the MPEG-2 standard. Spatial and temporal resolutions range from the low level to the high level. Standard television systems such as the NTSC conform to the main level, and Japanese HDTV to the high level. The maximum bit-rates applicable to each level and profile are listed in Table 2.8. These values reflect the upper limits of decoding capabilities, and in scalable profiles, reflect the total bit-rate. With regards to compatibility, top levels are capable of translating the bit streams of lower levels, and any profile to the right of another profile (in Table 2.8) is fully back-compatible with the lower profiles.

Table 2.8   MPEG-2 Profiles and Levels

Level \ Profile	Simple 4: 2: 0 I, P picture	Main 4: 2: 0 I, P, B picture	SNR scalable 4: 2: 0 I, P, B picture	Spatial scalable 4: 2: 0 I, P, B picture	High 4: 2: 0/4: 2: 2 I, P, B picture
High (HDTV-30M) Horizontal pixels : 1920 Lines : 1152 Frames : 60	×	80 Mbit/s	×	×	100 Mbit/s
High-1440 (HDTV-20M) Horizontal pixels : 1440 Lines : 1152 Frames : 60	×	60 Mbit/s	×	60 Mbit/s	80 Mbit/s
Main (Standard TV) Horizontal pixels : 720 Lines : 576 Frames : 30	15 Mbit/s	15 Mbit/s	15 Mbit/s	×	20 Mbit/s
Low (MPEG-1) Horizontal pixels : 352 Lines : 288 Frames : 30	×	4 Mbit/s	4 Mbit/s	×	×

2.3 Transmission and Reception Techniques

2.3.1 Transmitting Equipment

Component Configurations

Block diagrams depicting the components of conventional analog and digital broadcast transmitters are shown in fig. 2.19. In analog broadcast systems, the analog video and audio signals are directly applied to analog modulators prior to amplification and transmission. However, digital broadcasting requires that the video and audio signals be converted from analog to digital, then encoded, multiplexed, and forward error correction (FEC) codes added, all through digital processes, prior to digital modulation of the now digital signal. A brief description of each of the functional components which accomplish this process follows:

(a) High efficiency coding: High efficiency coding is also referred to as source coding, or simply compression. As discussed previously, the video signal is assembled from information derived from closely spatially-related pixels and from the correlation characteristics of sequential frames of pictures. These characteristics are analyzed for elimination of redundancies, which reduces the volume of data in the signal. In digital broadcasting, the MPEG-2 video standard specifies the coding parameters.

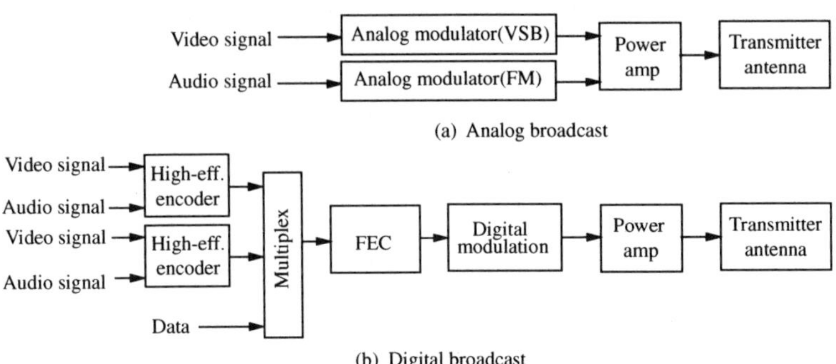

Fig. 2.19   TV Broadcast Systems (Analog & Digital)

(b) Multiplexing: The encoded video and audio signal sequence (bit stream) and the supplementary data bit streams are processed in groups of bits called *packets*. The various packets are combined into a single packet by the multiplexer. The multiplexing process also adds time information and other data as required by the system. Also, depending on the type of program service, packet streams are added to accommodate multiple program services. The multiplex procedure is also specified by the ISO/IEC MPEG-2 standard.

(c) Forward error correction coding: Generally, in the time that the signal leaves the transmitter, passes through the propagation path, and arrives at the receiver, it is subjected to noise and other influences that corrupt the coded signal. These errors can be corrected by inserting a small number of redundant bits in the signal sequence at the transmitter for use by the receiver in determining whether or not the received signal is error-free, which assists in the proper recovery of the original information. This procedure is referred to as forward error correction (FEC) coding. The error correction methods used in digital broadcasting vary depending on the transmission mode (Table 2.9). The quality of the channel also influences error correction requirements; for example, with the high-grade cables used in CATV systems, only the Reed-Solomon (RS) code is required. On the other hand, in terrestrial broadcasts the signal must contend with noise, multipath, and other propagation path conditions that lead to errors in the coded signal. In this case, convolutional coding using a Reed-Solomon code provides a very robust error correction system. A more detailed discussion on error correction coding will be presented in Chapter 3.

(d) Digital modulation: Modulation is the process of modifying the amplitude or phase of a carrier wave with the source information in order to transmit the information to a distant receiver. (The basic concepts of modulation are presented in Chapter 3.) As for error correction coding, the modulation systems used in digital broadcasting are selected for optimum performance in a specific transmission mode, and several different systems are available.

(e) Power amplifier: The power amplifier functions to amplify the level of the modulated signal. Amplifier design does not differ significantly between analog and digital systems, but since the OFDM (orthogonal frequency-division multiplexing) modulator is a multicarrier design, under actual terrestrial receiving conditions, amplifier nonlinearities, combined with the difficulty in maintaining orthogonality between carriers (which leads to intersymbol interference) means that the OFDM system cannot be used with existing analog broadcast power amplifiers without modifications. In such cases, the amplifier must be backed off so it is operating in the linear region. (More details in Chapter 4.)

(f) Transmitter antenna: The transmitter antenna functions to radiate the modulated and amplified signal into the transmission medium as the broadcast signal. As illustrated in Fig. 2.20, the super turnstile and super gain antennas are used for terrestrial broadcasts, while the parabolic antenna is used for satellite broadcasting purposes. The primary differences in antenna designs depend on the frequency band over which the broadcast signal is carried, and the same antenna can be used for both analog and digital broadcasts. In CATV, the signal is transmitted between the transmitter and the subscribers using coaxial cables or optical fiber cable, so antennas are not required.

Table 2.9   Transmission Media and Forward Error Correction Methods

Medium	Error correction codes
Satellite	Inner coding: Convolutional
Terrestrial	Outer code : Reed-Solomon
Cable	Reed-Solomon

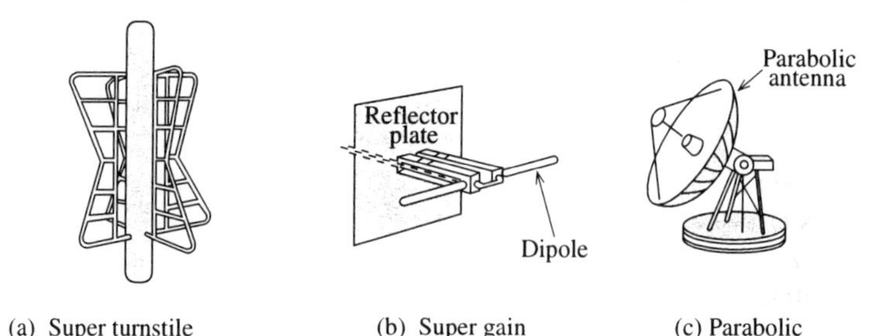

Fig. 2.20   Transmitter Antennas

Service Area – Cliff Effect

In conventional analog television broadcasts, as the strength of the received signal drops at the receiver, the quality of the picture degrades accordingly. However, as was discussed in Section 1.3.2, in digital broadcasts the picture quality remains constant up to the point a given threshold is reached, and then reception suddenly becomes impossible. This is called the cliff effect.

The service area of a terrestrial or a satellite television broadcast is defined as the area in which the radio waves in the transmission medium are strong enough to produce acceptable reception. Because of the cliff effect, the definition of the service area for analog broadcasts will not serve for defining the digital broadcast service area. For terrestrial analog broadcasts, the service area is defined to include those locations (of receivers) within an area in which the signal is strong enough to provide reception 50% or more of the time, to 50% or more of the receiver locations. In the VHF band, this is determined to be provided by a signal with an electrical field strength of 54 dBμV/m. However, if this definition is applied to a digital broadcast, at the edges of the service area it is possible that the signal strength could fall below the required level for reception 50% of the time. Given the cliff effect, this means that the subscriber could have no reception at all for up to 50% of the time. Also, with the "50% of localities" requirement, one of two subscribers (possibly neighbors) would not have reception. For digital broadcasts, the time and locality requirements must be set to a higher value, say 90 to 99% in both parameters. In 1999, the "99% of the time" parameter was provisionally decided.

For satellite broadcasts, FM modulation is employed for video, but digital modulation is used for the audio signal. Hence, the cliff effect exists with the sound signal. The service time percentage in satellite broadcasting has consequently been specified as 99%. This specification will continue to be effective, even after the video modulation systems are switched from analog to digital. Furthermore, since the channel between the satellite and receiver is virtually always a direct line, locality specifications are not required.

2.3.2 Receiving Equipment

Component Configurations

Block diagrams depicting the components of conventional analog and digital broadcast receivers are shown in fig. 2.21. In digital receiving equipment, the received signal is converted to a digital signal and then demodulated. The error correction codes are then decoded, followed by demultiplexing and decoding of the video/audio signals. Note that the process here is essentially the reverse of the operations conducted at the transmitter. A brief description of the system components is given below.

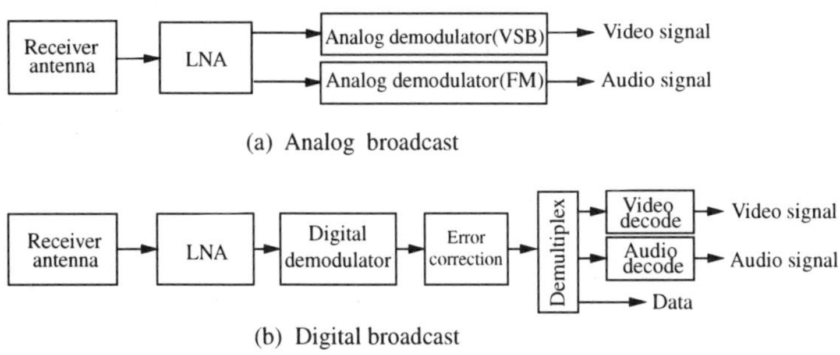

(a) Analog broadcast

(b) Digital broadcast

Fig. 2.21　Receiver Configurations

(a) Receiver antenna: The receiver antenna functions to convert the radio signals transmitted through space to an electrical signal. Antenna designs are basically the same for both analog and digital broadcasts, but the whip and the rabbit ears antennas (Fig. 2.22) are low gain antennas designed for portable and mobile receiver sets used to pick up digital broadcasts.

(b) Low noise amplifier (LNA): The electrical signals fed from the antenna into the amplifier are extremely weak, and the purpose of the low noise amplifier (LNA) is to boost the signal strength to a high enough level for demodulation. The noise figure (NF) specifies the "low noise" characteristics of the LNA, which indicates the C/N ratio between the input and output of the amplifier, expressed as a difference in decibels. (Refer to Fig. 2.2.3.) For example, given a signal with a C/N of 20 dB at input, and an amplifier NF of 3 dB, following amplification the C/N at output will be 17 dB. In terrestrial broadcasts, the NF of the receiver is 5 ~ 10 dB, and NF is 0.5 ~ 2 dB for satellite broadcasts.

(c) Digital demodulation: Demodulation is the process of recovering the original signal information from the amplitude and/or phase of the modulated carrier. The basic concepts of demodulation are explained in Chapter 3. As Fig. 2.24 illustrates, because of carrier synchronization and the symbol timing required to accurately regenerate the signal, the demodulation circuit for digital signals is much more complex and larger in scale than the analog demodulator.

(d) Error correction: The purpose of the error correction function in the receiver is to decode the error correction code(s) applied to the signal in the transmitter. Decoding the correction code restores the integrity of the original signal by removing the errors in the coded signal caused by noise and other error-generators in the transmission channel. Error correction is a function unique to digital signal systems. Transmission circuit quality varies depending on the transmission mode, so FEC procedures are accordingly slightly different. However, since receivers are generally similar, usually the outer code (which is the first code applied in an outer/inner convolutional code) is a Reed-Solomon RS(204, 188) code, and this is generally standard for all three transmission modes. More details on FEC will be given in Chapter 3.

(e) Demultiplexer: The demultiplexer functions to separate the received video and audio signals from the single transport packet signal into their original bit streams. This component is also unique to digital systems. Multiplex and demultiplex operations for digital broadcast systems are specified in the ISO/IEC MPEG-2 standard.

(f) Information source decoder: Decoding consists of retrieving the original information contained in the source-coded signal. Decoding is also unique to digital systems. Encoding procedures are specified in the MPEG-2 standard, as are the decoding procedures at the receiver end.

Conditional Access

Transmitted programs intended for conditional access are scrambled (by rearranging the coded data sequence, or through mathematical processing) so that only specified

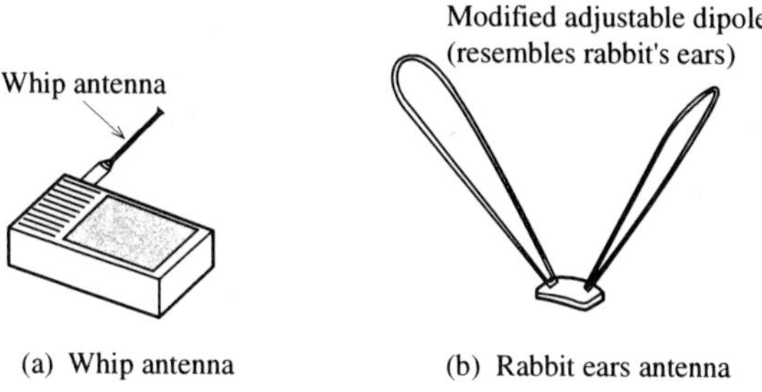

(a) Whip antenna　　　　　(b) Rabbit ears antenna

Fig. 2.22　Simple Receiver Antennas

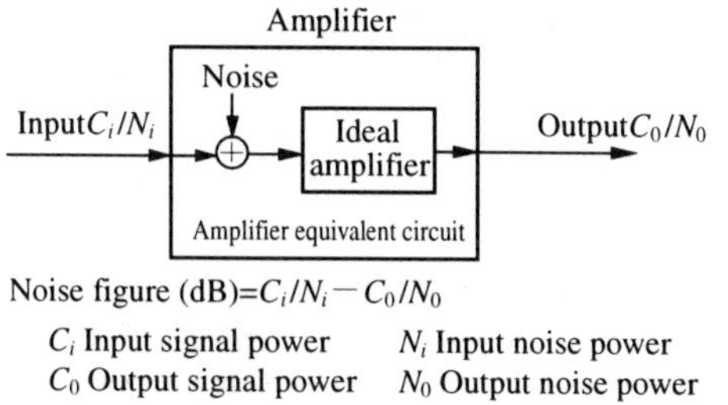

Noise figure (dB)=$C_i/N_i - C_0/N_0$

C_i Input signal power　　　N_i Input noise power
C_0 Output signal power　　　N_0 Output noise power

Fig. 2.23　Defining the LNA Noise Figure

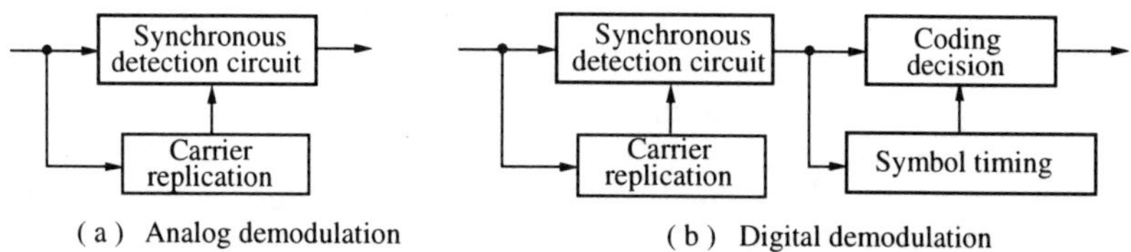

(a)　Analog demodulation　　　　(b)　Digital demodulation

Fig. 2.24　Demodulation Circuit Configurations

subscribers can receive the program. Conditional access is directed at specific subscribers, for subscribers paying special fees, or for subscribers living within a specific area.

A "key" is required to descramble the broadcast signal, and two types of keys are used. The first is the "secret key," a system which does not allow either the scrambling end (transmitter) or the descrambling end (receiver) to pass the key to a third party. The second system is called the "public key" system, in which the scrambling algorithm is in the public domain, but the subscribers use a secret key to descramble the received signal. Using this method, even though one of the keys is public knowledge, the subscriber key is an encryption code that is very difficult to break. (It would take a super computer hundreds of years to decipher this code!) The key is also periodically changed as an added safety factor [ISO, 1994]. When the key is changed, the updated information can be passed to the subscriber multiplexed on the broadcast signal (Fig. 2.25).

Receiver Classes and Required *C/Ns*

There are three classes of television receivers: the fixed receiver, portable receiver, and the mobile receiver. Because of the quality of reception, the fixed receiver has been traditionally considered the "conventional" television receiver, but with the

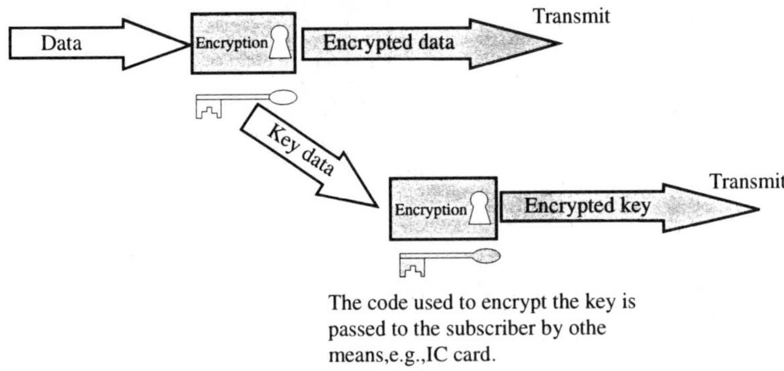

Fig. 2.25   Transmitting the Key

advent of digital broadcasts, digital television's immunity to ghost images (delayed waves) and fading (caused by time-related variations in the strength of the received signal) have opened up the possibility of increased usage of TV sets installed in automobiles, buses, and other mobile applications. The features of the various classes of receivers are summarized below:

(a) Fixed receiver: The receiver antenna and the television receiver remains in a "fixed" location. The fixed receiver can receive terrestrial, satellite, or CATV programs. A directional antenna can be used with a fixed receiver, and since fading is not a factor, the required *C/N* does not have to be particularly high.

(b) Portable receiver: The antenna and the receiver comes as a set, and the television set can be moved from place to place (but not while operating). Because of the difficulties in adjusting the antenna, the reception of satellite broadcasts is out of question – only terrestrial broadcasts are possible. A directional antenna cannot be used, and when the set is used indoors, the surrounding building also attenuates the strength of the received signal. A fairly high *C/N* is therefore required.

(c) Mobile receiver: Mobile receivers are capable of reception in a moving environment. Since mobile receivers use radio waves as the transmission medium, they are suited for both terrestrial and satellite broadcasts. Reductions in signal strength due to the "lay of the land," multipath fading, and low antenna gains require that for terrestrial broadcasts, *C/N* must be 20 ~ 30 dB greater than in fixed receivers [Okumura, 1986}. For satellite broadcasts, antenna controls must be provide which will allow the accurate tracking of the satellite. There is also the problem of obstruction shadowing (buildings, utility poles, trees, bridges, etc.), but this problem does not require a greater *C/N* than that of fixed receivers.

Selling Colorful Antennas for Color Television

NTSC color television programming began in Japan in 1960, but the receivers for color TV require a stronger signal from the antenna than did the older black and white models. When broadcasting stations attempted to raise transmitter power to meet those signal strength requirements, however, interference between nearby stations broadcasting at the same frequency became a problem, so that did not provide a satisfactory answer. The solution was found in getting better performance from the receiver, and introducing antennas with a higher gain. Antenna manufacturers did their part by increasing the number of elements on the antenna. Since these elements were brightly colored with different colors, the antennas were advertized as "color antennas for color TV." A product of this campaign was the now well-known Yagi-Uda antenna. The interesting part of this story is that coloring the antenna elements has no effect at all on performance, but coloring the elements accomplished its purpose with regards to sales.

2.3.3 Transmission Media and Modulation Systems

The transmission media (or modes) currently used for television broadcasts can be broadly divided into three categories: ground-wave terrestrial, satellite, and CATV. The proposals for modulator designs intended for use in digital broadcast applications are generally centered on taking advantage of the characteristics of a specific channel medium (Table 2.10). A brief description of the modulation systems used with the various transmission media follows:

Table 2.10　　Transmission Media – Modulation Systems

Media	Modulation
Satellite	QPSK TC8PSK
Terrestrial	OFDM 8-VSB (USA)
Cable	64 QAM 16-VSB (USA)

Terrestrial Broadcasts

In terrestrial broadcasts, the OFDM (orthogonal frequency-division multiplex) system is currently being investigated by the various European countries and in Japan because of its robust performance in combating ghost images and delayed wave interference in mobile receiver applications. In the United States, development of the future-generation Advanced TV (ATV) centers on using a single carrier VSB modulation scheme for terrestrial broadcasting purposes. The OFDM system has special features that have heretofore not been tested in use in communications and broadcast systems, and a more detailed description of the system will be presented in Chapter 4.

Satellite Broadcasts

Phase-shift keying (PSK) modulation systems such as QPSK (Quadrature Phase-Shift Keying) and 8-PSK are used in satellite broadcast applications. Since in PSK, the carrier amplitude is not modulated with signal information, the amplitude distortion typically found in satellite channels (when traveling-wave tube amplifiers are used) is not a problem with PSK. However, the spectral efficiency of BPSK (Binary PSK) and QPSK is low, and transmitting high volumes of information requires a wide bandwidth. The satellite broadcast bandwidth is 27 MHz, versus the 6 MHz bandwidth of terrestrial television systems. PSK systems are nevertheless suited for satellite television service.

CATV

The connecting cables used in CATV provide a high-quality transmission channel, and the system is thus well suited for 64-QAM (Quadrature Amplitude Modulation) because of its good spectrum efficiency. In QAM, two carriers, offset by 90 degrees (orthogonally oriented), each independently carry the amplitude-modulated information. BPSK and QPSK operate on binary source digits ($M = 2$ and 4, respectively). For higher multiple phase-shift keying, the modulation system is normally referred to as M-ary QAM.

The bit error rate (BER) characteristics for several M-ary QAM systems, QPSK, and BPSK are illustrated in Fig. 2.26 [Murota, 1985]. Note that for the same BER, 64-QAM is about 13 dB lower in *C/N* than QPSK. This is reflected in the fact that in order to obtain the same level of BER, 64-QAM requires about 20-times more power than QPSK. However, as noted earlier, since the transmission channel of CATV is so much better than space-propagated channels, *M*-ary QAM is a practical CATV modulation system.

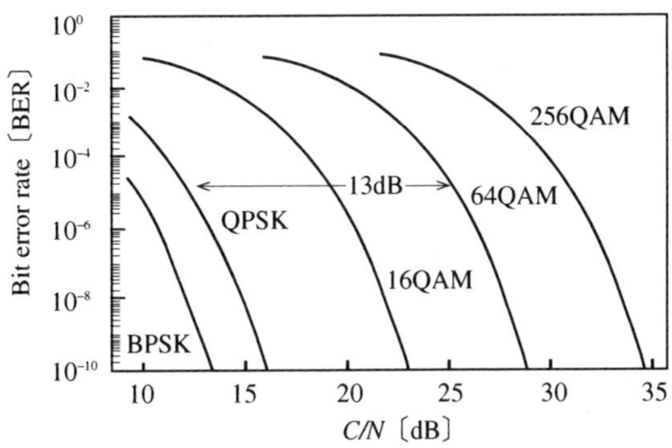

Fig. 2.26   BER Characteristics

2.4 Conclusion

An overview of the fundamental techniques of digital broadcasting has been presented in this chapter. The general concepts of the video (picture) signal, and high-efficiency coding as specified in the MPEG compression standards were discussed, along with the receiver technologies of digital television systems. For those readers desiring a more comprehensive coverage of picture coding, we invite you to refer to the [Kuroda, 1996] and [Fujiwara, 1994] references.

Chapter Questions

Q1: Compare progressive scanning and interlaced scanning, and list two advantages of the interlace technique.

Q2: Describe the basic configuration of the high-efficiency picture coding system.

Q3: Give an actual example of a transform coding operation.

Q4: List the three types of pictures that are coded in the MPEG standard.

Q5: Why is the service area defined differently for analog and digital television?

Q6: Why are different modulation systems used for different transmission media?

Q7: Describe the cliff effect.

Chapter References

[Fujiwara, 1994] H. Fujiwara kanshuu: "Saishin MPEG Kyoukasho", Asukii Shuppankyoku (1994) (in Japanese)

[ISO, 1994] IOS : "MULTI 2 Encryption Algorithm," ISO-9979/0009 (1994)

[Kuroda, 1996] H. Kuroda: "Gazoufugouka Gijutsu", Shoukoudou (1996) (in Japanese)

[Matsuo, 1997] K. Matsuo: Deijitaru Housou Gijutsu", Tokyo Denki Daigaku Shuppankyoku (1997) (in Japanese)

[Murakami, 1993] J. Murakami: "Gazoufugouka Gijutsu to Kokusai Hyoujunka Doukou", Denshi Jouhou Tsuushin Gakkaishi, Vol. 76, No. 12, pp. 1314-1325 (1993) (in Japanese)

[Murotani, 1985] Murotani, Yamamoto: "Deijitaru Musen Tsuushin", pp. 32-40, Sangyou Tosho (1985) (in Japanese)

[Netravali, 1994] A. N. Netravali, et al.,: "Digital Pictures", Plenum Press, New York (1994)

[Netravali, 1998] A. N. Netravali, et al.: "Digital Pictures," Plenum Press, New York (1988)

[NHK Housou Gijutsu Kenkyuujo, 1994] NHK Housou Gijutsu Kenkyuujohen: "Maruchimedeia Jidai no Deijitaru Housou Gijutsu Jiten", Maruzen (1994) (in Japanese)

[Okumura, 1986] Okumura, Shinji: "Idoutsuushin no Kiso", Denshijouhou Tsuushin Gakkaihen, Koronasha (1986) (in Japanese)

[Terebijyon Gakkai, 1995] "Tokushuu MPEG", Terebijyon Gakkaishi, Vol. 49, No. 4, pp. 408-534 (1995) (in Japanese)

[Yasuda, 1991] H. Yasuda Henchou: "Maruchimedeia Fugouka no Kokusai Hyoujun", Maruzen (1991) (in Japanese)

[Watanabe, 1996] K. Watanabe: Oodeio Shingou no Kounouritsufugouka", Denshi Jouhou Tsuushin Gakkaishi, Vol. 79, No. 8, pp. 793-796 (1996) (in Japanese)

Chapter 3
Digital Modulation and Error Correcting Codes

An overview of the basic concepts of digital modulation and methods used for evaluating the performance of modulation systems is presented in this chapter. Error correcting codes, an important design component of digital systems, are also discussed as a technique employed to improve the quality of digital signal transmission. Coded modulation, which combines the modulation and error correcting coding functions are also described.

3.1 Digital Modulation Fundamentals

The modulation and demodulation systems operate on a sequence of digital signals representing the source information, passed over a narrow frequency band in a finite, bandlimited channel. This signal is referred to as the baseband signal. On the transmitter, the baseband signal is modulated, passed through a band-pass filter, and sent to the transmission channel. At the receiver, the received signal is filtered to reduce the noise picked up in the channel, and demodulated to recover the original baseband signal. Modulation systems are selected based on transmission channel and bandlimiting characteristics. Modulator performance also varies, depending on the type of demodulation system used.

3.1.1 What is Digital Modulation?
Modulation Concepts
The transmitted message (source) signal consists of the modulating signal (which is the baseband signal) and the carrier wave on which the source information is impressed. The modulated signal is a product of modifying one (or more) of the parameters of the carrier wave with the modulating signal. A single frequency sine wave is normally used as the carrier.

Given time, t, the general form of the signal waveform, $s(t)$, is expressed in the following equation:

$$s(t) = A\cos(2\pi ft + \phi). \tag{3.1}$$

In the above equation, A, f, and ϕ represent amplitude, frequency, and phase, respectively. Where frequency is a known value, the above equation can be represented using vectors with A, ϕ, and the x_1 and y_1 components shown in Fig. 3.1.

The carrier parameters varied in order for the signal to carry the source information are amplitude, frequency, and phase, and the modulation systems which perform these operations are called amplitude modulation (AM), frequency modulation (FM), and phase modulation (PM) systems, respectively.

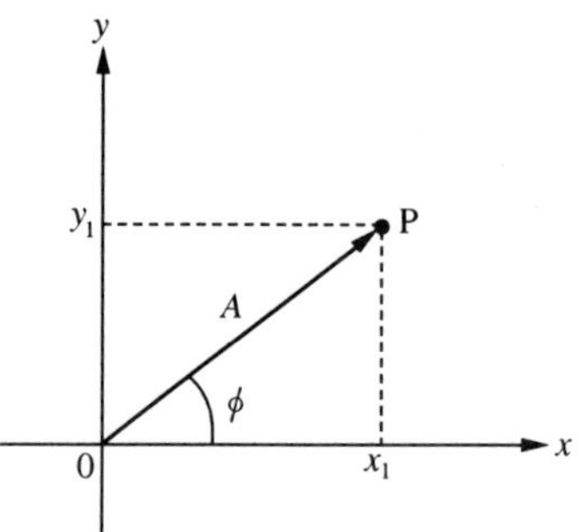

Fig. 3.1   Vector Representation of the Signal Waveform

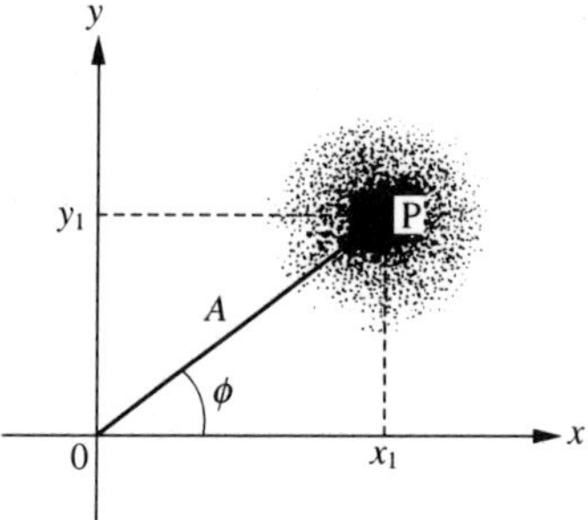

Fig. 3.2   Appearance of Noise in the Signal Waveform

In conventional analog modulation systems, an analog signal containing analog information (e.g., audio, etc.) is used as the modulating signal. In digital modulation systems, the baseband signal, consisting of a sequence of bits having discrete values serves the modulating signal function. In other words, the analog signals from the audio or video source are converted to digital signals represented in binary codes consisting of sequences of 0s and 1s. This is referred to as digital modulation, and the digital signal reflecting the source information after modulation is represented by changes in the amplitude, frequency, or phase of the carrier wave. In the demodulation process, the modified parameter of the received signal is tested against a given threshold, and the digital information is reconstructed in a decision process.

When the signal is transmitted over a radio channel, the signal is corrupted by noise, which is distributed around the signal point as shown in Fig. 3.2. As will be elaborated later, if the noise is Gaussian in nature, its distribution will be centered around the signal point. For analog modulated signals, the quality of the channel degrades in proportion to the noise corruption of the signal. However, in a digital transmission, if zero is established as the baseline level, the decision process simply determines whether a 1 or a −1 value was the transmitted bit. That being the case, if the symbol is a 1, it will remain a 1 unless it crosses the zero level into the negative region. Thus, even in the presence of moderate noise, the quality of the transmission channel does not change. Using error correcting codes provide added assurance of a high-quality signal. (Error correcting codes will be discussed later in the chapter.)

Modulation using Orthogonal Signals

Using a set of basis functions (in an orthogonal space) allows a set of signals to be combined for more efficient transmission of information. When the inner-product of any two independent functions is zero, the function is said to form an orthogonal set. In such cases, the signals are mutually independent, and can be carried in separate channels. For example, cosine waves and sine waves are orthogonally oriented, and as carriers, can be independently modulated as the in-phase (cosine) component, and the quadrature (sine) component. The signal space diagram is used to represent the signal point orientation of such systems. The signal space diagram is a method of representing the x and y (orthogonal) axis vectors shown in Fig. 3.1. Figure 3.3 shows the normal representation of the in-phase component (called the I-channel) and the orthogonal component (called the quadrature or Q-channel). In Fig. 3.3, the two signal points are shown as 1 and −1 in the I-channel. The geometrical (Euclidean) distance between the two signal points is called the Euclidean distance.

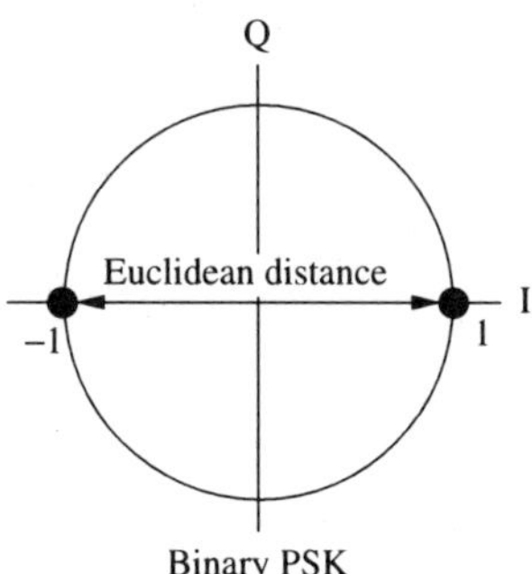

Fig. 3.3   Signal Space Diagram

3.1.2 Basic Modulation Systems

The fundamental purpose of the digital modulator is to place the information contained in the source onto the carrier signal by modifying the carrier's phase or frequency characteristics with the modulating signal containing the source information.

Phase Modulation Systems

The most basic phase modulation system is the binary phase-shift keying (BPSK) modulator. For BPSK, in response to modulating data digits of 1 and 0, the amplitude of the carrier waveform becomes 1 and −1, respectively, and the phase of the waveform shifts (or keys) between the 0 and π states. In eq. (3.1), A is equal to 1. Figure 3.3 is representative of the signal space diagram for BPSK. To demodulate a BPSK signal, the carrier component is regenerated using the received signal, and the modulating signal information is extracted through a multiplication process.

Frequency Modulation Systems

In FSK (frequency-shift keying) signaling, the carrier is shifted to different frequencies to reflect the information transmitted. Orthogonal signal frequencies, f_1 and f_2, represent symbol inputs of 0 and 1, respectively. A simple system consisting of two signal envelope-matching filters can be used for demodulating the FSK signal, selecting the frequency having the highest power for output as the detected signal.

The modulated signal waveforms for BPSK and FSK are shown in Fig. 3.4. For BPSK, binary digits of 0 and 1 (i.e., one bit of information) are converted to a carrier having an amplitude of −1 and 1, and phase shifts between π and 0. In FSK, the same information is represented by changes in the carrier frequency.

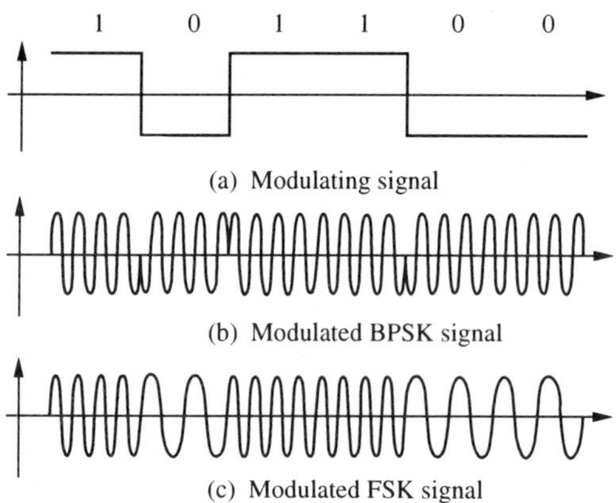

Fig. 3.4   Digitally Modulated Signal Waveforms (Time-transition)

M-ary PSK

In binary PSK, a source bit can only be transmitted over the I-channel or the Q-channel, but not both. A more efficient modulation system is be possible if both orthogonally-oriented channels (I and Q) are used in a multiple-phase configuration.

In QPSK (quadriphase or 4-ary PSK) the same carrier frequency, f_c, is used independently in the orthogonally-oriented I-channel and Q-channel to transmit a binary signal. This means that a total of four source digits, or two bits of information can be transmitted. The digital data in a one-time phase shift of the modulated signal is called a symbol. For example, in binary PSK, one symbol contains one bit of information ($M = 2^n$, $n = 1$ bit), and in QPSK, one symbol contains two bits of information ($M = 2^n$, $n = 2$ bits).

An example of a QPSK modulator is illustrated in Fig. 3.5. The digital signal input is divided into two bit sequences by the serial/parallel converter, and as in BPSK operations, the 0 and 1 data digits are converted to waveforms with amplitudes of −1 and 1. These signals are then multiplied (heterodyned) with the orthogonal carriers (cosine and sine waves) to become the modulated signal. The signal space diagrams for these systems are shown in Fig. 3.6.

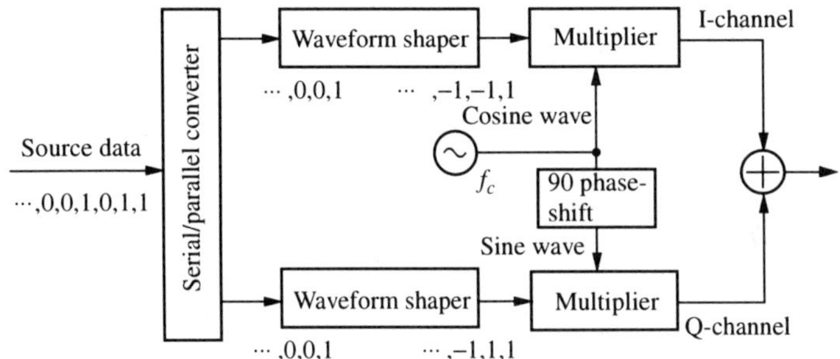

Fig. 3.5 Orthogonal (Quadrature) Amplitude Modulator

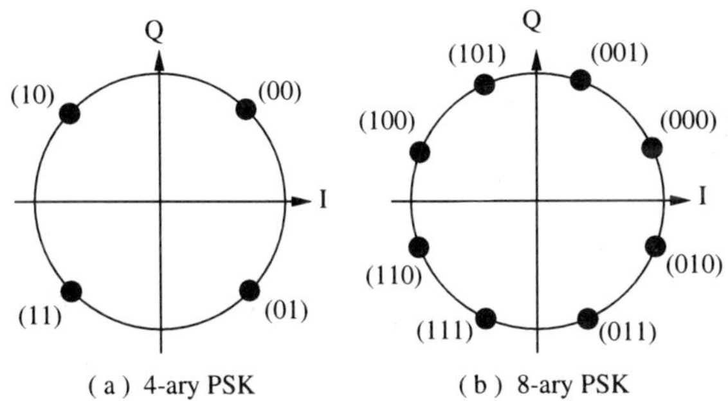

Fig. 3.6 MPSK Signal Space Diagram

As the size of the symbol set increases (higher M-ary systems), n bits of data are simultaneously modulated by the MPSK system, and this permits a greater amount of data to be carried over a comparatively narrower bandwidth. Note however, that the resulting decrease in Euclidean distance in higher M-ary systems does cause a degradation of BER values when signal power is equal. For a system with a given throughput (transmission capacity), selection of the modulation system involves a trade-off between signal power and bandwidth for an acceptable BER. Figure 3.6(b) illustrates the signal space diagram for an 8-ary PSK (or 8-PSK) modulator. In order to improve the BER in *M*-ary systems, the signal points are arranged so that the difference between neighboring points is only one bit. This is referred to as Gray coding.

π/4-Shifted QPSK

π/4-shifted QPSK is a special case of QPSK, in which the amount of change in the envelope is reduced by avoiding the radical phase shifts of normal QPSK. This also results in a reduction in the nonlinear distortion in the power amplifier.

As illustrated in Fig. 3.7, the input signal, a_n, is converted to two bit sequences by the serial/parallel converter, and phase is rotated in accordance with the values produced by the differential encoder. The signals are then modulated in their respective orthogonal channels. The bandlimiting filters shown in the diagram are a required part of the communications system, and are described later.

The phase transitions in response to inputs X_k and Y_k are illustrated in Table 3.1. When the phase of the prior digit is ϕ_{n-1}, the phase of the digit transmitted next will be $\phi_n = \phi_{n-1} + \theta$. For example, in the previous symbol time, where at the π/4 signal point we have $X_k = 1$ and $Y_k = 1$, the transition will be $-3\pi/4$, which means that the point to which the transition is made will be $\pi/4 + (-3\pi/4) = -2\pi/4$. As Fig. 3.8 illustrates, the transitions alternate between the point sets marked by the empty circles and the solid dots, as shown by the arrows. The transition diagram makes this system look like 8-PSK, but because of the way phase transitions are made, transmission efficiency and performance indicates that π/4-shifted QPSK is actually a modified QPSK system.

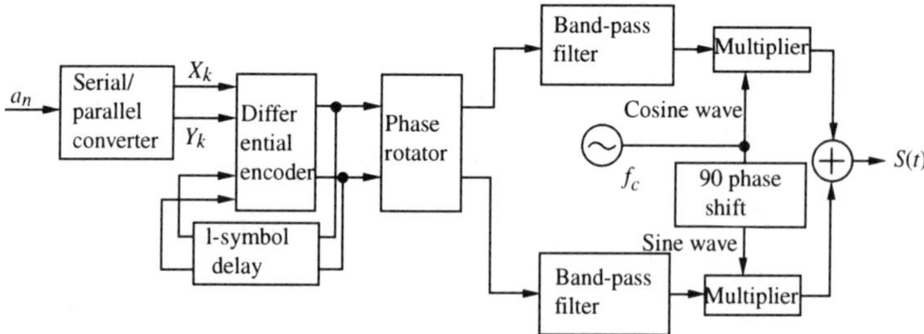

Fig. 3.7 π/4-shifted QPSK Modulator

Table 3.1   Phase Transitions

X_k	Y_k	θ
1	1	$-3\pi/4$
0	1	$3\pi/4$
0	0	$\pi/4$
1	0	$-\pi/4$

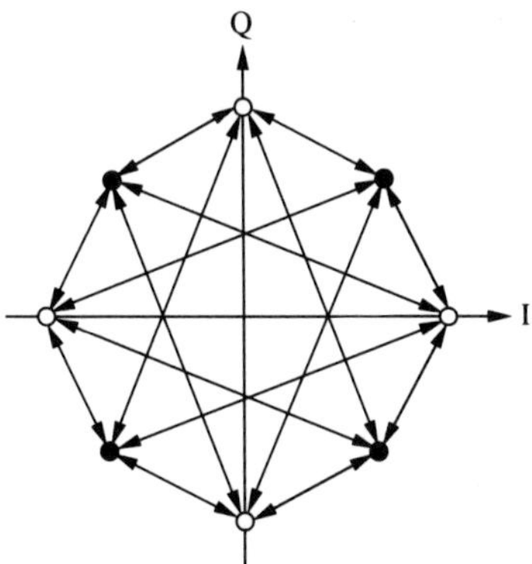

Fig. 3.8 π/4-shifted QPSK signal point Transitions

At the receiver , since the phase difference can be detected and differentially decoded, simple methods such as a differential coherent detector can be used to demodulate the signal. In some QPSK systems, the QPSK signals remain at the same phase, which is not the case with π/4-shifted QPSK because the π/4 phase is usually additive. Also, in other QPSK systems the carrier phase of the pre-filtered QPSK signal is rotated by π/4 for each symbol, which permits the use of a coherent detector (absolute-phase coherent detector) in the receiver.

3.1.3 Spectrum Distribution and Bandlimiting

Transmission wave Spectrum

An example of the power spectrum of a QPSK modulated wave is shown in Fig. 3.9. Where the symbol time duration (i.e.,the inverse of symbol transmission rate, or the number of symbols sent per unit of time) is T_s, the largest concentration of spectral density is distributed across a width of $1/T_s$ on either side of the center-frequency of the carrier. The central $2/T_s$ part of the spectrum is called the main lobe, while the outer lobes are called the side lobes. This diagram also indicates

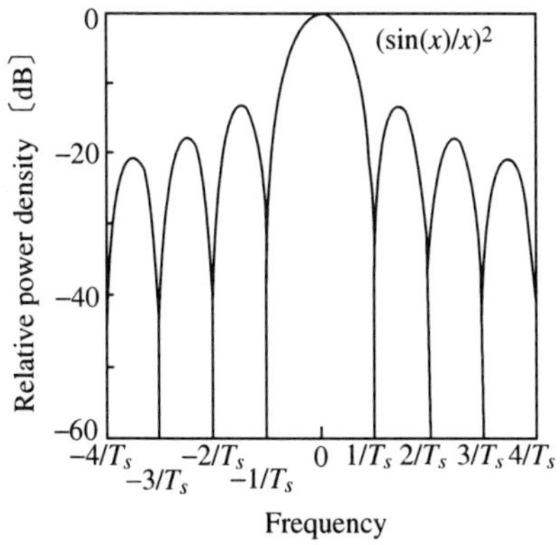

Fig. 3.9   Example Power Spectrum (normalized on center-frequency)

how the unconstrained modulated wave of a digital signal can expand beyond the designated bandwidth, which is the reason that bandlimiting must be used in most communications systems.

Bandlimiting the Channels

As noted in Fig. 3.4(a), the digital signal is normally shown in its "ideal" form, which is as a rectangular wave. However, as we observe in Fig. 3.9, since the power radiates outside the specified bandwidth, this is not a realistic depiction. In practical systems, filters are used to eliminate the extraneous side lobes, thereby shaping the waveform. This is called bandlimiting. By using strictly band-limited channels, more channels can be allocated within a given frequency band, which improves the spectral efficiency of the system.

Figure 3.10 shows the eye pattern and power spectrum when Gaussian filters are used for bandlimiting the channel. The eye pattern is formed by repeatedly tracing the baseband signal at integer multiples of T_s. The arrow in each diagram indicates the sampling points, and the wider the eye opens, the greater the noise immunity of the channel. Figure 3.10(b) illustrates a highly band-limited channel. In this case, the signal is interfered with by the leading and/or following signals (which tends to change the symbol sequence), and this results in the arrow growing shorter (a closing of the eye). This is referred to as intersymbol interference (ISI).

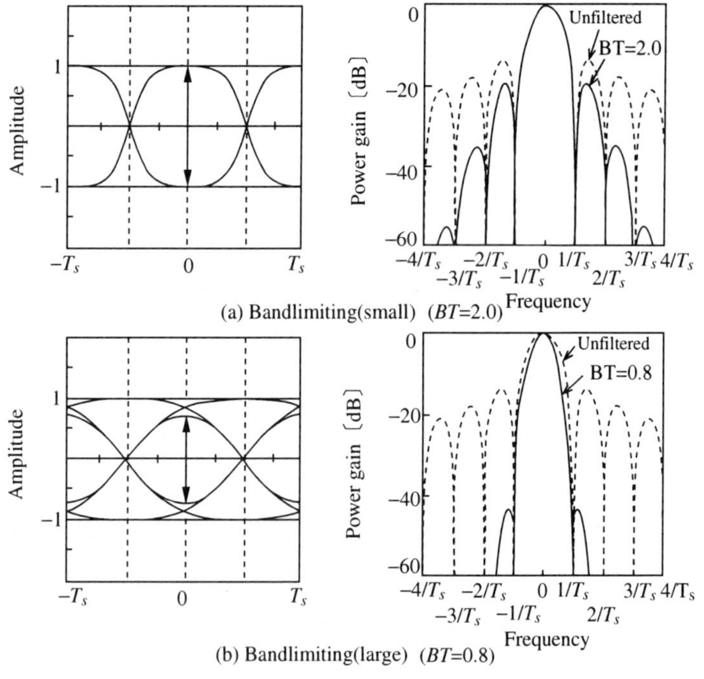

(a) Bandlimiting(small) (*BT*=2.0)

(b) Bandlimiting(large) (*BT*=0.8)

Fig. 3.10   Bandlimiting Using Gaussian Filters

The most suitable filter is a filter which provides a maximum *SNR* (signal-to-noise ratio) at output, and is called a matched filter. The impulse response of a matched filter is a time-inverted waveform of the input signal [Proakis, 1995]. Here, impulse response is defined as the filter's frequency response in the time-domain (measured at filter output) to an infinitely narrow pulse of finite power, presented at the filter input.

When the channel is severely band-limited, the trade-off is between reducing ISI and noise suppression of the receiver filter, and filter characteristics must be selected accordingly. ISI-induced degradation can be avoided by using combinations of transmitter/receiver filters optimized so that ISI does not occur at the sampling point in the receiver. Where the transfer function is $H_T(\omega)$ and $H_R(\omega)$ for the transmitter and receiver, respectively, the filter transfer function between the transmitter/receiver of $H(\omega) = H_T(\omega) H_R(\omega)$ is given by the following equation:

$$H(f) = \begin{cases} 1 & : 0 \leq |f| \leq \dfrac{1-\alpha}{2T_s} \\[2ex] \cos^2\left[\dfrac{\pi T_s}{2\alpha}\left(f - \dfrac{1-\alpha}{2T_s}\right)\right] & : \dfrac{1-\alpha}{2T_s} < |f| \leq \dfrac{1+\alpha}{2T_s} \\[2ex] 0 & : \text{everywhere else} \end{cases} \qquad (3.2)$$

where T_s represents symbol duration. α is the roll-off factor, and is 0 for a rectangular filter, and 1 for a raised cosine filter. Also, since frequency is one-half of the $1/T_s$ frequency, the voltage gain of the filter is 0.5. Filters that ISI dose not occur at the sampling point by pulse shaping are referred to as Nyquist filters.

Graphical representations of the transfer function and the impulse response of a Nyquist filter are illustrated in Fig. 3.11. As we can see in Fig. 3.11(b), no interference components exist within the T_s interval of the sampling point.

In a linear channel, the system transfer function characteristics are a product of the transfer functions of the transmitter and receiver filters. Generally in such cases, these characteristics are a square root distribution between the transmitter and

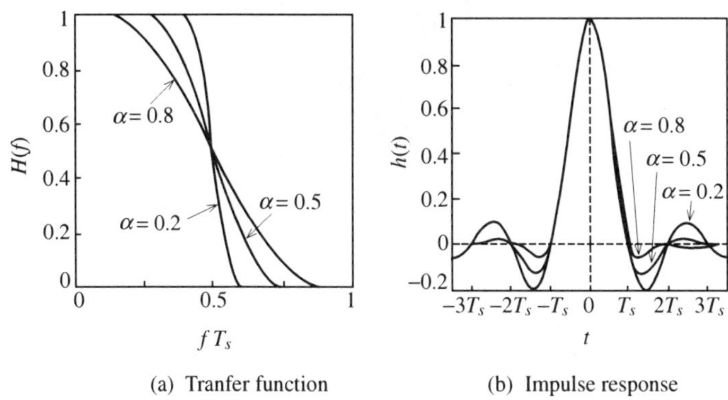

(a) Tranfer function   (b) Impulse response

Fig. 3.11 Nyquist Filter Characteristics

and receiver filters. When this is the case, the filter's impulse responses are symmetical in time, hence their name, matched filters.

When we have transfer function characteristics distributed as described above, Fig. 3.12 illustrates the eye pattern following demodulation for $\alpha = 0.8$ and 0.2, respectively. The points of the arrows are between 1 and -1 (although a number of other sample points are taken), but it is evident that the eye is always open. Also, when bandlimiting is severe, the variation becomes greater, which means that the effects of timing error at the decision points are greater. (With a small timing error, decision errors are caused by interference.) Figure 3.13 shows the trace of the signal points in the I- and Q-channels of a QPSK signal when $\alpha = 0.5$. Here, we should see only the four points in transition at a fixed amplitude, but bandlimiting has caused the trace to vary from its normal track. Thus, use of an amplifier with nonlinear characteristics causes the bandwidth to expand.

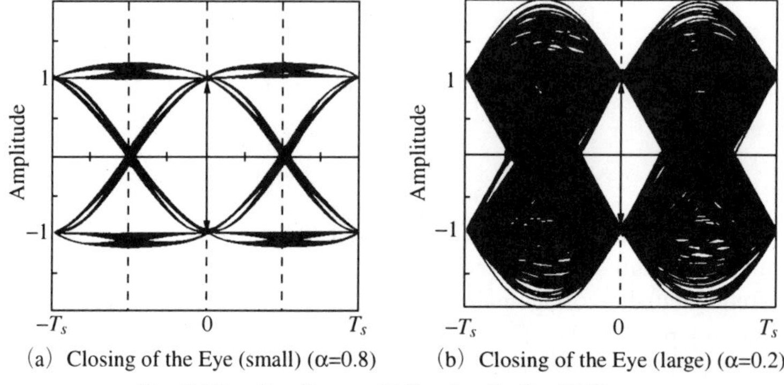

<table>
<tr><td>(a) Closing of the Eye (small) (α=0.8)</td><td>(b) Closing of the Eye (large) (α=0.2)</td></tr>
</table>

Fig. 3.12   Eye Pattern Following Roll-off Filter

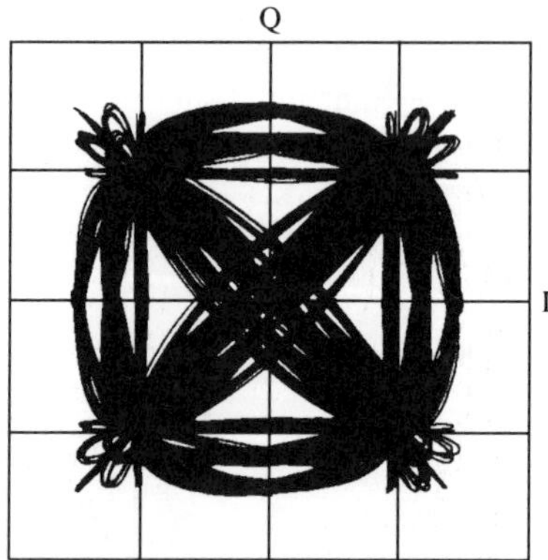

Fig. 3.13   QPSK Signal Trace ($\alpha = 0.5$) Following Bandlimiting

3.1.4 Demodulation Systems

In this section, the demodulation process is briefly described based on the perfomance of the well-known PSK system.

FSK Demodulation

Demodulation of the FSK signal is a rather simple procedure, consisting of using two band-pass filters (BPF) centered on frequencies f_1 and f_2. The specific filter output at which the received signal appears is made by the decision process, and the transmitted data is estimated using a noncoherent detection system.

PSK Demodulation

Coherent detection is typically used to demodulate the PSK signal. The differential coherent detector (featuring a simple circuit design) is also used. These two systems will be described below.

(a) Coherent detection

Coherent detection requires regeneration the carrier and a synchronous reference signal. Methods of extracting the source information include reconstructing the carrier phase by sending a known bit sequence (absolute phase coherent detection), and in the case of MPSK, by partitioning the signal space into M regions, and using the decision rule of the detector to decide which signal was transmitted based on which region the received signal falls in. Figure 3.14 illustrates a block diagram of the BPSK coherent detector. The reference signal obtained from the carrier regeneration circuit is described by $s_i(t) = \cos(2\pi f_c t)$.

Subsequently, multiplying the received signal, $s_r(t) = A\cos(2\pi f_c t + \phi_n) + n(t)$, with the reference signal yields

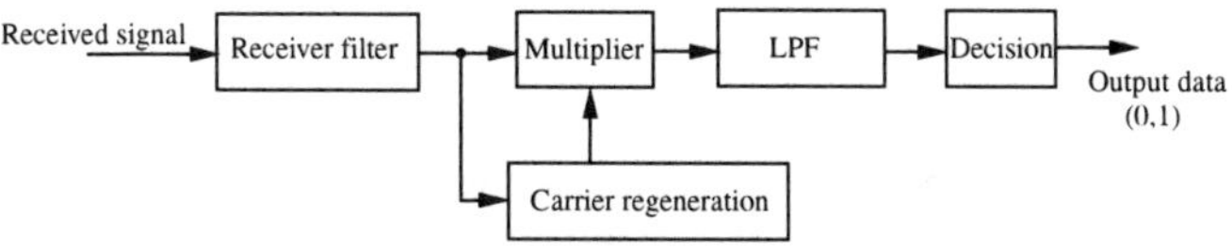

Fig. 3.14   BPSK Coherent Detection System

$$s(t) \cdot s_i(t) = \{A\cos(2\pi f_c t + \phi_n) + n(t)\}\cos(2\pi f_c t)$$

$$= \frac{A}{2}\{A\cos(4\pi f_c t + \phi_n) + \cos(\phi_n)\} + n(t)\cos(2\pi f_c t) \ , (3.3)$$

where $n(t)$ is the noise component. The LPF suppresses the carrier harmonics which are higher than second-order, and the signal is then presented to detector output. For BPSK, as was discussed in Section 3.1.2, the transmitted data is estimated by deciding polarity of the signal at output.

(b) Differential coherent detection

Differential coherent detection does not require carrier regeneration, making the circuit design comparatively less complex than that of the coherent detection system. However, the BER characteristics of differential coherent detection are poorer than in coherent detection.

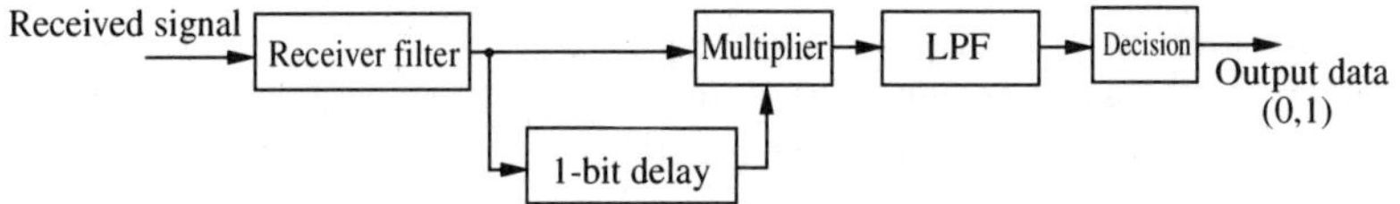

Fig. 3.15   BPSK Differential Coherent Detection System

The differential coherent detection demodulation system for BPSK is illustrated in Fig. 3.15. For differential coherent detection, we use the premise that all channel variations other than for the present and the previously transmitted symbol can be ignored. Detector output consists of the signal phase difference between the present symbol and the symbol received immediately prior to the present symbol. Any changes in phase are detectable. At the transmitter end, where the nth phase is ϕ_n, the symbol is differentially coded as

$$\phi_n' = \phi_{n-1}' + \phi_n \quad (\text{mod } 2\pi). \tag{3.4}$$

At the receiver end, the product of the present and previous symbol is computed by

$$s_r'(t) = A^2 \cos(\phi_n' - \phi_{n-1}')$$
$$= A^2 \cos\{(\phi_{n-1}' + \phi_n') - \phi_{n-1}'\} = A^2 \cos(\phi_n) \tag{3.5}$$

for detection of ϕ_n. Following the decision step, the data sequence is added to the previous sequence using modulo-2 arithmetic to regenerate the original data sequence.

3.1.5 Characteristics Evaluation

Evaluation Methods

The basic standard used for evaluating digital modulation system performance in an additive white Gaussian noise (AWGN) channel is the bit error rate (BER). BER is a measure of the number of bit errors generated in the channel versus the total number of bits transmitted. Other evaluation measures are also important, depending on the particular system, and the most important ones used to evaluate and compare system perfomance and radio communication system designs are briefly described in the following:

(a) *C/N* ratio or CNR (carrier power-to-noise power ratio)

CNR is closely related to received signal power, and as such, is an important parameter with which to compare the overall perfomance of modulation systems.

(b) *S/N* ratio or SNR (signal power-to-noise power ratio)

SNR is relatively easy to measure, and is thus a widely-used parameter for determining signal power versus noise power in the post-detected baseband signal. Where CNR is a high value, SNR can be easily converted to other evaluation measures. Note however, that when making comparisons with other systems, one must indicate from what point in the system power is being measured.

(c) E_b/N_0 (energy per bit-to-noise power spectral density ratio)

The E_b/N_0 value is the basic performance standard by which modulation systems are compared. E_b/N_0 can be converted from SNR and from band-pass filter characteristics values.

For more details on evaluation parameters, the reader is invited to refer to the [Saitoh, 1996] and other references.

Bit Error Probability

We shall begin by discussing noise as a factor in error generation in digital systems. Noise in a radio channel can be statistically modeled as a Gaussian distribution, the equation of which is given below. When noise is designated as x, the probability density function, $p(x)$, is expressed as

$$p(x) = \frac{1}{\sqrt{2\pi\sigma^2}} \exp\left(-\frac{x^2}{2\sigma^2}\right), \tag{3.6}$$

where σ^2 is the variance of noise.

As an example illustrated in Fig. 3.16, let A and $-A$ be the amplitudes of two amplitude modulated signals, with noise added at the receiver end. The graphs in the diagram indicate the probability distribution of the received signals. In digital communications, the received signal is estimated based on a decision process, which itself is based on a threshold level acting as a "dividing line" between the two pulse signals. In this case, the threshold is set to 0, evenly dividing the positive and negative amplitudes. If a signal is transmitted at one amplitude (say, $-A$) is determined by the decision process to have been received on the opposite side of the amplitude threshold, an error has occurred. A signal transmitted at $-A$ has its normal distribution marked by the solid line, but should that signal be received at any point in the other region (the positive side, marked by the hatched area), it becomes an error. This is where we see the importance of signal Euclidean distance (in this case, the distance between A and $-A$, or $2A$). The further apart the two signal points are spaced, the lower the probability of error occurrence.

Referring again to Fig. 3.16, in BPSK coherent detection, what constitutes an error is when a signal pulse is transmitted at $-A$, and is received at any point in

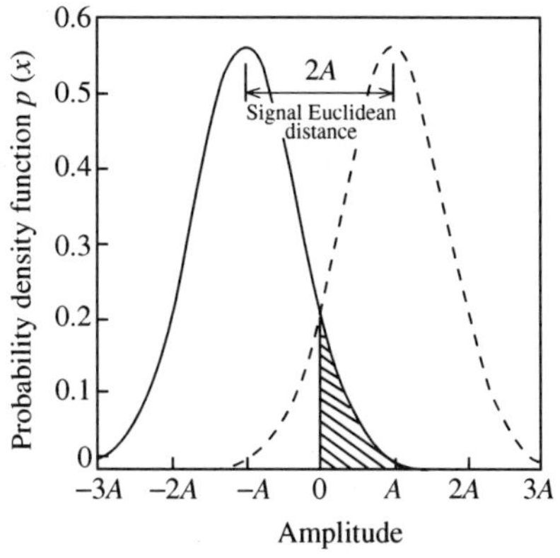

Fig. 3.16   Distribution of Received Signals

the distribution curve between zero and ∞, or between zero and $-\infty$ for a $+A$ signal. The probability of a symbol error (equivalent to a bit in BPSK) is therefore given by

$$P_e = \int_0^\infty \frac{1}{\sigma\sqrt{2\pi}} \exp\left[-\frac{(x+A)^2}{2\sigma}\right] dx$$

$$= \frac{1}{2}\mathrm{erfc}(\sqrt{\gamma}), \tag{3.7}$$

where σ is the standard deviation for noise, and $\gamma = A^2/2\sigma^2$. γ for BPSK is equivalent to E_b/N_0. $\mathrm{erfc}(x)$ is called the complementary error function, and is defined as

$$\mathrm{erfc}(x) = \int_x^\infty \frac{2}{\sqrt{\pi}} \exp(-t^2)dt . \tag{3.8}$$

The error probability given in this case is applicable to absolute phase coherent detection. In a standard demodulation system with a carrier regeneration circuit, since there is an ambiguity with regards to phase in the differential coding/decoding process, BER is approximately two-times greater than in absolute phase coherent detection.

In QPSK signaling, signal power is uniformly distributed between the I- and Q-channels, and since each channel is independent, where γ takes the value of E_b/N_0, the probability of error becomes

$$P_e = \frac{1}{2}\mathrm{erfc}\left(\frac{\sqrt{\gamma}}{2}\right). \tag{3.9}$$

Given the same bandwidth, QPSK is also capable of transmitting two-times the amount of information of BPSK, so performance with respect to E_b/N_0 remains the same.

In differential coherent detection, part of the detection process consists of taking the product of the noisy received signals, which generates a component that is a product of the individual signal noise components. BER becomes more difficult to derive in this case, because the resulting noise component is no longer a Gaussian distribution. We will show the results here, without going through the derivation [Sampei, 1997].

Bit error probability for differential coherent detection in BPSK is given by

$$P_e = \frac{1}{2}\exp(-\gamma) . \tag{3.10}$$

Bit error probability for differential coherent detection in QPSK is

$$P_e = Q(a,b) - \frac{1}{2}\exp\left(-\frac{a^2+b^2}{2}\right)I_0(ab)$$

$$\begin{cases} a = \sqrt{2\gamma(1-1/\sqrt{2})} \\ b = \sqrt{2\gamma(1+1/\sqrt{2})} \end{cases}, \tag{3.11}$$

where $Q\,(a,\,b)$ is Marcum's Q-function, defined by

$$Q(x,y)=\int_{y}^{\infty}t\exp\!\left(-\frac{x^{2}+t^{2}}{2}\right)I_{0}(xt)dt,$$

where $I_{0}\,(\cdot)$ is a zero-order Bessel function of the first kind.

Figure 3.17 shows the plots for BER versus CNR as derived from the above theoretical equations. These graphs demonstrate the superiority of the coherent detection system in BER performance. Note also that in BER performance, $\pi/4$-shifted QPSK is the same as standard QPSK.

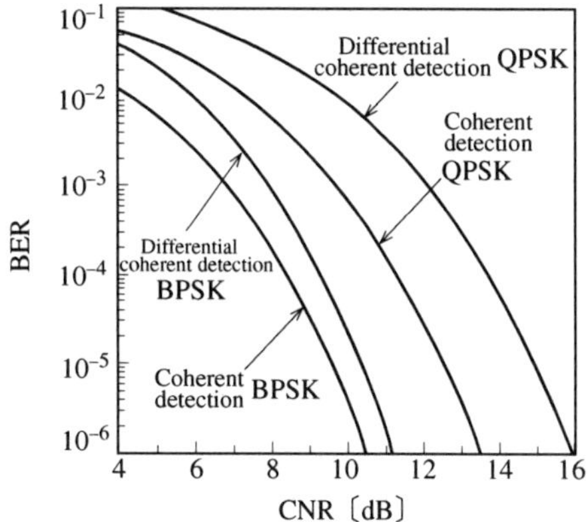

Fig. 3.17   BER Performance for BPSK and QPSK

3.2 The Basic Principles and Features of *M*-ary Modulation Systems

3.2.1 Quadrature Amplitude Modulation

In *M*-ary phase shift keying (MPSK), by increasing the binary alphabet in multiples of two (e.g., to 4-ary, 8-ary, 16-ary), we have an increase in the number of signal points, which means that more information can be transmitted during each symbol period. In this case, however, since amplitude remains constant, the Euclidean distance becomes smaller, resulting in degraded BER performance. As was noted in Section 3.1.1, quadrature amplitude modulation (QAM) is a system in which the I- and Q-channels are independently amplitude-modulated in quadrature. QAM features superior BER performance over phase modulation systems having the same *M*-ary symbol set size, as it uses both phase and amplitude information for better efficiency. However, the envelope does not remain constant (i.e., it is not constant-envelope modulation), so an amplifier with good linear characteristics must be used.

When the symbol set size is increased to four source digits per channel, a total of 16 source digits can be transmitted. This is called 16-QAM, and each symbol consists of four bits. The signal space diagram for this system is shown in Fig. 3.18. If we increase the symbol set size to eight source digits per symbol, 256 source digits can be transmitted, and the system is referred to a 256-QAM. The source digit arrangements within the parentheses in Fig. 3.18 are the *Gray coding* arrangement for 16-QAM.

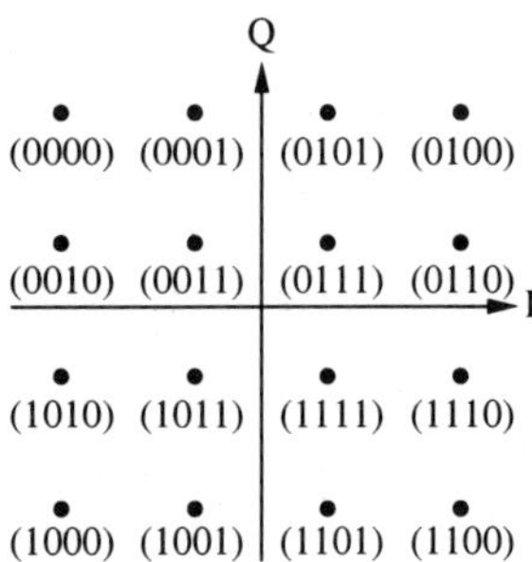

Fig. 3.18   Signal Space Diagram for 16-QAM

The approximations for average BER in a Gaussian noise channel, when $\gamma = E_b/N_0$, for Gray coded, absolute phase coherent detection 16-, 64-, and 256-QAM are given as follows [Saitoh, 1996], [Sampei, 1997]:

16-QAM

$$P_e \approx \frac{3}{8}\operatorname{erfc}\left(\sqrt{\frac{2}{5}\gamma}\right), \tag{3.12}$$

64-QAM

$$P_e \approx \frac{7}{24}\operatorname{erfc}\left(\sqrt{\frac{1}{7}\gamma}\right), \tag{3.13}$$

256-QAM

$$P_e \approx \frac{15}{64}\operatorname{erfc}\left(\sqrt{\frac{4}{85}\gamma}\right). \tag{3.14}$$

The bit error rate (BER) performance for the various QAM systems are shown in Fig. 3.19. As is evident by these plots, BER degrades corresponding to higher M-ary values, which means that ISI and other degradation factors must be compensated by waveform equalization circuits. The increased noise due to the reduction in signal spacing distance can be countered by boosting transmission power.

3.2.2 M-ary Vestigial Sideband (VSB) Transmission

Transmitting with one full sideband and one partial sideband is termed vestigial sideband (VSB) transmission. The concept of VSB is briefly described in this section. An amplitude-modulated signal is shown in Fig. 3.20(a). Moving to the (b) diagram, we see that the power spectrum of the modulated signal is distributed evenly on both sides of the carrier frequency. This is therefore called double sideband (DSB) transmission. In single sideband (SSB) transmission, one complete

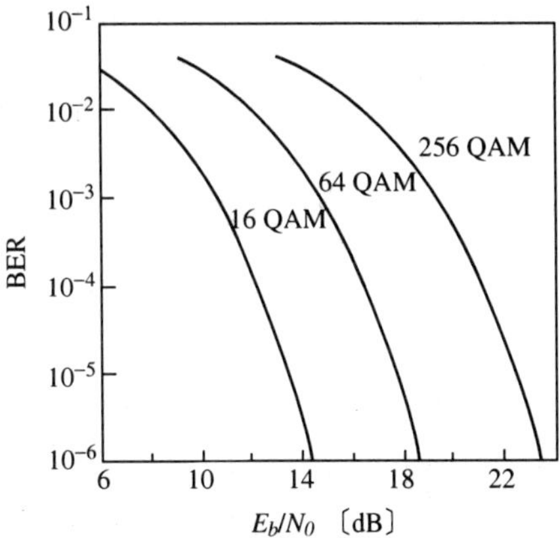

Fig. 3.19   BER Performance in QAM Systems

sideband is removed, and the information is transmitted over only one side of the signal spectrum. The removal of a sideband is accomplished by using a band-pass filter (BPF), but fully eliminating one sideband is difficult, requiring the frequency attenuation "edge" of the filter to be very sharp at the center frequency of the carrier. In VSB, the frequency characteristics at the edges of the filter are less sharp, and as shown in Fig. 3.20(c), a portion (vestige) of the spectrum on one side remains.

In order to obtain spectral efficiency levels equal to orthogonal modulation systems, one sideband must be completely suppressed by filtering. However, in a practical system, part of the carrier frequency still exists at the dividing line

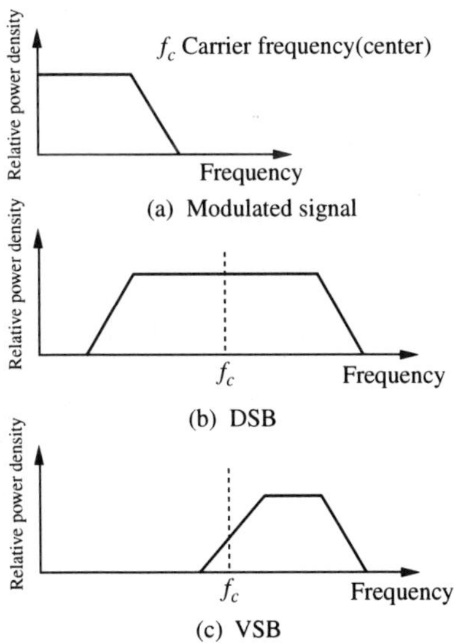

Fig. 3.20   Power Spectrum for DSB and VSB (normalized on the power density at f_c)

between sidebands. To completely eliminate one sideband without increasing ISI would require a filter having a perfectly rectangular edge, and such a filter cannot be realized.

In M-ary VSB, the signal waveform is sampled, and the results are transmitted over the vestigial sideband as a multi-valued pulse signal. To accomplish this, a filter with roll-off characteristics meeting the Nyquist criterion is used, which removes most of one sideband, but its bandlimiting slope is gradual enough near the carrier frequency edge that part of the spectrum remains on the vestigial side. Because M-ary VSB occupies a somewhat larger bandwidth (by virtue of the vestigial side), it is slightly less spectrally efficient than QAM. The advantages of M-ary VSB versus QAM are listed below:

1: In QAM, any presence of phase distortion results in interference between the orthogonal signals, which degrades performance. This is not possible in an M-ary VSB transmission.

2: In QAM, a waveform equalization circuit is required between the pulse sequences in the I- and Q-channels. With M-ary VSB, only one pulse sequence exists, so only one waveform equalization circuit is required.

3.3 Forward Error Correction

Forward error correction (FEC) coding is used in the transmitter end of a digital system, enabling the receiver to correct the errors generated in the transmission channel. FEC is typically implemented using redundancy bits for error detection and correction, and provides the most economical and efficient (with regards to the number of transmitted bits) means of error correction.

In the field of radio communications, forward error correction is one of the techniques unique to digital systems, used following filtering, diversity reception, and equalization methods to fulfill the quality-of-service requirements of digital transmissions. Research continues in the field of error correction and related techniques in an effort to upgrade signal quality. Even so, note that the improvement gained using error correction codes in systems having a large amount of errors can still be less than desired.

FEC is most effective in digital satellite communications, where the primary source of disturbance to the channel is thermal noise. FEC is also very advantageous in orthogonal frequency-division multiplexing (OFDM) applications, as countermeasures to burst errors (concentrated groups of errors caused by multipath fading due to reflections and refracted radio waves) can be effectively exploited.

Forward error correction can be generally divided into block coding and convolutional coding mthods. Block coding is a structurally simple system, and has thus been around a long time. With the discovery of the Viterbi algorithm, convolutional coding has also become relatively easy, and features a high coding gain. FEC finds its most practical use in satellite communications.

3.3.1 Error Detection, Error Correction, and Hamming Distance

Information data with redundancy bits added to it following a specific set of rules is called a code word. Since information bits and code words have a one-to-one correspondence, it is possible to distinguish between code words. Where k represents the number of information bits, and n the number of code word bits, then k/n becomes the coding rate.

To simplify matters, we shall here consider the code as a repeating sequence of transmitted bits. Thus, in Fig. 3.21(a), sending a 0 results in a 00 code, and sending a 1 gives a 11 code. These code words have two differing bits, and the sum of the different bits is what comprises the Hamming distance. The horizontal axis in Fig. 3.21 represents the Hamming distance (amount of separation), and in this case the Hamming distance is two. Consequently, the point a, differing from code A by one bit (e.g., changing 00 to 01) is the same as the point which differs from code B by one bit. In this case, an error can be detected, but whether the error stems from A or from B cannot be determined.

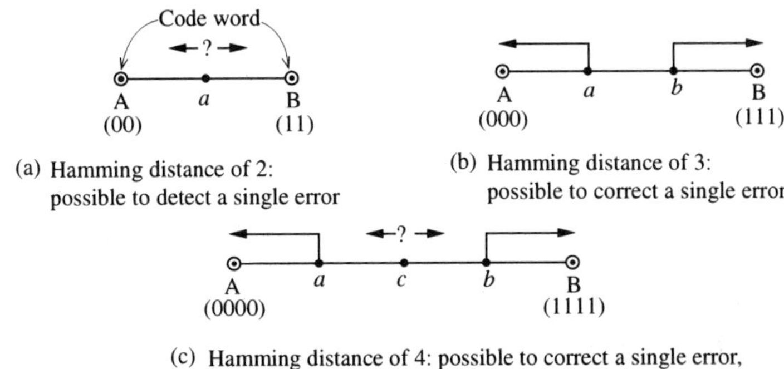

(a) Hamming distance of 2:
possible to detect a single error

(b) Hamming distance of 3:
possible to correct a single error

(c) Hamming distance of 4: possible to correct a single error,
and detect a double error

Fig. 3.21   Hamming Distance and the Relationship Between Error Detection and Error Correction

In Fig. 3.21(b), the Hamming distance between the codes 000 and 111 is three. In this case, the point a separated by one bit from A is not the same as the point b separated by one bit from B. Therefore, when a 1-bit error exists, there is a high probability that the point a will be corrected to A, and point b will be corrected to B. Now, consider the case where the error consistis of two bits. The error point 2-bits distant from A is the same as the error point 1-bit distant from B. Therefore, it is possible to correct the 1-bit error, but many errors of over 2-bits will remain uncorrected.

Moving from the Fig. 3.21(b) case to diagram (c), we can see that a 1-bit error is correctable. Also, for a 2-bit error, if we consider the (a) case, a 2-bit error (double error) is detectable. In other words, $d = 2t + 1$ defines the relationship between Hamming distance d and the number of correctable bits t.

From the above discussion, we can conclude that the greater the Hamming distance, the greater the capability of the code to correct and detect errors. At the same time, greater Hamming distances require a greater number of redundancy bits in the code, and the most efficient means for extending Hamming distance must be considered.

One method of evaluating the performance of a channel in which error correction codes are used is through its coding gain. Figure 3.22 compares the bit-error-rate (BER) versus E_b/N_0 performances of a system not using error correction and a system using a rate 1/2 correcting code. In this case, to make the comparison at the same transmission rate requires a doubling of the power, and a conversion procedure must be used to account for that. As the graph in Fig. 3.22 illustrates, characteristics are good where BER is low. The performance differences in the two systems are illustrated by the differences in E_b/N_0 (or SNR) for a given BER value, and this difference is referred to as the coding gain. The point of comparison in the graph shows that for a BER of 10^{-4}, the coding gain is 1.4 dB.

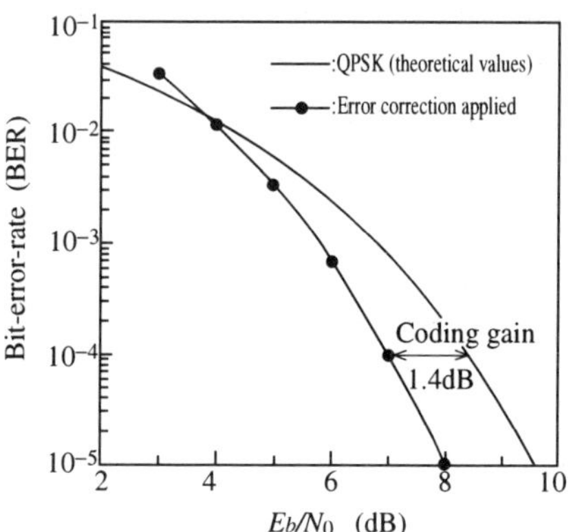

Fig. 3.22   Coding Gain Illustrated

3.3.2 Block Codes

A block code consists of an information code of k bits, along with $n - k$ redundancy bits used for error corrections ($n > k$, where n is the number of code word bits). The block code concept is illustrated in Fig. 3.23. This type of code is referred to as a (n, k) code. The minimum Hamming distance, or minimum distance is the minimum value of the Hamming distance (defined above) between two different code words. Including minimum distance d as a parameter, the code then becomes a (n, k, d) code. Cyclic redundancy checks (CRC) using cyclic codes, Hamming codes, and Bose-Chaudhuri-Hocquengham (BCH) codes are typical block codes used in random error corrections, and the Reed-Solomon (RS) code is a block code used for symbol error correction.

Because the Hamming code is simple from the hardware implementation standpoint, it was developed early, and is widely used. However, it is capable of only correcting one bit error. Also, the RS code, which is a special case of the BCH code, is capable of correcting errors in multiple symbols (which consist of several bits).

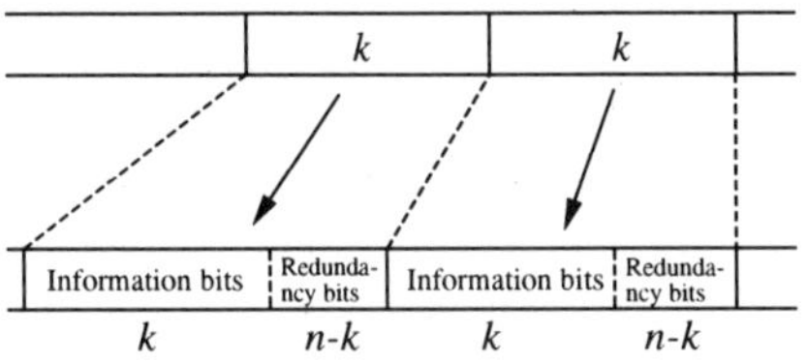

Fig. 3.23   Block Coding Concept

General Discussion

Before proceeding with the description of block codes, we shall first discuss some of the basic principles on which block codes are based.

(a) Cyclic codes

A cyclic code is a cyclic shift (or end-around shift) code in which the bits are shifted to the right or left as the code word is processed. Since the code vectors can be expressed in polynomial form, it is widely used. For example, for a word in the cyclic code that originally reads $w = (w_{n-1}, w_{n-2}, \cdots, w_1, w_0)$, following a left-shift, now reads $w' = (w_{n-2}, \cdots, w_1, w_0, w_{n-1})$. Coding of cyclic codes is easily accomplished using shift registers.

(b) Expressing codes using polynomials

Polynomials are a convenient means of expressing code vectors. For example, consider the 6-bit code 101011. The left-most digit can be expressed as x to the 5th power, the next digit as the 4th power of x, and so on as x^n. Thus, the polynomial for the above code becomes $1 \times x^5 + 0 \times x^4 + 1 \times x^3 + 0 \times x^2 + 1 \times x^1 + 1 \times x^0 = x^5 + x^3 + x^1 + 1$. Here, the degree of x represents the bit position in the code, and the coefficient of x indicates whether the bit is a 1 or a 0. The general expression for a code of n-bits is a $(n - 1)$th degree equation.

$$a_{n-1} x^{n-1} + a_{n-2} x^{n-2} + \cdots + a_2 x^2 + a_1 x^1 + a_0 x^0, \tag{3.15}$$

where: $a_{n-1}, a_{n-2}, \cdots, a_2, a_1, a_0$ represent the value of 0 or 1.

(c) Galois field structuring

Block codes are developed using mathematical expressions, and requires the ability to structure Galois fields. The description here is very brief. For more details, the reader is invited to refer to the [Imai, 1990] and [Lin, 1983] references.

The finite field constructed using the four fundamental rules of arithmetic on a p (prime) set of elements is referred to as a Galois field, expressed by $GF(p)$. $GF(p)$ consists of a commutative ring structured using modulo p arithmetic on the elements 0, 1, 2, ... , $p - 1$. For example, the Galois field $GF(2)$ contains a set of two elements, 0 and 1, and the computations are as shown in Table 3.2. Note that the addition and subtraction operations are the same.

In the case discussed above, a Galois field can only be constructed using a prime p. However, we would like to formulate polynomials having $GF(q)$ elements as coefficients, and from that set of polynomials, construct a Galois field from all polynomials having degrees smaller than a certain value. The degree m field extension of $GF(2)$ is expressed as $GF(2^m)$, and arithmetic operations are performed on the element set of 0, 1, a, a^2, ..., a^{q-2} (where $q = 2^m$). An extended field $GF(2^m)$ containing $GF(2)$ is termed "the extension field of $GF(2)$," and operations in such cases are termed, "over $GF(2^m)$." For example, the polynomial over $\{0, 1\}$ is $\{0, 1, x, x + 1, x^2, x^2 + 1, \}$. From this set, a Galois field can be constructed from the set of polynomials having degrees smaller than $m - 1$.

Table 3.2 $GF(2)$ Operations Table

(a) Addition   (b) Multiplication

+	0	1
0	0	1
1	1	0

·	0	1
0	0	0
1	0	1

The roots of a degree n primitive polynomial are the primitive elements of $GF(2^m)$ (the first element that makes a^{q-1} a one when root a is multiplied repeatedly). Here, the primitive polynomial $G(x)$ is an irreducible polynomial in which the degree n term cannot be factored, and is moreover a special type of polynomial where e is the smallest integer for which $x^e + 1 = G(x) M(x)$. In this case, e also satisfies the relationship $e = 2^m - 1$, and $M(x)$ is any polynomial. For example, the degree 3 irreducible polynomial $G(x) = x^3 + x + 1$,where $e = 7$, can be factored as $x^7 + 1 = (x^3 + x + 1) (x^3 + x^2 + 1) (x + 1)$which satisfies the relationship $7 = 2^3 - 1$, and makes $G(x)$ a primitive polynomial. Assuming one root of $G(x)$ is a, finding $x^7 + 1 = 0$ satisfies the relationship $a^7 = 1$.

Since in the computation of the Galois field we have $\alpha^{p^{m-1}} = 1$, it is more convenient to perform the multiplication/division operations using power expressions for $\alpha^i \cdot \alpha^j = \alpha^{(i + j)\mathrm{mod}(p^m - 1)}$ and $\alpha^i / \alpha^j = \alpha^{(i - j)\mathrm{mod}(p^m - 1)}$, and to use polynomial base expressions (expanded with α^{m-1}, $\alpha^{m-2}, \cdots \alpha$,1) or vector expressions to perform the addition and subtraction operations. Dividing x^3 by $x^3 + x + 1$ yields a remainder of $x + 1$, so substituting the roots a of $x^3 + x + 1$, the term other than the remainder is 0, thus we have $\alpha^3 = \alpha + 1$. Expressions derived from the polynomial base can be obtained in the same manner, and have the relationships shown in Table 3.3. The relationships found in the table are $\alpha^3 \alpha^5 = \alpha^8 = \alpha$, $\alpha^2 + \alpha^4 = (\alpha^2) + (\alpha^2 + \alpha) = \alpha$, $\alpha^2 + \alpha^2 = 0$, and $-a = 0 - \alpha = \alpha$.

Hamming Codes

The Hamming code is a cyclic code having a block length of $n = 2^m - 1$ that uses the primitive element a of $GF(2^m)$ to develop a minimal polynomial, which is a nonzero polynomial of minimum degree with a root of a. This polynomial is used as the generator polynomial.

Assume that a 7-bit code sequence consists of four information bits and three redundancy bits, n. For $n = 3$, we have the (7, 4) Hamming code, which can correct one error bit in a 7-bit word. To determine the redundancy bits, divide the information polynomial by the generator polynomial to find the remainder, and affix

Table 3.3   Power and Vector Expressions of $GF(2^3)$ Elements ($x^3 + x + 1$)

Power	Polynomial expansion	Vector
0	0	0 0 0
$\alpha^0 = 1$	1	0 0 1
α^1	α	0 1 0
α^2	α^2	1 0 0
α^3	$\alpha + 1$	0 1 1
α^4	$\alpha^2 + \alpha$	1 1 0
α^5	$\alpha^2 + \alpha + 1$	1 1 1
α^6	$\alpha^2 \quad + 1$	1 0 1

the remainder as the last three bits of the code word. In this way, the code can always be divided by the generator polynomial. When there is a 1-bit error, and division operations of the generator polynomial $G(x)$ produce the three remainder bits (000 indicates no error), the 7-bit pattern containing the 3-bit remainder becomes the primitive polynomial with a one-to-one correspondence to the error position.

Selecting $G(x) = x^3 + x + 1$ as the generator polynomial, a 3-bit left-shift of the information sequence 1001 produces 1001000 (polynomial $x^6 + x^3$). Dividing by the generator polynomial gives a remainder of 110 (polynomial $x^2 + x$). Subtracting the remainder (same as mod 2 addition on each bit) gives a transmitted code sequence of 1001110 (polynomial $x^6 + x^3 + x^2 + x$). At the receiver end, the transmitted sequence is divided by the generator polynomial. If the remainder is zero, there is no error. If a remainder results, one error bit in the remainder pattern can be corrected. Also, as will be discussed in BCH codes (next section), corrections can be made through the computations of syndromes.

The minimum Hamming distance between the transmitted code sequences is three, but since only one error bit can be corrected, if there are two or more error bits, the process will be in fault.

Bose-Chaudhuri-Hocquenghem (BCH) Codes
In the Hamming code, only one primitive element can be used. In general form, we write

$$a, a^3, \cdots, a^{2t-1}. \tag{3.16}$$

The Bose-Chaudhuri-Hocquenghem (BCH) code can be considered as a cyclic code applicable to a word length n, and at the same time, a minimum degree generator polynomial with t elements as roots. Given $t = 1$, the BCH code is a Hamming code.

Here, we shall consider the decoding process for a code of length n. The received signal polynomial arrives at the receiver as $Y(x) = y_{n-1}\, x^{n-1} + \cdots y_1 x + y_0$, and the decoder produces a *syndrome* of

$$S_i = Y(\alpha^i) \quad (i = 1, 2, 3, \cdots, 2t). \tag{3.17}$$

As a result of the relationship $Y(x^2) = \{Y(x)\}^2$ we have $S_{2i} = (S_i)^2$ which permits the even-numbered syndromes to be computed from the other syndromes.

Assume that l errors occur in the $j_1, j_2, \cdots, j_l$ positions. In this case, $l \leq t$ and $n-1 \geq j_1 > j_2 > \cdots > j_l \geq 0$. Thus, we have

$$S_i = a^{ij_1} + a^{ij_2} + \cdots + a^{ij_l}. \tag{3.18}$$

Decoding operations use this syndrome to locate error positions $j_1, j_2, \cdots, j_l$. However, since a direct solution is difficult, the following degree l polynomial is first solved:

$$\sigma(z) = (1 - \alpha^{j_1}z)\,(1 - \alpha^{j_2}z) \cdots (1 - \alpha^{j_l}z). \tag{3.19}$$

Equation (3.19) is referred to as the error locator polynomial. The roots of $\sigma(z)$ are $a^{-j_1}, a^{-j_2}, \cdots, a^{-j_l}$, and these are used to locate the error positions $j_1, j_2, \cdots, j_l$. Where the operations fail to derive the error locator polynomial, or find its roots, the error is uncorrectable.

Here, we will use as an example a BCH (15, 7, 5) code, with a minimum Hamming distance of 5, and a capability of correcting two bit errors. We use the product of the two polynomials, $G_1(x) = x^4 + x + 1$ and $G_2(x) = x^4 + x^3 + x^2 + x + 1$, which yields the generator polynomial $G(x) = G_1(x)\,G_2(x) = x^8 + x^7 + x^6 + x^4 + 1$. When two errors exist, the odd-number syndromes of S_1 and S_3 become $S_1 = a^{j_1} + a^{j_2}$ and $S_3 = a^{3j_1} + a^{3j_2}$ respectively. From eq. (3.19), the error locator polynomial in this case is

$$\sigma(z) = (1 - \alpha^{j_1}z)\,(1 - \alpha^{j_2}z) = 1 + (\alpha^{j_1} + \alpha^{j_2})z + \alpha^{j_1}\alpha^{j_2}z^2. \tag{3.20}$$

To represent the polynomial coefficients are $(\alpha^{j_1} + \alpha^{j_2})$ and $a^{j_1}a^{j_2}$, with S_1 and S_3, the cube of S_1 is taked. Then, we have

$$S_1^3 = (\alpha^{j_1} + \alpha^{j_2})^3 = \alpha^{3j_1} + \alpha^{3j_2} + \alpha^{j_1}\alpha^{j_2}(\alpha^{j_1} + \alpha^{j_2})$$
$$= S_3 + \alpha^{j_1}\alpha^{j_2}S_1. \tag{3.21}$$

We then have the relationship $\alpha^{j_1}\alpha^{j_2} = (S_1^3 + S_3)\,S_1^{-1}$ hence, $\sigma(z)$ is expressed as

$$\sigma(z) = 1 + S_1 z + (S_1^3 + S_3)S_1^{-1}z^2. \tag{3.22}$$

We see that $\sigma(z)$ from eq. (3.22) is computed from the syndromes S_1 and S_3, and its roots are located. If only one error is present, the coefficient of the z^2 term in eq. (3.20) becomes zero, and the same equation can be used as the error locator polynomial.

Let us assume that the code word $y = (000000101000001)$ has arrived at the receiver end. Where the polynomial is expressed as $Y(x) = x^8 + x^6 + 1$, the syndromes are

$$S_1 = Y(\alpha) = \alpha^8 + \alpha^6 + 1 = \alpha^3 \tag{3.23}$$
$$S_3 = Y(\alpha^3) = \alpha^{24} + \alpha^{18} + 1 = \alpha^4. \tag{3.24}$$

Here, the expression is from the polynomial base of $GF(2^4)$. Since a is the root of $x^4 + x + 1$, we have the relationships shown in Table 3.4.

Solving for $\sigma(z)$, we have

$$\sigma(z) = 1 + \alpha^3 z + (\alpha^9 + \alpha^4)\alpha^{-3}z^2$$
$$= 1 + \alpha^3 z + \alpha^{11}z^2. \tag{3.25}$$

When $\sigma(z)$ is tested for zero by successively substituting $z = 1, \alpha, \alpha^2, \cdots$, we find the roots are $\alpha^8 = \alpha^{-7}$ and $\alpha^{11} = \alpha^{-4}$ Thus, when the root is α^{-j}, we can predict that the errors (j) are at positions $j_1 = 7$, and $j_2 = 4$. Correcting these, the transmitted word is predicted as $\hat{x} = (000000111010001)$.

When t exceeds four or five bits, expressing $\sigma(z)$ using syndromes is not a simple matter. Although procedures do exist for computing error locator polynomials from syndromes that are capable of handling more errors, we shall omit their details here. When the number of error bits is small, the syndrome pattern and related error positions can be preprogrammed into a ROM table, which makes the error correction procedure extremely fast.

Reed-Solomon Codes

The Reed-Solomon code (RS) is a code that is capable of correcting errors in units of symbols (also called bytes). Because of its ability to correct several error bits

Table 3.4   Power and Vector Expressions of $GF(2^4)$ Elements

Power	Polynomial expansion	Vector
0	0	0 0 0 0
$\alpha^0 = 1$	1	0 0 0 1
α^1	α	0 0 1 0
α^2	α^2	0 1 0 0
α^3	α^3	1 0 0 0
α^4	$\alpha + 1$	0 0 1 1
α^5	$\alpha^2 + \alpha$	0 1 1 0
α^6	$\alpha^3 + \alpha^2$	1 1 0 0
α^7	$\alpha^3 + \alpha + 1$	1 0 1 1
α^8	$\alpha^2 + 1$	0 1 0 1
α^9	$\alpha^3 + \alpha$	1 0 1 0
α^{10}	$\alpha^2 + \alpha + 1$	0 1 1 1
α^{11}	$\alpha^3 + \alpha^2 + \alpha$	1 1 1 0
α^{12}	$\alpha^3 + \alpha^2 + \alpha + 1$	1 1 1 1
α^{13}	$\alpha^3 + \alpha^2 + 1$	1 1 0 1
α^{14}	$\alpha^3 + 1$	1 0 0 1

simultaneously, the RS code is very useful in correcting burst errors. The RS encoder codes a symbol from elements consisting of m bits (a binary sequence of 2^m variations of m bits). The roots and generator polynomial over $GF(2^m)$ are encoded as for the BCH code. In fact, we can consider the RS code as a special type of BCH code.

For correcting up to t symbol errors, the RS code generator polynomial $G(x)$ is given by

$$G(x) = (x - 1)(x - \alpha) \cdots (x - \alpha^{2t-1}).\tag{3.26}$$

The code length for the RS code is $n = 2^m - 1$, and the number of information symbols is $k = n - 2t$.

The RS code is capable of correcting errors in multiple symbols, but to simplify matters, we shall now consider a generator polynomial, $G(x) = (x - 1)(x - \alpha)$, capable of correcting only one symbol. In this case, three bits make up one symbol, and the two redundancy symbols, C_1 and C_2 are added to the three information symbols, A_1, A_2, and A_3. The code polynomial for correcting one symbol over $GF(2^3)$ is given as

$$Y(x) = A_1 x^4 + A_2 x^3 + A_3 x^2 + C_1 x + C_2.\tag{3.27}$$

Assuming that a is one of the roots of the primitive polynomial $G(x) = x^3 + x + 1$, C_1 and C_2 are selected so as to satisfy the two following equations by substituting the roots of the generator polynomial and setting to zero:

$$Y(1) = A_1 + A_2 + A_3 + C_1 + C_2 = 0\tag{3.28}$$

$$Y(a) = \alpha^4 A_1 + \alpha^3 A_2 + \alpha^2 A_3 + \alpha C_1 + C_2 = 0.\tag{3.29}$$

Using eqs. (3.28), (3.29), and Table 3.3, C_1 and C_2 are given by

$$C_1 = \alpha^2 A_1 + \alpha^5 A_2 + \alpha^3 A_3\tag{3.30}$$

$$C_2 = \alpha^6 A_1 + \alpha^4 A_2 + \alpha A_3.\tag{3.31}$$

For example, $A_1 = (100)$, $A_2 = (010)$, and $A_3 = (001)$ corresponds respectively to $A_1 = \alpha^2$, $A_2 = \alpha$, and $A_3 = 1$, which results in $C_1 = 0$ and $C_2 = \alpha^5$. Now, defining syndromes S_0 and S_1 as

$$S_0 = A_1 + A_2 + A_3 + C_1 + C_2 \text{ and}\tag{3.32}$$

$$S_1 = \alpha^4 A_1 + \alpha^3 A_2 + \alpha^2 A_3 + \alpha C_1 + C_2,\tag{3.33}$$

we find $S_0 = S_1 = 0$ when there is no error. When one symbol error exists, from the relationship $\alpha^{5-i} = S_1/S_0$ we can predict the symbol errors in A_1, A_2, A_3, C_1, C_2 that correspond to $i = 1, 2, 3, 4, 5$. By adding S_0 to the error symbol, then one symbol error is corrected.

As an example, assume that A_3 is in error, and is received as (100). Syndromes S_0 and S_1 become $S_0 = \alpha^6$ and $S_1 = \alpha = \alpha^8$, which sets up the relationship. By this, we know that A_3 is in error, and the correct A_3 should be $(100) + S_0 = (001)$. Thus, when all error bits are concentrated in one symbol, the RS code allows all bits to be corrected simultaneously. We therefore see that the BCH decoding procedure for correcting a symbol error value (the value corresponding to the erroneous bit pattern) has been added to the RS decoding procedure. Furthermore, adding more information symbols (i.e., A_4, A_5) also forms an RS(7, 5) code over $GF(2^3)$.

Figure 3.24 illustrates a computer simulation of the graphs of BCH(15, 7), BCH(31, 16), and RS(15, 7) codes over $GF(2^4)$ in a Gaussian noise channel using

BPSK signaling. Note that the characteristics of the RS(15, 7) code are still good, even though error corrections are in symbol units. This is due to the fact that the RS(15, 7) code consists of four bits per symbol, and operations on 60-bit code lengths are very efficient. The BCH(31, 16) code is more efficient in error correction than the BCH(15, 7) code, but because it is capable of correcting more error bits, the code is more complex. In selecting the desired correction code, one must consider the source of generated errors, coding efficiency factors, and the complexity of the code.

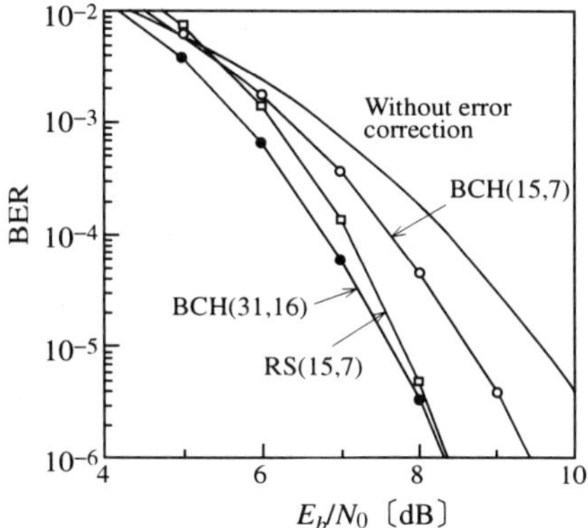

Fig. 3.24   Block Code Performance Graphs

Shortened Codes and Extended Codes
In an actual system implementation, the ratio between the total system bit transfer and information bit transfer is about 2: 1, for which a (7, 3) code is unsuited. The code, however, can be expanded to an (8, 4) code. Thus, where the desired coding rate cannot be obtained in the design of the system, the coding rate can be adjusted by employing a shortened or an expanded code.

Encoding using a shortened code consists of filling a part of the code allocated for information bits with zeros, with the parts of the code other than the zero strings forming the shortened code. The decoding procedure is the same as for normal decoding, except for the 0 bits. Note that the shortening of the code does not change minimum Hamming distance, which results in a lower coding rate.

An extended code consists of a longer minimum Hamming distance $(n + 1, k)$ code with a (n, k) code added as a check symbol. Since there is no general procedure for constructing an extended code, the number of symbols which can be extended varies with each different code.

3.3.3 Difference Set Cyclic Codes
The difference set cyclic code is a code capable of error correction using majority logic circuits. The decoding circuit for the difference set cyclic code is somewhat simpler than the BCH decoder. The difference set cyclic decoder can also be

adjusted to increase the number of bit corrections to exceed its specified error correction performance (specified error correction performance means the number of bits correctable without fail), which is an advantage.

One of the difference set cyclic codes, (273, 191), can be shortened by one bit to a (272, 191) code, which is used for character transmission.

Perfect Difference Set

The difference set cyclic code works on the difference set concept, a brief description of which follows. Given a set P, and a difference of the elements D, we have

$$P = \{l_0, l_1, \cdots, q\} \quad (0 \leq l_1 < l_2 \cdots < q \leq q(q + 1)) \tag{3.34}$$

$$D_{i,j} = l_i - l_j \quad (i \neq j; i, j = 1, \cdots, n, q = 1, \cdots n), \tag{3.35}$$

where n is the number of elements. The perfect difference set P must satisfy the three following conditions:

1: In the $D_{i,j}$ set, all positive values must be different.

2: In the $D_{i,j}$ set, all negative values must be different.

3: When $D_{i,j}$ is negative, $q(q + 1) + 1 + D_{i,j}$ is not in agreement with other $D_{i,j}$ values.

For example, when $q = 3$, then the set $\{0, 1, 3\}$, as is shown in Table 3.5, becomes $q(q + 1) + 1 = 13$. This satisfies the above conditions.

Table 3.5   Perfect Difference Set Example

j \ i	0	1	3
0	–	−1	−3
1	1	–	−2
3	3	2	–

Difference Set Cyclic Code

For a binary code, the perfect difference set is constructed using the relationship $q = 2s$ (s: all positive integers). Using the set elements as powers in the terms of the polynomial $Z(x)$, we have

$$Z(x) = 1 + X^{l_1} + X^{l_2} + \cdots + X^{l_q}, \tag{3.36}$$

and the generator polynomial, $g(x)$, for the difference set cyclic code is given by

$$h(x) = \text{GCD}\,\{Z(x), X^n - 1\} = 1 + h_1X + h_1X^2 + ... + h_{k-1}X^{k-1} + X^k, \tag{3.37}$$

which yields

$$g(x) = \frac{X^n - 1}{h(x)}. \tag{3.38}$$

Here, GCD is the greatest common divisor polynomial. In this case, for a code length of $n = 2^{2s} + 2^s + 1$, the code has a minimum signal distance of $d = 2^s + 2$. Each bit in this code can be provided with an orthogonally-oriented parity check sum of q elements.

Let us now consider the inner-product resulting from the received $y = w + e$, and the ith codewords of the code v (consisting of all vectors that in combination with all codewords in code w yield an inner-product of zero) in the relation of dual for the code w. Using the fact that when there is no error, the inner-product of v and w is zero, the parity check sum, s_i, is given as

$$s_i = v_{i,\,n-1}\, e_{n-1} + \cdots + v_{i,\,0}\, e_0, \tag{3.39}$$

where e $(e_{n-1}, \cdots, e_1, e_0)$ is the error vector corresponding to each coded bit. (No error, 0; error expressed by a component of 1.) In the orthogonal parity check sum, the kth component, e_k, of the error vector e is included in all s_k, but all other error vector components are only included in one s_k.

Using the above information for error correction, e_k is estimated in eq. (3.40), and for an $e_k = 1$, the position k bit is inverted for correction.

$$e_k = \mathrm{maj}\,\{s_1, s_2, \cdots, s_q, 0\}. \tag{3.40}$$

Here, maj $(\cdot)$ implies operations on the majority of the values included in each component. In other words, when position k is not in error ($e_k = 0$), then where the number of 1s included in $s_1, \cdots, s_q$ are designated w, we have $w \leq t \leq \lfloor J/2 \rfloor$, and eq. (3.40) becomes zero. On the other hand, where $e_k = 1$, e_k is included in all s_i, and if there are no other errors, all s_i become 1. If there are other errors, then a $t - 1$ number of s_i become 0. Since now $w \geq J - t + 1 > \lfloor J/2 \rfloor$, the majority value between $s_1, \cdots, s_J$ and 0 becomes 1, and eq. (3.40) yields a 1. Here, $\lfloor x \rfloor$ represents the largest integer below x, and $\lceil x \rceil$ represents the smallest integer above x. This procedure is referred to as majority logic decoding.

A measure that can be used in the decoding process to improve error correction performance is to set the majority logic decoding threshold level $((q/2) + 1$ in the above discussion) to a level higher than the original setting. After the first decoding process is completed, if there are any syndrome bits other than 0, the threshold value is lowered by one step, and the decoding process is repeated. This process is continued, testing the syndrome bits for all zeroes as necessary, until the original threshold value is reached. Adjusting the decoder thresholds normally results in an improvement in the performance of the error correction function.

More information with regards to implementation and the performance of the (272, 190) code can be found in the [Etoh, 1996] reference.

3.3.4 Convolutional Codes

Convolutional Coding

A convolutional code is coded based on previously input information data. As Fig. 3.25 illustrates, the encoder consists of shift-registers and mod 2 adders, and the code is generated using the input bits and data contained in the shift-registers. The encoder yields a three bit output of $d_{k,1}$, $d_{k,2}$, and $d_{k,3}$ in response to the kth 1-bit input, a_k. Figure 3.26 illustrates the trellis diagram using the encoder shown in Fig. 3.25. The trellis diagram depicts the various states and how encoder output influences the transitions between states. The numbers in the circles represent the numerical values (states) of the shift-registers. The dashed lines show the transitions for an input of 0,

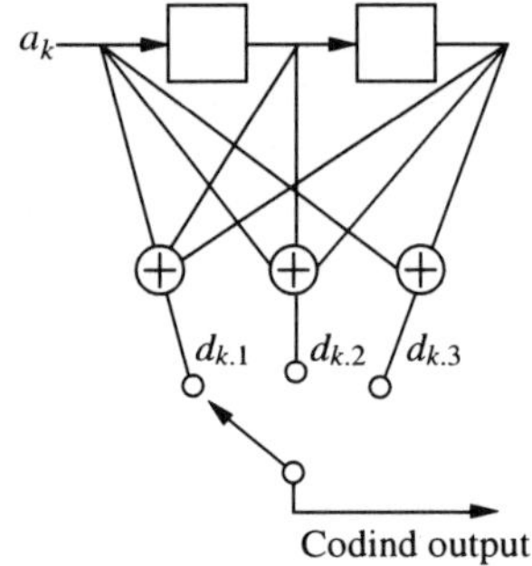

Fig. 3.25   Encoder for the Convolution Code

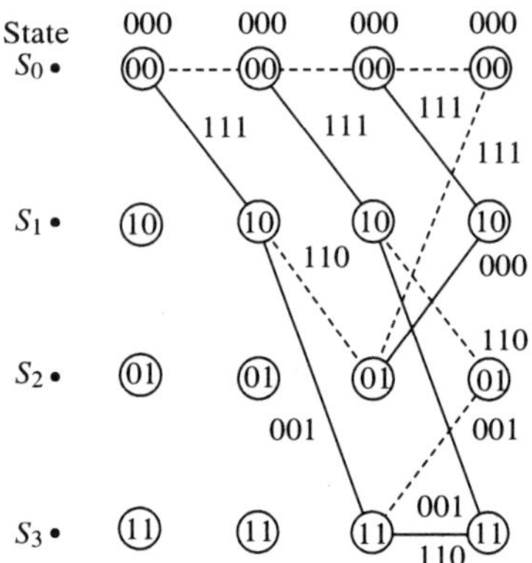

Fig. 3.26   Trellis Diagram

and the solid lines show the transitions for 1. The extent of influence of the previous information is called the constraint length, and in Fig. 3.25, constraint length is three.

Convolutional Code Decoders and Performance

In maximum likelihood (ML) decoding, the currently received data sequence is compared with all other sequences possibly transmitted. To decode a binary sequence of length L, it compares the likelihood of 2^L different code words (out of X received sequences, the conditional probability that Y was transmitted), and selects the one code word with the maximum likelihood of being transmitted.

In ML decoding, the path in the trellis diagram is chosen based on the computation of a cumulative metric. However, as the sequence increases in length, it becomes impossible to search all paths. The process therefore drops a path as soon as it determines that the metric eliminates it as a possibility, and selects the path remaining (the survivor) at the end of the process. The early abandonment of unlikely paths is the only feature that distinguishes maximum likelihood decoding from similar decoding systems that search all paths. Viterbi and sequential decoding

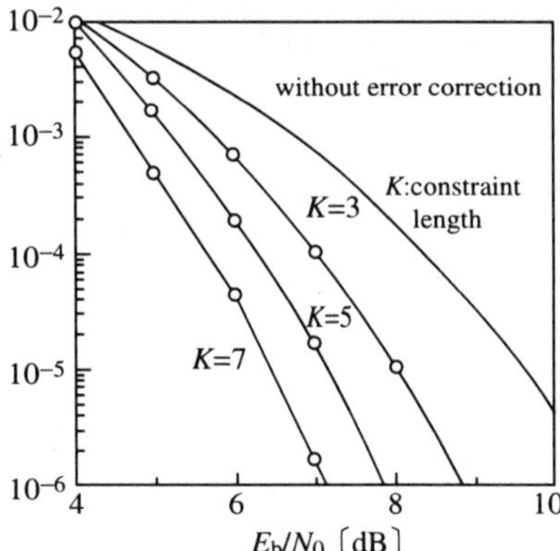

Fig. 3.27   Convolutional coding/Viterbi Decoding Performance

are representative of ML decoding methods.

The encoder connections shown in Fig. 3.25 determine the code structure and the constraint length, which results in a difference in performance for each configuration. Generally, an increase in the constraint length results in a greater symbol spacing distance, which improves bit error rate (BER). Figure 3.27 shows the BER performance by computer simulation in a Gaussian noise channel for a rate 1/2 encoder, and as constraint lengths are varied. Viterbi decoding (described below) was used in this simulation.

Viterbi Decoding

The algorithm proposed by Andrew J. Viterbi is used in the Viterbi decoding process. The algorithm operates on a maximum likelihood rule, periodically shifting the path connections in the trellis diagram until the surviving path is determined. Viterbi decoding is a very efficient process.

Define K as constraint length, and $S_i^{(n)}$ as the states of S_i, where $i = 1, 2, \cdots, 2^{k-1}$, and time n as the branch points. Here, the states are indicated as values stored in the encoder shift registers. For binary code, the two paths entering each state are checked, and the most probable path is selected. The selected path is then stored, while the other path is eliminated. This procedure is repeated for all states for each increment of time. The final determination regarding selection of the surviving path for decoding consists of terminating the trellis with a known information symbol having a constraint length of $K - 1$. After the last symbol has been received and decoded, only one surviving path remains in the trellis. This is used as the decoding path, as it corresponds to the most likely transmitted sequence.

The above operations are carried out for each state, so the complexity of the decoder is in direct proportion to the number of states, and increases exponentially with the constraint length of the code.

Consider the example illustrated in Fig. 3.25, where a convolutional code is transmitted with $K = 3$, and coded at rate 1/3. With the bits added to terminate state

0 and the 4-bit information sequence, we have a sequence of 1, 0, 1, 1, 0, 0. The corresponding code word was transmitted as 111, 110, 000, 001, 001, 111. Here, we shall assume that the 3rd, 8th, 15th, and 18th coded symbols are in error at the receiver. The received sequence is therefore 11$\underline{0}$, 110, 0$\underline{1}$0, 001, 00$\underline{0}$, 11$\underline{0}$, with the underlines representing the positions of the inverted bits (errors).

Referring to Fig. 3.28, the algorithm starts from the three points of origin and computes the eight path metrics, choosing between contenders at the four states until the path with the minimum Hamming distance (the surviving path) is selected and stored. The metrics are computed based on Hamming distances. The other paths (marked by dashed lines) are discarded. The two branches emanating from the survivor at time 3 extend to time 4, with each branch metric computed at each time point. These results are added to the surviving path Hamming distance at time 3, which is then used to determine the surviving path at time 4. At each point where the paths merge, the surviving path is selected as described above. The surviving path (and losing contender) at time 4 is shown in Fig. 3.29. Occasions arise when there are two contender paths having the same cumulative metric in a branch at a given time. In such cases, the next received symbol will again set each contending path to the same value, so one path is arbitrarily selected to eliminate the continued uncertainty. At time 3 and time 4, we see that the contending paths arrive from states $S_3^{(3)}$, $S_1^{(4)}$ and $S_2^{(4)}$, respectively.

Since the last two code bits of the received sequence are 00, the decoder extends only the branch corresponding to 0. As shown in Fig. 3.30, this results in only one surviving path at time 7, and that path represents the decoded sequence. Tracing the path yields a decoded information sequence of 10110.

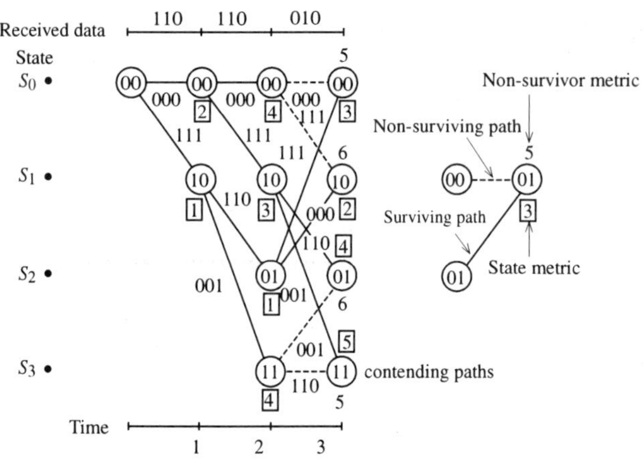

Fig. 3.28   Trellis Paths (1 of 2)

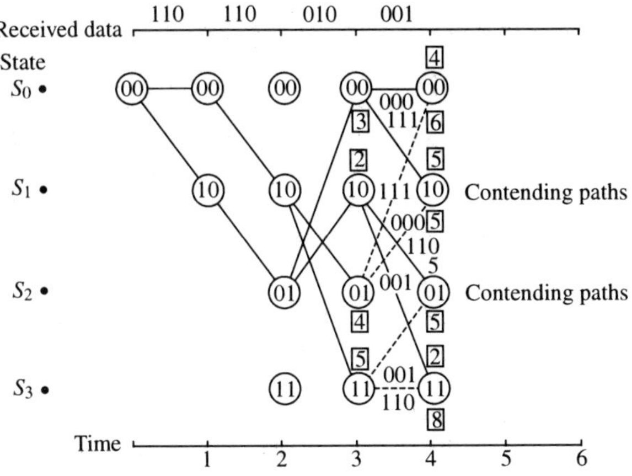

Fig. 3.29   Trellis Paths (2 of 2)

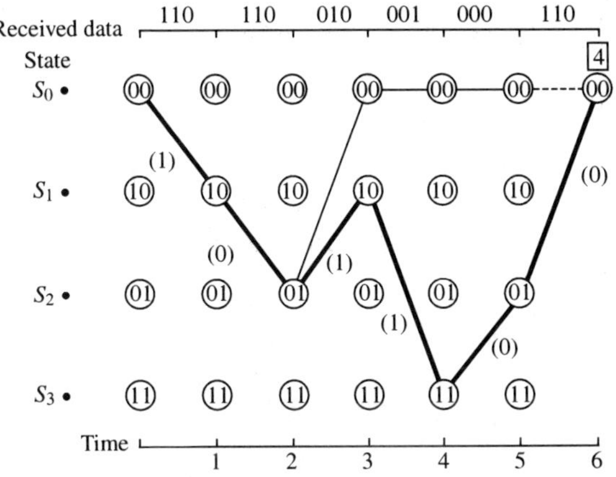

Fig. 3.30   Path Selection Process

Modification of the Algorithm

In the normal Viterbi decoding process, a memory is required to store the path information. Also, since the process must wait until the first symbol is output before decoding all information symbols, which determines the known termination symbol, the decoding process involves a time delay.

Since all surviving paths are not clearly delineated, tracing the time points in the surviving path from a given time point back to the starting point, the path passes through only one state. In other words, from the first branch to that given time point, only one surviving path exists, and that path will not be changed by the subsequent decoding process. Consequently, the decoded bit can be output without waiting for the final time point to be reached. For example, because the paths with $S_2^{(3)}$ and $S_3^{(3)}$ become the non-survivor path in Fig. 3.29, from the first branch to time point 2, the path only goes through states of $S_0^{(0)}$, $S_1^{(1)}$, and $S_2^{(2)}$, at which point the decoding bit can be output. Therefore, rather than the decoder having to store the complete path history, the history need only be recorded up to the common point of convergence at $S_2^{(2)}$ Should the common point of convergence fall after time M, the required memory for the path history of one state is M bits, and the total memory capacity becomes $M2^{K-1}$ bits. Determination of the information bits can be carried out by arbitrarily selecting one of the 2^{K-1} surviving paths, or by selecting the path with the smallest cumulative metric that has the oldest information bits.

In decoding, the oldest information bits are output with each time increment, or alternatively, by output of the oldest information bits from the surviving path with the largest metric. If memory capacity is available for 3- to 5-times the constraint length, any degradation in BER from an ideal decoder can be ignored.

Punctured Coding and Decoding

A punctured coding system is capable of eliminating some of the symbols from the basic convolutional code data, which results in an encoding process featuring a higher coding rate. Consider a convolutional code having a coding rate $R = 1/2$, from which is generated a punctured code with a higher coding rate $R = (n - 1)/n$, $(n = 3, 4, 5, \cdots)$. The information sequence first entered into the convolutional encoder is $x = x_1, x_2, x_3, \cdots$. After encoding (convolutionally) coding at rate R 1/2, the coded sequence becomes $y = y_{1,1}, y_{1,2}, y_{2,1}, y_{2,2}, \cdots$. One block of data consists of $y_{1,1}$, $y_{1,2}, \cdots, y_{n-1,1}, y_{n-1,2}$ in a continuing sequence taken $n - 1$ bits at a time, with symbols occupying specific positions within the block. With punctured coding, these symbols are periodically eliminated by block from their normal positions. Figure 3.31 shows a coding rate 2/3 encoder constructed from a coding rate 1/2 encoder. The bits marked by × s are not transmitted, resulting in an increased coding rate.

Since the punctured code is based on the convolutional code, the same trellis diagram can be used, and the Viterbi algorithm can be used for decoding. In decoding, the symbol positions eliminated during the coding process are filled with dummy symbols to reconstruct the original information. In Viterbi decoding of this translated sequence, the dummy symbols are not involved in the computation of the metric. (For details, refer to the [Cain, 1979] reference.)

Using the punctured code with a compatible encoder/decoder permits the

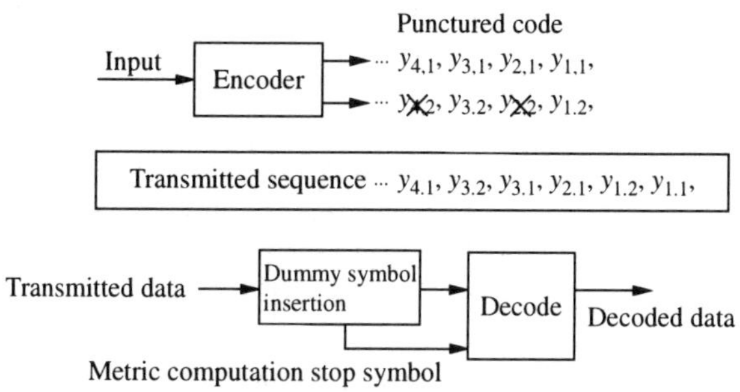

Fig. 3.31   Punctured Code Encoder/Decoder

realization of a variable-rate coding system.

3.3.5 Techniques Related to Error Correction

Interleaving

Burst errors are common in mobile communications applications and often prevent the full potential of a forward error correcting code from being realized. Interleaving is a technique that spreads out a concentrated error sequence so that the random errors are more easily corrected. Although several methods can be used for interleaving, here we shall discuss the method of switching the columns and rows of a matrix.

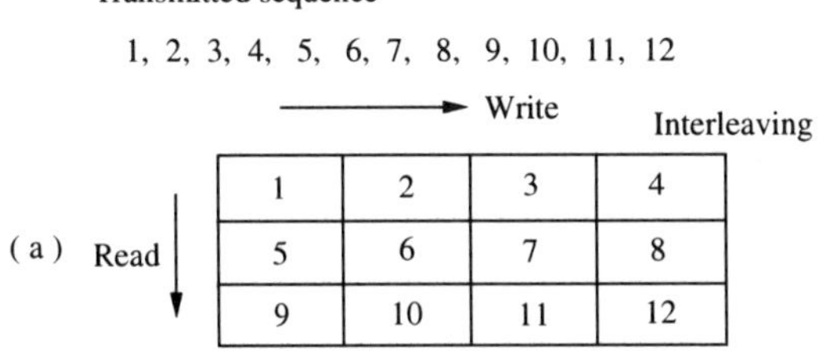

Fig. 3.32   Interleaving Process

As Fig. 3.32(a) illustrates, at the transmitter, the n-row, m-column matrix is written by row and read by column to form the transmission sequence. The reverse operation occurs at the receiver, which recovers the original data sequence. As illustrated in Fig. 3.32(b), this scheme spreads out the burst errors present in the received sequence. The numbers in the diagram represent the data sequence, and those marked with an × are in error.

This method of interleaving results in a time delay at processing, so for applications such as voice communications that require an immediate response time, the improvements gained using this particular method may be unsatisfactory.

Note that this description assumes that the interleaving is carried out in the time domain. However, interleaving can also be effectively implemented in the frequency domain. Coded-OFDM is one such example.

Soft Decision Decoding

The normal decoding process consists of applying the decision rule to modulated data, and operating on the resulting digital values. This is termed hard decision decoding. On the other hand, soft decision decoding carried out in a Gaussian noise channel uses side information (such as the envelope of the received signal) to improve BER with respect to E_b/N_0 by approximately 2 dB over hard decision decoding.

In block codes, methods such as generalized minimum distance (GMD) decoding and Chase decoding are available. These decoding methods append a reliability check to the received symbol, and through repeated erasure decoding, start eliminating the least reliable data until the desired level of correction has been obtained.

Soft decision Viterbi decoding (based on the Viterbi algorithm) uses the quantized demodulator output as the metric, and reliability information primarily consists of the received signal level. This data is used in the interleaving process, but must be deinterleaved in the decoding process.

3.3.6 Combinations of Codes

In systems where a relatively extended delay is permissible in the decoding process, using a combination of coding procedures can result in a robust error correction system, while at the same time keeping hardware requirements relatively simple. In transmission channels where random errors and burst errors are mixed, codes that can handle long burst errors are quite easily designed. However, when one compares a combination code with a code having a long code length, a combination code will not always result in the best performance as regards efficiency, coding rate, and inter-codeword distance.

Concatenated Codes

A concatenated code is a long code formed from several shorter codes. Longer codes generally increase the complexity of the system, but concatenated codes can use combinations of short code decoders, alleviating this. The code is divided into two parts called the inner code C_1, and the outer code C_2.

As illustrated in Fig. 3.33, the encoder first creates the outer code C_2, consisting of a (n_2, k_2) linear code. The symbols encoded as C_2 are then operated on (encoded) by the inner code (the C_1 code consisting of (n_1, k_1) linear symbols) to produce a code word of total length $n_1 \times n_2$. The composite code is a two-element $(n_1 n_2, k_1 k_2)$ code.

Considering that the concatenated code consists of an inner and outer code, the combined use of block codes and convolutional codes yields four different combination codes.

Decoding is normally a 2-step process that entails decoding of the inner code, followed by decoding of the outer code. Using the results of the decoded C_1 as

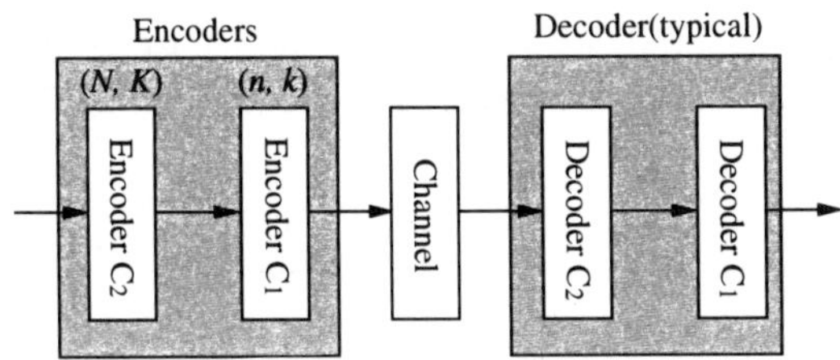

Fig. 3.33   Concatenated Code System

reliability information, generalized minimum distance (GMD) makes it possible to decode with soft decision processes.

Using Viterbi decoding, a wrong path selection is possible due to the errors remaining after decoding, and this can cause a burst error situation. It is therefore common to use combinations with the Reed-Solomon code to cope with this possibility.

Satellite communications applications and CS digital broadcasts typically use a convolutional code as the inner code and an RS code as the outer code. This combination is also being studied for use in terrestrial digital broadcasts.

Turbo Codes

Use of the turbo code achieves a coding performance that approaches the Shannon limit [Berrou, 1993]. The turbo code employs encoded interleaving, and is a parallel concatenated-convolutional code. For more details, refer to the [Ri, 1996] and [Schlegel, 1997] references.

The reliability information produced during decoding is used in repeated decoding operations, and in a novel method of exchanging information between the decoders assists in improving performance. Decoding is carried out by using decoders featuring a soft decision output based on the maximum a posteriori (MAP) algorithm [Bahl, 1974].

The structure of a typical encoder is shown in Fig. 3.34. The information bits are output without modification, while the redundancy bits are transmitted by alternating between encoder 1 and encoder 2. Each encoding operation is the same as the punctured code encoding process. Here, interleaving between encoder 1 and encoder 2 produces independently coded sequences. As illustrated by Fig. 3.35, the results from decoder 1 are output as the soft decision value of the kth information bit, X_k. Decoder 2 accepts this soft decision output and the redundancy bit data, decodes this information, and outputs the results as a soft decision value. Since the code sequences at the two decoders are independent, the noise-induced spread of reliability information is reduced at the decoders. The decoding process is repeated until the desired degree of performance is obtained. The delay shown in the diagram is due to the time adjustment between the next received input for decoding (X_k, Y_k) and the decoder soft decision output (decoding process delay D bit). According to the [Berrou, 1993] reference, under Gaussian noise conditions and at $E_b/N_0 = 0.7$ dB, a BER of 10^{-5} is achievable with this system.

The turbo code has only recently been proposed. Efforts to develop this code continue, but we expect it to prove an easy-to-use, high performing tool in the future.

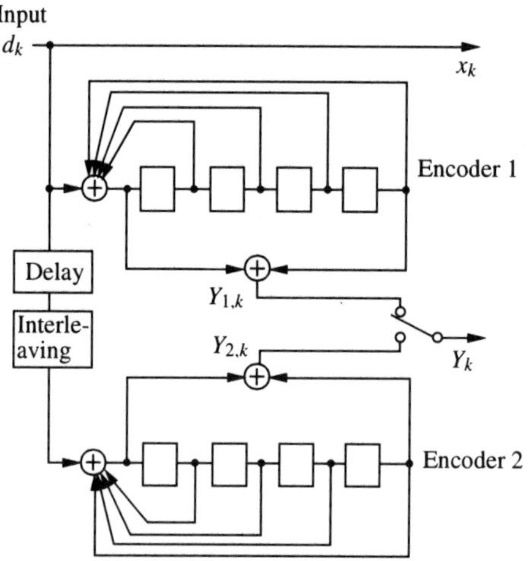

Fig. 3.34   Turbo Code Encoding

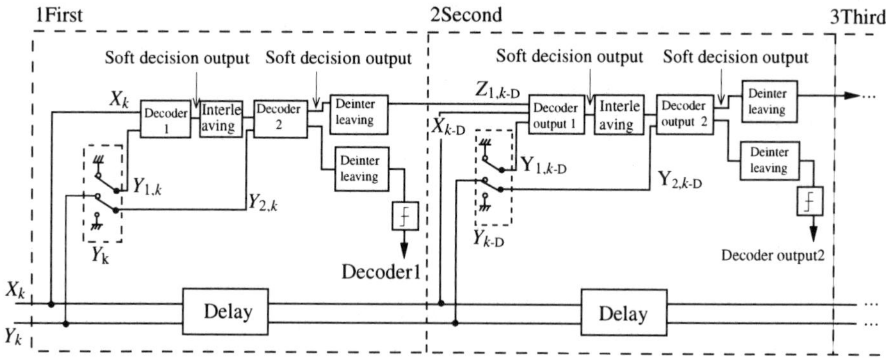

Fig. 3.35   Turbo Code Decoding

3.4 Trellis-Coded Modulation

Coded modulation systems work using the concept of Euclidean distance, which defines the physical distance between signal points. The modulation system and the FEC are contained in a single functional block. Coded modulation can use block codes, convolutional codes (such as trellis-coded modulation), or multi-dimensional coded modulation, using an increased number of orthogonal time bases [Kasahara, 1989], [Schlegel, 1997].

Trellis-coded modulation (TCM) has a combined modulation and error correcting function, which improves bit error performance in the channel. The COMETS project to be discussed in Ch. 5 uses this type of system. It is also being studied for application in the European OFDM system (as Coded-OFDM, COFDM) [Alard, 1987]. In this section, after discussing the relationship between distance and coding errors, and briefly explaining the trellis (state transition) diagram, we shall cover the basic principles of TCM and describe its perforrnance in actual application.

3.4.1 Basic Principles of Trellis-Coded Modulation
Encoding and Distance

Errors generated in the digital transmission process are a result of additive noise affecting the positions ofthe signal points between the transmitted signal and the received signal. As was related in Sec. 3.1.5, a greater signal Euclidean distance results in a system with a greater immunity to noise. During the encoding process, system performance can be improved by considering Euclidean distance, as well as Hamming distance. This is an important concept in coded modulation, as is illustrated in the following equation. Where the distance between signal points is increased from d_a to d_b, the gain G is given by

$$G = 20 \log (d_b/d_a). \tag{3.41}$$

The Trellis Diagram

A convolutional code is used in trellis-coded modulation. A typical coded modulator is shown in Fig. 3.36. Inputs are b_1 and b_2, and output for 8-PSK signaling consists of the values corresponding to 0(000) – 7(111). The trellis diagram (Fig. 3.37) indicates the states of the shift-register values, which is why this system is referred to as "trellis-coded modulation." This method can also be considered a convolutional code with a coding rate 2/3 encoder.

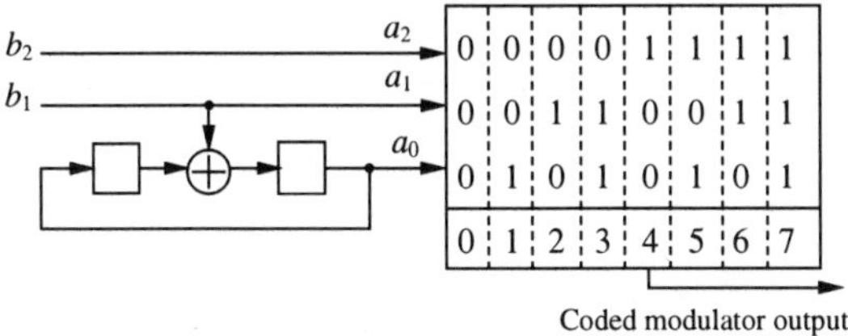

Fig. 3.36   Coded Modulator (typical)

3.4.2 Trellis-Coded Modulation

The description here applies to a trellis-coded 8-PSK (or TC 8-PSK) modulation system. Figure 3.38 shows the signal point constellation for 4-PSK and 8-PSK.

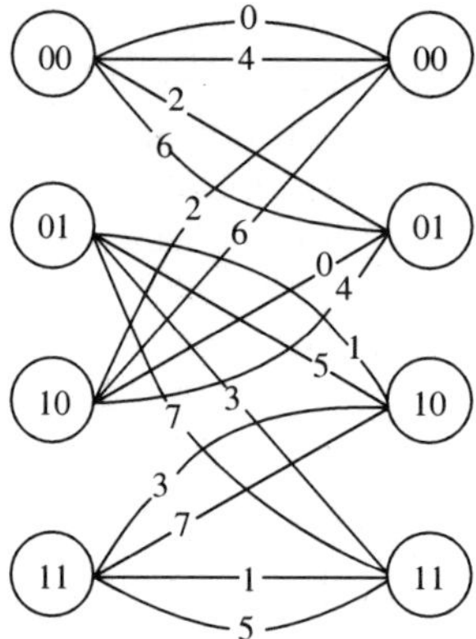

Fig. 3.37   State Transition Diagram (Trellis)

One of the important features of trellis-coded modulation is the division of signal points into several subsets, and to monotonically increase the minimum distance between the signal points included in each subset. This is known as mapping by set partitioning. Mapping by set partitioning of the 8-PSK signal is shown in Fig. 3.39. Of the eight signal points produced by 8-PSK signaling, a subset of four signals is selected from B_0 and B_1, and as accepted by the modulator input, one of the four signal points sets is selected for modulation. One of the advantageous features of trellis-coded modulation is that set partitioning makes designing the code easier.

Referring to Fig. 3.34, note that for 8-PSK, the minimum distance between signal points is $\Delta_0 = 2 \sin(\pi/8) = 0.765$. By further partitioning the signal points into additional sets, the minimum signal Euclidean distance is increased to $\Delta_1 = 1.414$. Another partition increases this value to $\Delta_2 = 2.0$. We also observe that there are uncoded areas in the diagram. These are called parallel transitions, where multiple paths lead to the same state. In parallel transitions, signals are received either from

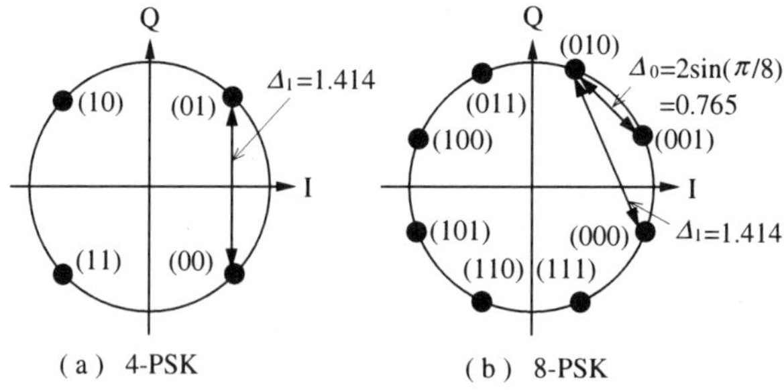

Fig. 3.38   MPSK Signal Spacing Distance

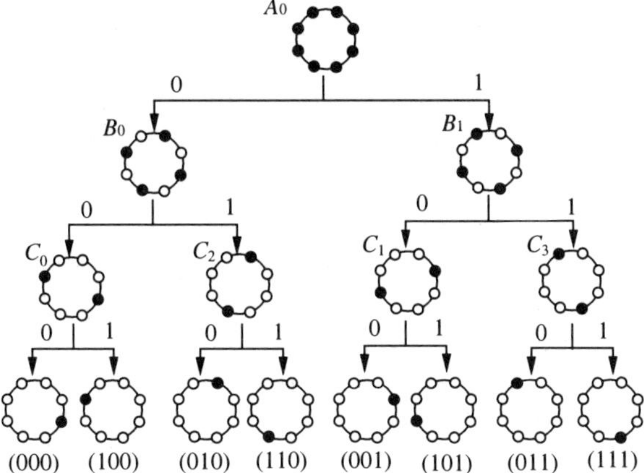

Fig. 3.39　Mapping by Set Partitioning

the subset C_0, or C_1, or C_2, or C_3 (which have a minimum distance between signal points). By partitioning in this manner, the signal points in parallel transition are assured of having a minimum separation distance of $\Delta_2 = 2.0$.

As Fig. 3.40 illustrates, the signal point separation distance at time 1 is Δ_1 (signal points 0(000) and 2(010) in the diagram), Δ_0 at time 2 (signal points 0(000) and 1(001)), and Δ_1 at time 3 (signal points 0 and 2). These signal points are transmitted in a time sequence as an orthogonal phase-modulated signal. Consequently, the distance between the two signal points (the distance between paths) carrying the sequences 0, 0, 0 and 2, 1, 2, since the signals are transmitted independently at separate time points, is the sum of the distances at each time point. Thus, we have $\sqrt{\Delta_1^2 + \Delta_0^2 + \Delta_1^2} > 2.0$, in which case the minimum distance between paths becomes the value for parallel transitions, $\Delta_2 = 2.0$. In the same manner, the four paths originating at state 00 and merging as a pair at time 3 will also be

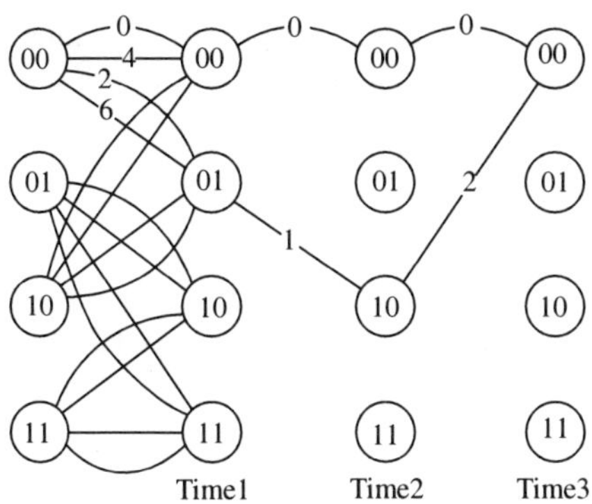

Fig. 3.40　Inter-Signal Distances

separated by $\sqrt{\Delta_1^2 + \Delta_0^2 + \Delta_1^2}$ >2.0. Consequently, $\Delta_2 = 2.0$ will be the minimum distance for these paths, which compared with the $\Delta_1 = 1.414$ minimum distance for an uncoded 4-PSK signal means that a trellis-coded 8-PSK system is 3 dB better in coding gain. Note however, that several path combinations are available.

Table 3.6   Coding Examples [Ungerboeck, 1987]

States	h^2	h^1	h^0	Asymptotic coding gain (dB)
4	–	2	5	3.01
8	04	02	11	3.60
16	16	04	23	4.13

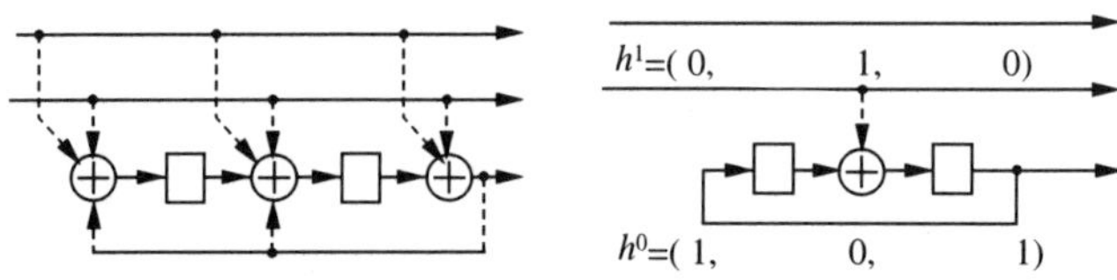

(a) Connection points for a 4-state encoder   (b) Wiring diagram for a 4-stateencoder

Fig. 3.41   Encoder Wiring Diagrams

Table 3.6 provides an example code [Ungerboeck, 1987]. The codes h show the encoder connection points in octal notation. Figure 3.41(a) illustrates the connection points for an encoder with four states. Where $h^0 = (1, 0, 1)$ and $h^1 = (0, 1, 0)$, h^0 represents the lower stage, and h^1 the upper stage. 1 means connected, and 0 means disconnected. Figure 3.41(b) shows the wiring diagram for the encoder. Codes designed to accommodate phase ambiguities are also being investigated [Ungerboeck, 1987].

3.4.3 Trellis-Coded Modulation Performance

The decoding error probability $P(e)$ for trellis-coded modulation in a Gaussian noise channel, and for a high SNR is given by the following equation [Ungerboeck, 1982]:

$$P(e) = N_F Q\left(\frac{D_F}{2\sigma}\right),   \tag{3.42}$$

where D_F is the minimum Euclidean distance between coded sequences, N_F is the number of sequences satisfying D_F, and σ is the standard deviation of the Gaussian noise. $Q(x)$ is the integration of the canonical normal distribution ($\sigma=1$ in eq. 3.6) from x to infinity, given by

$$Q(x) = \frac{1}{\sqrt{2\pi}} \int_x^\infty \exp\left(-\frac{t^2}{2}\right) dt.   \tag{3.43}$$

Figure 3.42 shows a computer simulation of the BER performance versus SNR at the sampling points of 4-PSK using absolute coherent detection, and 8-PSK system using the trellis coding scheme discussed above. The graphs indicate that in the same frequency band, BER performance of the thellis-coded 8-PSK system is superior. Note however, that the 8-ary signaling of 8-PSK makes it more sensitive to phase jitter than is 4-PSK [Ungerboeck, 1987].

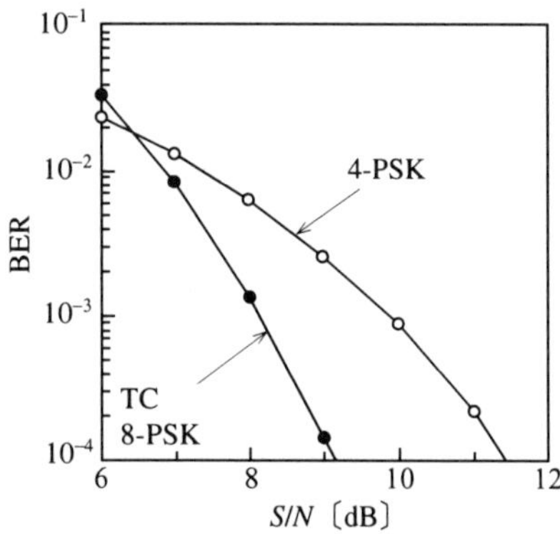

Fig. 3.42 BER Performance of TC 8-PSK

Is the Analog Generation in the Immediate Future?

Marconi's first radio transmission was a message sent using digital techniques (code), but its success had little to do with the use of radio waves. As radio techniques were subsequently developed, the transmission of voice signals ushered analog broadcasting into its Golden Age. Now, the recent improvements in digital techniques have resulted in remarkable achievements in the fields of data communications. The need for transmitting high volumes of information at high data transmission rates have also forced the development of high M-ary systems, and such systems are very rapidly approaching analog systems with regards to signal quality and system performance. Future research and development in error control techniques, coded modulation and demodulation techniques, added to the ability to compensate the degradation resulting from reduced signal spacing should quickly make M-ary systems a practical reality. This all means that analog conception may be needed in order with regards to radio transmission technologies.

3.4.4 Decoding of Trellis-Coded Modulation

We base the description of the decoding procedure of trellis-coded modulation on the system discussed in Sec. 3.4.2.

Step 1: Since there are uncoded regions, the decision rule is applied to determine which part of the code is being input. This process selects the subset ($C_0 \sim C_3$)

nearest the received signal point. The square of the distance is calculated, and the decision value a_2 is stored in the subset.

Step 2: The Euclidean distance is applied as the metric, and the most likely path is selected based on the Viterbi algorithm.

Step 3: Using a_2 from step 1, and the path selected in step 2, the transmitted data is estimated from the trellis diagram in Fig. 3.37.

The above procedure outlines the decoding process, but to cope with the usual fading-induced level variations caused by fading and encountered in mobile applications, interleaving is also typically used. This means that the SNR of each of the symbols will vary. The Viterbi decoding process features a weighting scheme based on channel side information (CSI) to deal with these variations in SNR values [Simon, 1988].

3.5 Conclusion

A general description was given of the digital modulation systems and the FEC techniques used to improve signal quality in digital communication systems. Coded modulation, which combines modulation and error-correcting functions in one system, was also discussed. For readers desiring a more comprehensive treatment of these subjects, the [Saitoh, 1996] and [Sampei, 1997] references are recommended for modulation systems, and the [Lin, 1983], [Schlegel, 1997], [Imai, 1990] and [Etoh, 1996] references are recommended for error-correction coding.

Chapter Questions

Q1: Describe a typical procedure used to transform CNR and E_b/N_0 for a roll-off filter.

Q2: Describe the features of digital modulation systems.

Q3: Draw the signal space diagram for a Gray coded 64-QAM system.

Q4: Consider a Reed-Solomon code over $GF(2^4)$ capable of correcting one error bit. Use eqs. (3.28) and (3.29) to encode three information symbols, and determine C_1 and C_2 for $A_1 = (0001)$, $A_2 = (1001)$, and $A_3 = (1100)$.

Q5: Consider a code generated by the encoder shown in Fig. 3.25 using $d_{k,1}$ and $d_{k,3}$, and a coding rate of 1/2. As described in Sec. 3.3.4, decode (using Viterbi decoding) the encoded output for an input sequence of 1, 0, 1, 1, 0, 0. Check to make sure your answer conforms to the input.

Q6: Design a concatenated code for a coding rate 2/3 encoder.

Q7: Use the 8-PSK system to explain the differences between Hamming distance and Euclidean distance.

Q8: Describe the process of mapping by set partitioning as used with 16-QAM.

Chapter References

[Alard, 1987] M. Alard et al. : "Principles of modulation and channel coding for digital broadcasting for mobile receivers," EBU Review Technical, No. 224, pp. 168-190 (Aug. 1987)

[Bahl, 1974] L. Bahl, J. Cocke, F. Jelinek, and J. Raviv : "Optimal decoding of linear codes for minimizing symbol error rate," IEEE Trans. IT., Vol. 20, pp. 284-287 (March 1974)

[Berrou, 1993] C. Berrou, A. Glavieux and P. Thitimajshima: "NER SHANNON LIMIT ERROR-CORRECTING CODING AND DECODING : TURBO-CODES (1)," in Proc. ICC'93, pp. 1064-1070 (May 1993)

[Cain, 1979] J.B. Cain, G.C. Clark, Jr., and J.M. Geist: "Punctured Convolutional Codes of Rate (n − 1)/n and Simplified Maximum Likelihood Decoding," IEEE Trans. Inf. Theory, IT-25, pp. 97-100 (1979)

[Divsalar, 1988] D. Divsalar and M. D. Simon : "The design of trellis coded MPSK for fading channels : Performance Criteria," IEEE Trans. Commun. Vol. COM-36, pp. 1004-1011 (Sept. 1988)

[Etoh, 1996] Y. Etoh, T. Kaneko, eds.: "Ayamari Teisei Fugou to sono Ouyou", Ohmusha (1996) (in Japanese)

[Imai, 1977] H. Imai and S. Hirakawa : "A New Multilevel Coding Method Using Error-Correcting Codes," IEEE Trans. Inf. Theory, Vol. IT-23, pp.371-377 (May 1977)

[Imai, 1990] H. Imai: "Fugou Riron", Denshi Jouhou Tsuushin Gakkai (1990) (in Japanese)

[Kasahara, 1989] M. Kasahara: "Fugouka Henchou Houshiki I ~ III", Denshi Jouhou Tsuushin Gakkaishi, Vol. 72, No. 1 ~ 3 (1989) (in Japanese)

[Lin, 1983] S. Lin and D.J. Costello: "Error Control Coding: Fundamentals and Applications," Printice-Hall (1983)

[Proakis, 1995] J.G. Proakis: "Digital Communications," 3rd ed., McGraw-Hill (1995)

[Ri, 1996] T. Ri, O.Y. Takeshita, H. Imai: "Ta-bo Fugou ni tsuite", Denshi Jouhou Tsuushin Gakkai, Jouhou Riron Kenkyuukai, IT 96-45 (1996) (in Japanese)

[Saitoh, 1996] Y. Saitoh: "Dhijitaru Musentsuushin no Henfukuchou", Denshi Jouhou Tsuushin Gakkai (1996) (in Japanese)

[Sasaoka, 2000] Y. Sasaoka, editor : "Mobile Communications", Ohmsha (2000)

[Schlegel, 1997] C. Schlegel: "Trellis Coding," IEEE Press (1997)

[Simon, 1988] M. K. Simon and D. Divsalar : "The Performance of Trellis Coded Multilevel DPSK on a Fading Mobile Satellite Channel," IEEE Trans. Vech. Technol., VT-37, 2, pp. 78-91 (May 1988)

[Ungerboeck, 1982] G. Ungerboeck : "Channel coding with multi-level/phase signals," IEEE Trans. Inf. Theory, Vol. IT-28, 1, pp. 55-67 (Jan. 1982)

[Ungerboeck, 1987] G. Ungerboeck : "Trellis-Coded Modulation with Redundant Signal Sets Part II : State of Art," IEEE Commun. Magazine, 25, 2, pp. 12-21 (Feb. 1987)

Chapter 4
Terrestrial Broadcasts

In this chapter we describe the systems used in digital broadcast television in the terrestrial environment. Much of the discussion centers on orthogonal frequency-division multiplexing (OFDM), a recently developed system designed for terrestrial digital television, and heretofore not employed in other communications applications. We shall also discuss the phenomenon of delayed wave interference (producing what are called "ghost images" in analog TV), which is a problem unique to the terrestrial television environment. The subject of mobile receivers, as they relate to terrestrial broadcasting, will also be covered in this chapter.

4.1 Terrestrial Digital Television

4.1.1 Broadcasts using Ground-Wave Signals

Television broadcast service started in Japan in 1953, using radio waves in the VHF band (called surface- or ground-waves) as the transmission medium. Since that initial start, the number of TV stations using the ground-wave medium have gradually increased to over 14,900 as of 1997. As an indication of the importance of terrestrial television service, in the half-century since TV programming began, broadcasts using ground-waves cover every region of the country, and have become an indispensable part of our daily lives. The points listed below have each played a large role in the proliferation of terrestrial television broadcasting.

1) Relatively simple receiver antennas.

   Acceptable reception is often possible using "rabbit ears" or whip antennas (Sec. 2.3.2) in all but a few areas where delayed wave interference is a problem. The development of the Yagi-Uda antenna and low noise signal booster amplifiers permit reception even in areas where signals are weak.

2) Portable/Mobile receivers.

   Reception over portable or mobile receivers is possible in areas where CATV is not available.

3) Television signal repeaters.

   The television signal can be transmitted by ground-waves from the main station to more remote locations through repeaters (called on-air relay). Other transmission media are not required.

4) Wide frequency bands allocated for television signal use.
   The VHF band allocated for terrestrial TV service is 70 MHz, while the UHF band is 300 MHz. (As we shall discuss later, even this wide frequency band allocation is becoming crowded.)
   The items noted above are applicable to the general field of terrestrial broadcast television service. As was noted in Sec. 1.3, the following points are specifically applicable to digital TV in the terrestrial broadcast environment.
5) Using high-efficiency signal coding techniques, a large number of channels can be transmitted over a comparatively narrower frequency bandwidth.
6) Digital modulation is very immune to interference.
7) High-efficiency coding techniques permit the transmission of HDTV programming.
8) Since digital modulation has a strong immunity to noise, transmission power requirements are on the order of ten-times to several hundred-times smaller than the power required to transmit analog signals.
9) Common standards used for digitizing the video, audio, and data signals makes the manipulation of information easier.
10) Signal scrambling makes it possible to direct programming to specific subscribers.
   The above positive points notwithstanding, terrestrial broadcasting still faces some problems, which we shall discuss in the following section.

4.1.2 Problems with Ground-Wave Broadcasts
Delayed Wave Interference
The rapid increase in the number of large buildings in urban areas has been accompanied by a growing problem with interference caused by multiple carrier signals, which is one of the primary factors behind poor picture reception. Multiple carrier signals are referred to as multipath in the general communications field, and cause ghost images in analog TV. However, in digital TV, since this does not result in a double image problem (i.e., a ghost image), we will henceforth use the term delayed waves to indicate that phenomenon in digital television. According to a 1988 survey conducted by BTA (Broadcast Technology Association, currently the Association of Radio Industries and Businesses), 75.4% of all urban households (in Japan) suffered some level of ghost image interference in television reception [BTA, 1988]. In order to advance an effective countermeasure to ghost interference, the scope of the problem must be accurately defined, and solutions positively implemented. The subject of delayed waves as they affect digital television will be further discussed in Secs. 4.3.2 and 4.3.3.

Relaying Broadcast Signals
Currently in Japan and Europe, orthogonal frequency-division multiplexing (OFDM) appears to be the choice in modulation systems for digital television. The use of OFDM permits the configuration of a single frequency network (SFN). In SFN, repeaters for neighboring TV stations can be assigned the same channel

frequencies. In regular analog broadcast TV, different channel frequencies must be assigned in order to prevent interchannel crosstalk (Fig. 4.1), but since digital systems strongly reject interchannel interference, the SFN is a worthwhile concept to consider. Note however, that SFN does have some problems, which we will briefly list and discuss below.

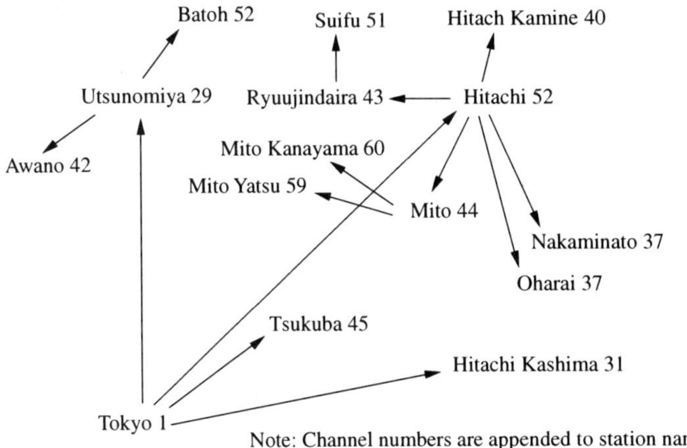

Fig. 4.1   TV Station Repeater Network and Channel Frequencies (by channel number)

1) The existing repeaters cannot be used in SFN. (The degree of signal isolation between the receiving and transmitting antennas is insufficient to prevent the repeater-transmitted signal from turning around and being picked up by the repeater. Also, where signal isolation is insufficient, attempting to amplify the signal without modifying its frequency results in ringing (the system going into a state of oscillation).)

2) Abandoning the repeater concept in favor of alternative transmission media (e.g., cables or microwave circuits) is too expensive to be practical.

3) Delay time compensation is required to reduce delayed wave interference.

4) Where long distances separate the stations, long guard intervals are required. (The guard interval is a measure used in OFDM to reduce delayed wave interference. Further discussion in Sec. 4.2.4.)

As a means of circumventing these problems, the double frequency network (DFN), which uses two signal frequencies alternatively, has been proposed [Tsuzuku, 1995]. The SFN and DFN concepts are illustrated in Fig. 4.2, and the features of the DFN system are listed below.

1) Provides a practical on-air relay function.

2) The existing transmission network remains usable.

3) Less delayed wave interference than in SFN.

4) Requires two frequencies per channel, which makes it less spectrally efficient than SFN, but more efficient than conventional repeater networks.

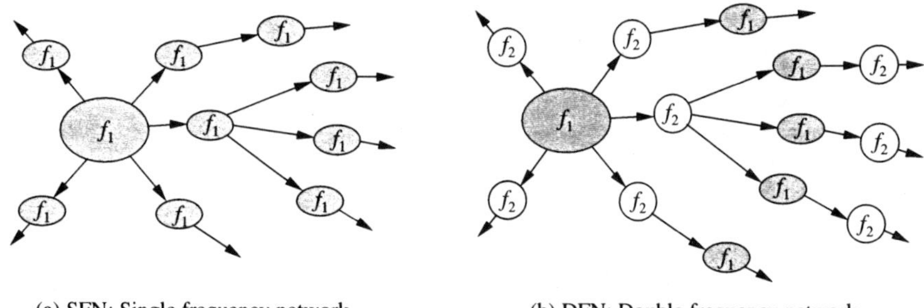

(a) SFN: Single frequency network   (b) DFN: Double frequency network

Fig. 4.2   Television Signal Repeater Networks

The Effects of Channel Nonlinearities

The problems with OFDM as regards intermodulation due to channel nonlinearities has been well-documented [Nagatsuka, 1994]. This problem results from nonlinearities in the channel, which cause the two sets of carrier waves to modulate each other and produce unwanted harmonics in the channel. This point will be elaborated in Sec. 4.2.5.

The Pressures on Broadcast Frequency Allocations

In Japan, there are currently 12 VHF channels and 50 UHF channels (for a total bandwidth of 370 MHz) allocated to television broadcasts using the ground-wave medium. As of March, 1997, about 14,900 stations (including repeater (translator) stations) are broadcasting over that frequency allocation (Fig. 4.3). Consequently, there are numerous stations assigned the same channel frequency that must work together so as to avoid interfering with each other's signal. There are approximately 600 stations in all of Japan sharing the channel 51 frequency, for example.

Given this scenario, introducing digital television as a new service will require strategically assigning channel frequencies (channel numbers) so that the existing analog service is not interfered with. Where OFDM is employed to modulate the terrestrial digital television signal, the spectrum will have flat frequency characteristics, which means that any interference to an analog broadcast can be treated the same as white noise. Thus, the electric-field strength ratio (called the interference protection ratio) which will prevent the digital broadcast from interfering with the analog broadcast must be on the order of 40 dB [ITU, 1996]. The interference protection ratio is illustrated in Fig. 4.4, where distance is given on the horizontal axis and electric-field strength is represented on the vertical axis. Channel frequencies must be assigned so that the interference protection ratio is maintained throughout the complete service area of the analog broadcast. (As illustrated in the diagram, however, the edges of the analog service area are where the problems lie.) What the above explanation boils down to is that if a digital broadcast station is located near an existing analog TV station, the digital station cannot be assigned the same channel frequency as the analog station, as it will interfere with the analog signal transmitted to the subscribers in the original service area.

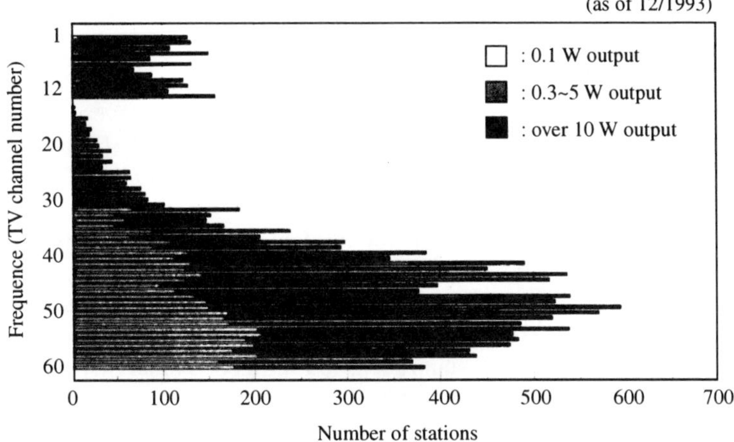

Fig. 4.3   TV Broadcast Stations and Assigned Channels

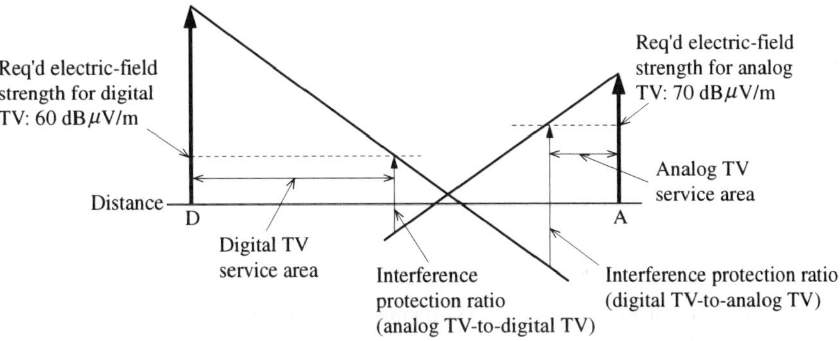

Fig. 4.4   Interference Protection Ratio

4.2 OFDM Modulation System

The digitalization of terrestrial television service is making rapid progress, particularly in Europe and the United States. The modulation system chosen for use in the US is the VSB (Vestigial SideBand) system (Sec. 3.2.2), while the system selected by the European Union and Japan is the multi-carrier OFDM (Orthogonal Frequency-Division Multiplexing) system, which is still under development. Even though the transmission channel of the OFDM system is afflicted with the delayed wave problem, experiments have confirmed that it has good transfer characteristics, and in Europe and Canada, its practicality as a modulation system has been demonstrated in the DAB (Digital Audio Broadcasting) system, which is designed for mobile applications. A more detailed description of the OFDM system is provided in this section.

4.2.1 The History of OFDM

The recent events that have brought OFDM once again to the forefront has been its use in the DAB audio system, as noted above. OFDM does have a long history however, as it was originally proposed in the 1950s [Chang, 1966]. The theoretical viability of the system was generally confirmed by the latter 1960s [Chang, 1968], and a US patent was granted for the OFDM system in 1966 [Cimini, 1985]. In the early 70s, the system was proposed for implementation in a receiver, used in conjunction with a discrete Fourier transform (DFT) [Weinstein, 1971]. By the early 1980s, OFDM was being investigated as an M-ary modulation system in QAM signaling [Hirosaki, 1980], and in 1985, its use in mobile communications applications was documented [Sistanizadeh, 1993]. Reports of its first application in broadcasting were in 1987 [Alard, 1987], when it was being proposed as a coded modulation system designed for robust performance under difficult channel conditions. The system was subsequently called COFDM (Coded-OFDM), which became the common title for DAB [Rault, 1989].

Simulcast Frequencies

Given the present state of frequency band occupancy, it is unreasonable to expect that digital TV should start service with a frequency allocation completely separate from the analog TV frequencies. It has therefore become necessary to modify the frequency band for digital broadcasts, and share it with the existing analog system. One of the primary purposes of introducing digital television has been to efficiently utilize the frequency resources allocated for broadcasting, but since it is impossible to make the transition overnight, for the time being both analog and digital systems must share frequency bands in what has become known as the simulcast period. During this period, digital TV must occupy the spaces existing in the analog frequency band, but those "gaps" are quite scarce. (Currently there are about 15,000 analog TV stations operating in 62 channels.) Settling this dilemma has become one of digital television's greatest challenges.

4.2.2 The Principles of OFDM

The primary problem confronting design engineers in the terrestrial (ground-wave) environment is the interference caused by delayed waves in the channel. The problem is particularly acute in the large cities of Japan, where there are many high-rise structures. When planning a new terrestrial digital broadcast system, devising countermeasures to the effects of delayed waves is thus a very important part of the total task. One of the measures used to reduce the effects of delayed waves is extending the symbol length (Fig. 4.5), which lessens the overall intersymbol interference (ISI) by cutting the degree of overlap between adjacent symbols. (For a theoretical analysis of this problem, refer to Sec. 4.3.3.) The information sequence is also divided into a number of frequency branches (i.e., FDM) for transmission, which cuts down on the problem of frequency selective fading. (With FDM, there is less chance of fading affecting all frequency branches simultaneously.) Also, using the multicarrier scheme means that if the bandwidth of one wave becomes smaller, then fading will affect all frequency components equally (flat fading), which is easier to cope with.

Even though OFDM belongs to the family of frequency-division modulators (FDM), all carriers are synchronous when modulated, so using orthogonal basis functions assures a minimum spacing between carriers. The level of spectrum efficiency of the OFDM system is thus almost equal to the single carrier systems.

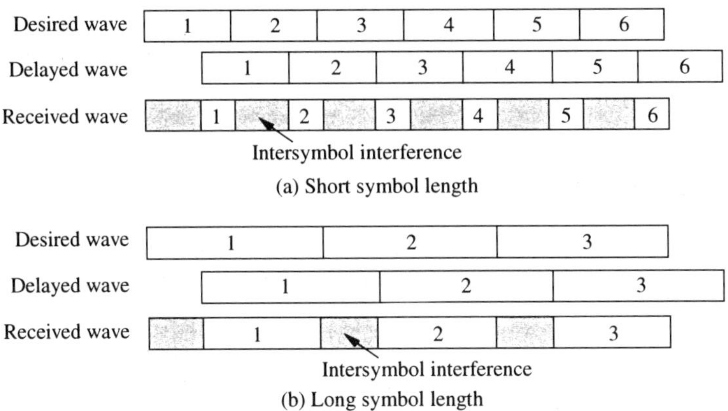

Fig. 4.5   Symbol Length Effect on Delayed Wave Interference

The Basics of OFDM

The orthogonality of the set of functions defining the signal space is basic to the description of OFDM, and we shall start there. The orthogonality of a complex function having a finite energy is defined as

$$\int_{-\infty}^{\infty} f_n(t) \cdot f_m{}^*(t)dt = \begin{cases} 0 & n \neq m \\ \text{finite value}, & n = m, \end{cases} \tag{4.1}$$

where m and n are integers, and $*$ is the complex conjugate operator. Where the function f_n in the above equation is a trigonometric function, orthogonality is expressed by

$$\int_{0}^{1/f} a \cdot \sin(2\pi nft) \cdot b \cdot \sin(2\pi mft)dt = \begin{cases} 0 & n \neq m \\ \dfrac{1}{2f} \cdot ab & n = m. \end{cases} \tag{4.2}$$

In the above equation, when the frequencies of the trigonometric functions differ ($n \neq m$), and moreover, when that difference is an integer multiple of the fundamental wave frequency, the integral yields a zero, independent of the values of a and b. In other words, there is no interference between the respective waves.

Designating symbol length (symbol duration) as T_s, where frequency spacing Δf is taken as $1/T^s$, the above equation indicates that the frequencies of the $n\Delta f$ and $m\Delta f$ sine waves are orthogonally oriented. The transmitted signal is comprised of these two additive waves, and the resultant composite signal waveform, $x(t)$, is expressed by

$$x(t) = \sum_k \left\{ a_k \cos(2\pi f_k t) + b_k \sin(2\pi f_k t) \right\}. \tag{4.3}$$

The situation described above is depicted in Fig. 4.6. In eq. (4.3), a_k represents the in-phase component (real part) of the transmitted symbol, and b_k the orthogonal component (imaginary part). Also, f_k is the frequency of the kth sine wave, and $f_k = k/T_s$. Where the number of sine waves added to form the composite wave is N, and the symbol duration T_s is partitioned into N equal parts (T_s/N is the sampling interval ΔT), the signal amplitude at each sampling point, $t = n\Delta T$, is written

$$x(n\Delta T) = \sum_{k=0}^{N-1} \left\{ d_k e^{j2\pi nk/N} \right\}, \tag{4.4}$$

where $d_k = a_k - jb_k$. An inspection of the above equation reveals it to be an inverse Fourier transform of the complex number d_k, and performing an inverse Fourier transform on the transmitted data indicates that it can be modulated by the OFDM system.

Let us next extract, or demodulate, an arbitrary frequency component from the composite signal. Applying to an integral the transmitted signal $x(t)$ from whose sine wave we wish to extract a frequency, and integrating from 0 to T_s, from the orthogonality of the function demonstrated in eq. (4.2), we have

$$\int_0^{T_s} x(t) \cdot \cos(2\pi f_i t) dt = \frac{1}{2f_i} \cdot a_i, \tag{4.5}$$

which yields the information a_i that was modulated and placed on the carrier at frequency f_i. Where the received signal is $x(n\Delta T)$, the demodulated data is expressed in the following discrete Fourier transform:

$$d_k = \sum_{n=0}^{N-1} \left\{ x(n\Delta T) e^{-j2\pi nk/N} \right\}. \tag{4.6}$$

Note that the information carried by the signal can be analog (where d_k is a continuous value) or digital (where d_k is a discrete value), and the process is compatible with several types of modulation systems. For example, in terrestrial digital broadcasting, M-ary QAM signaling is being investigated for applications with fixed receivers, and the use of QPSK is being considered for use in mobile receivers. (To distinguish normal single-carrier modulation systems from modulation systems using multi-carrier OFDM, the acronyms are generally combined, e.g., QPSK-OFDM.)

The Characteristics of OFDM Signals
We shall start by considering the instantaneous power distribution of the OFDM

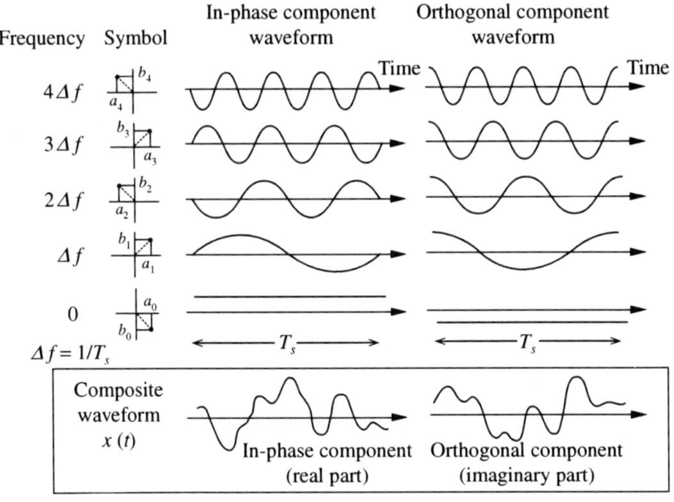

Fig. 4.6   OFDM Signal Waveforms

signal. Assuming that the modulation system and power is equal for all carrier waves, we can consider the in-phase and orthogonal components of the signals to consist of an additive set of sine waves having various amplitudes and phases. Thus, if the number of multiplexed channels is sufficiently large, then by the central-limit theorem, each has its own normal distribution. Since instantaneous power is the square of the sum of these two channels, we have a chi-square distribution with two degrees of freedom (Fig. 4.7). This indicates a very large difference between the average value and the peak value of instantaneous power. As an example, if there are 1,024 carriers, then the theoretical peak value will exceed the average value by approximately 30 dB.

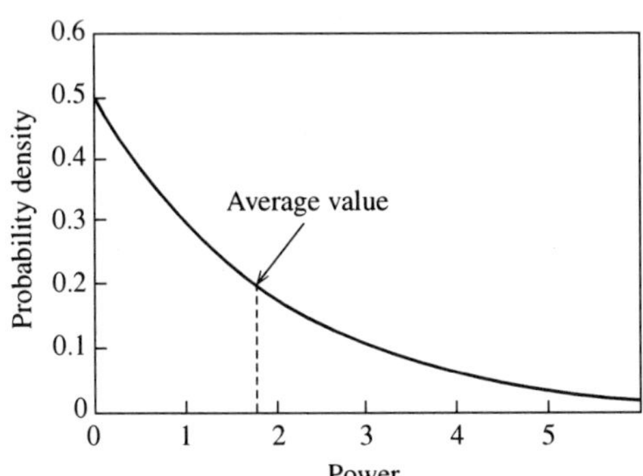

Fig. 4.7 χ^2 Distribution (two degrees of freedom)

OFDM Circuit Configuration

A block diagram of an OFDM signaling system employing the Fourier transform function is shown in Fig. 4.8. At the transmitter end, the input symbol sequence is converted from series to parallel, and the resulting low-rate symbol pulses are mapped into their respective channels to become the information carried by the set of carrier waves. The mapping circuit carries out this operation. The mapping circuit functions to modify the amplitude and phase of the carriers, which is essentially the modulation process. In the QPSK-OFDM case, for example, the digits in the binary signal produced by the series-parallel converter are taken as paired bits, and converted to a corresponding complex signal, the components of which are represented by a_k and b_k in Fig. 4.6 and eq. (4.3). The signals are next passed through the inverse Fourier transform, the parallel-series-converter, and are then converted by the digital-analog converter to the baseband signal, which is a time-sequential signal. From this point, the process resembles normal modulation. The signal is passed through the frequency converter ("up-converted") and a band-pass filter, then placed on the transmission channel.

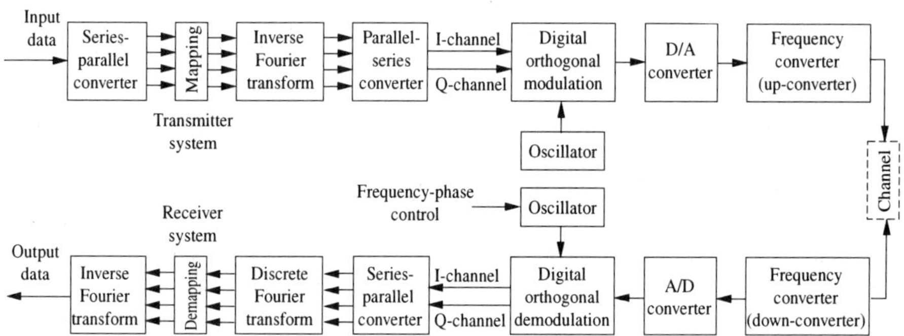

Fig. 4.8   OFDM Circuit Block Diagram

At the receiver end, the signal goes through the opposite process as at the transmitter, finally resulting in a replica of the transmitted signal. In OFDM, the larger the number of carriers the system uses, the closer the roll-off factor approaches zero, and the closer the multiplexed spectrum resembles an "ideal" rectangular wave. Assuming that the two systems have the same capacity, this gives the OFDM system an advantage in spectrum efficiency over single-carrier digital modulation systems. Figure 4.9 shows the results of a test conducted by the Communications Research Laboratory (CRL) on a test device having 795 carriers. The graphs illustrate the OFDM signal waveform and its frequency spectrum, and the square wave response of the system.

The Differences in OFDM, FDM, and CDM

OFDM is being treated in this text as a type of modulation system, but by virtue of the fact that it also functions as an orthogonal frequency-division multiplexing system, it can also be considered as a type of multiplexing system. The multiplexing systems we have available include TDM (Time-Division Multiplexing), FDM

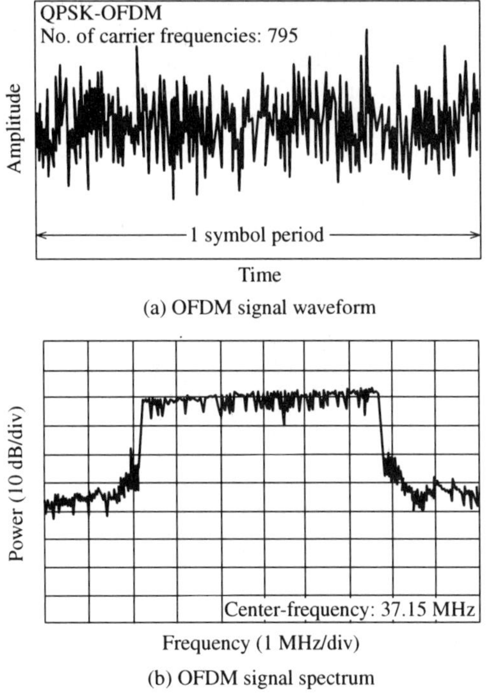

(a) OFDM signal waveform

(b) OFDM signal spectrum

Fig. 4.9   OFDM Signal Waveform and Spectrum

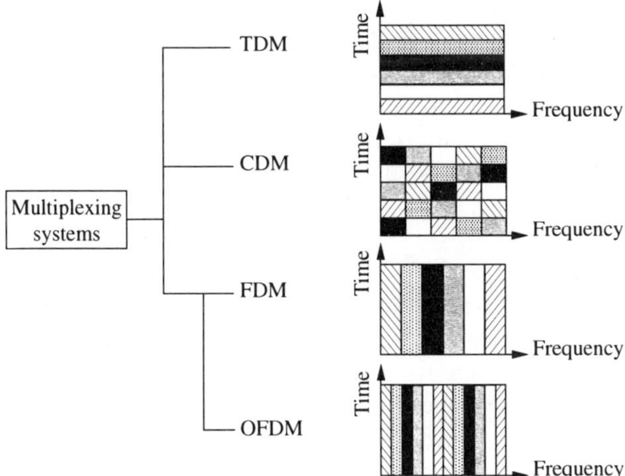

Fig. 4.10   The Various Multiplexing Systems

(Frequency-Division Multiplexing), and CDM (Coded-Division Multiplexing). The illustrations in Fig. 4.10 depict the basic differences in these systems.

From its title, it is evident that OFDM is a special type of frequency-division multiplexing system, and given the fact that ISI is virtually nonexistent between carrier frequencies, theoretically the width of the channels can be made narrower. (Symbol length is the inverse of frequency spacing.) FDM requires roll-off filters and guard bands between frequencies in order to prevent interchannel interference, and FDM is therefore not as spectrally efficient as OFDM.(see. Fig. 4.11)

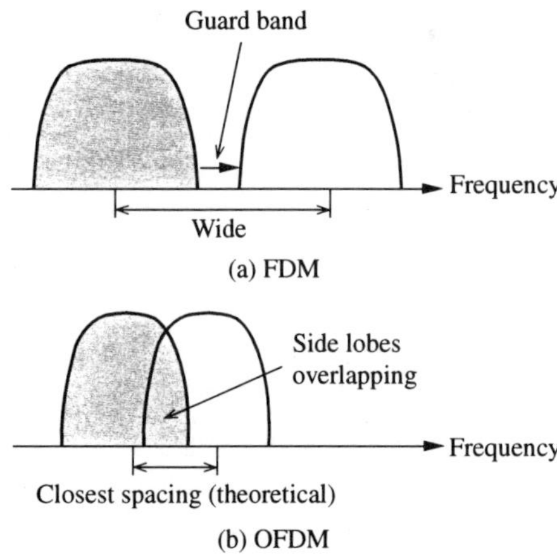

Fig. 4.11   Difference in FDM and OFDM

CDM is a spread-spectrum transmission system, which uses a low cross-correlation spreading code to spread the bandwidth of the transmitted signal. If we make the assumption that sine waves in orthogonal (differing periods) are used as the spreading code, then CDM and OFDM are the same system. (Orthogonally oriented sine waves (i.e., non-cross-correlated sine waves) were discussed previously in this section.) Thus, OFDM can also be considered as a special case of code-division multiplexing.

4.2.3 OFDM Features

In OFDM, the source information is divided into multiple carrier frequencies (referred to as subchannels) for transmission. OFDM lists the following points as advantages:

1) Since numerous subchannels are used, the system is immune to frequency-selective fading caused by delayed waves in certain subchannels.
2) Both time interleave and frequency interleave can be exploited to enhance the effectiveness of forward error correction (FEC).
3) Long symbol intervals and guard intervals (the period over which a part of the signal sequence is retransmitted) reduce the effects of delayed wave interference.
4) Closely spaced subchannels (a dense spectrum) result in high spectrum

efficiency. (Better than single-carrier systems. The greater the number of subchannels, the better the system meets Nyquist minimum bandwidth requirements.)

5) Information can be arbitrarily assigned to specific subchannels. Subchannels where interference is expected can thus be avoided (called a "carrier hole"), adding more flexibility to the system. (Carrier holes are further discussed in Sec. 4.2.4.)

6) The modulation of each subchannel and the ability of modify power makes hierarchical layering of the information easier.

Of the advantages listed above, the robustness of OFDM to delayed wave interference is the most important, and this is the reason that OFDM is now employed in the DAB system [Rau, 1990]. It was during the investigative process of DAB that the advantages offered by the OFDM system became widely understood, and in that process, the OFDM system was deemed suitable for use in fixed and mobile receiver applications where reflected waves are present in the transmission channel.

However, as no system is "perfect," OFDM also has the following disadvantages:

1) Since OFDM is a multicarrier system, nonlinearities in the transmission channel result in cross-modulation, which degrades the overall performance of the system.

2) Preserving the orthogonal orientation of the many carriers requires a complex transmitter/receiver structure. In particular, special measures must be implemented in the receiver in order to assure accurate regeneration of the signal in the face of fluctuations in channel conditions.

3) While a high number of carriers enhance other advantages of the system, even signals (such as audio) where a low transmission rate will suffice are multiplexed for transmission over most) subchannels, which requires a wide bandwidth for all applications.

Proposals for solving the complexity issue include using an OFDM wave generator, and a DFT section in the receiver. Also, the recent progress made in developing the digital signal processor (DSP) provides another promising solution. A more detailed discussion of the problems facing OFDM is given in Sec. 4.2.5.

4.2.4 Techniques Used to Improve OFDM Performance

The immunity of OFDM to delayed wave interference has previously been noted, but employing the techniques described in this section should provide additional improvements in performance.

Interleaving

In channels suffering mutually dependent signal transmission impairments, the time-sequential data in the subchannels are interleaved in both the frequency- and time-domains as illustrated by the arrows in Fig. 4.12. Interleaving the data spreads out any concentrations (bursts) of channel errors so that they become random in nature. Note however, that long interleaving jumps in the time-plane can themselves result in a time delay, which can create additional problems.

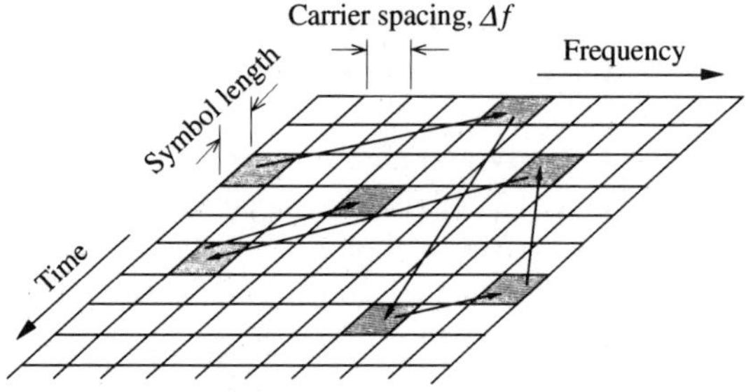

Fig. 4.12   Interleaving

Can OFDM also be Employed in Communications Applications?

As was noted in the advantages listed in Sec. 4.2.3, OFDM offers a very spectrally efficient system that also has good immunity to delayed wave interference. These are features that should make it effective in both TV broadcasting and voice communications. So, why not use OFDM in communications also? The answer to that question lies with the problems that OFDM has with symbol synchronization. As the name OFDM implies, the symbols must be orthogonally oriented, which means that the timing of the symbols is offset. When used in broadcasting systems, the symbols are modulated synchronized "en masse," which averts any problems with timing. However, in communications systems, in order to assure the preservation of the orthogonality of the signals arriving randomly at the base station from the various terminals located at differing distances from the base station, control of the signal timing becomes an insurmountable problem. Also, the circuit nonlinearities that cause problems in broadcasting are also a major problem in communication systems. Therefore, the amplifier linearity that must be sacrificed in order to keep communications devices small in size (most voice communications systems use class C power amps) makes the use of OFDM unfeasible in communication systems.

Guard Intervals

In investigating OFDM, it was discovered that without the presence of nonlinearities in the transmission channel, orthogonality between subchannels could be preserved, even with departures from the "ideal" of frequency characteristics in the channel (which also includes filters, etc.). In other words, the ability to replicate the source information has been experimentally confirmed [Chang, 1966]. This means that orthogonality can be maintained between subchannels carrying signals of the same nature, even when we have a delayed signal added to a directly received signal. In the transmission of digital signals, however, where different symbol information is added to a symbol with a long delay time, the transmission channel can no longer be considered linear, and even depending on equalization will not prevent errors from occurring. In developing DAB, establishing guard intervals in the time-domain was proposed as a solution to this problem [Alard, 1987]. The use of guard intervals allows the guard interval length to be adjusted based on expected delay times for a particular symbol length, and this avoids having to change the subchannel frequency spacing. At the receiver end, data in the section of the guard interval where delayed wave-induced ISI is projected are ignored, and the remaining data are OFDM demodulated. In such a case, the breakdown of orthogonality between the subchannel signals can normally be expected, but as illustrated by the received signal in Fig. 4.13, a time window is established (an operation in which the effective part of the symbol data are extracted from the received signal, and only the guard interval portion of the data are discarded), and by using the effective portion of the received signal, a condition equivalent to orthogonality is maintained. Also, the guard interval portion of transmitted power does not become a factor in the signal demodulation process, so the use of guard intervals does reduce the power efficiency of the system. The length of the guard interval must therefore be established after considering the expected delay times of the delayed waves in the channel, power efficiency, and other applicable factors.

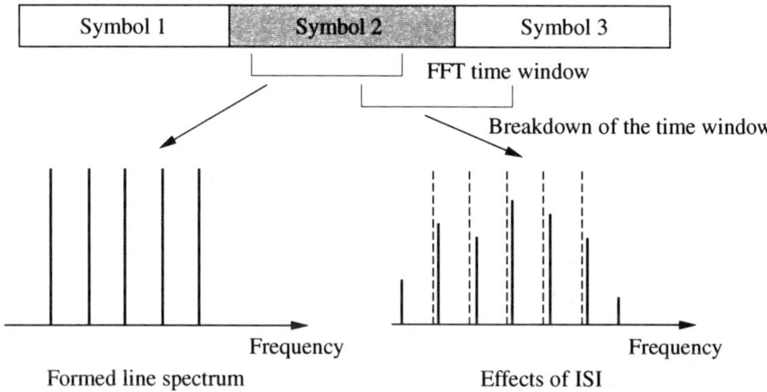

Fig. 4.13   FFT Time Window and Spectrum

Referring to Fig. 4.14(a) and (b), we shall now take a closer look at the effects of the guard interval. In Fig. 4.14(a), since there is not a guard interval present, the channel transmission characteristics include, in addition to the direct wave, three delayed waves having different delay times. In this case, the operational time of the Fourier transform on symbol j is over the range T_s. Note however, that in the area marked by A, a mixing occurs between the components of the previous symbols, and the orthogonality between subchannels cannot be maintained. The same condition develops during the demodulation of symbol k, where we see a mixing of undesired symbol components in area C.

In Fig. 4.14(b), a guard interval, Δ , is inserted at the head of each symbol. Note here however, that we would also like to transmit the symbol information present in the area covered by Δ . If, in this instance, we assume that all subchannels have the same impulse response characteristics, then while the Fourier transform is

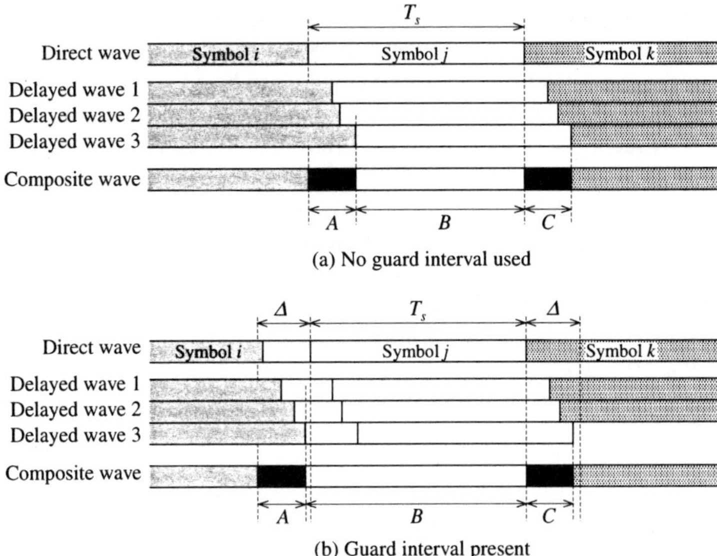

(a) No guard interval used

(b) Guard interval present

Fig. 4.14   Effects of Using a Guard Interval

operating on symbol j, the mixing of previous symbols in area A falls within the guard interval, and a Fourier transform over the range of T_s will accomplish the demodulation process without bringing about any intersymbol interference.

Not only does the use of guard intervals reduce delayed wave interference, as was noted in Fig. 4.13, a time window is established in which a timing margin is created so that the symbol switching point is not included within the guard interval (see Fig. 4.15). Consequently, a part of the guard interval includes the trailing part of the signal itself. Thus, no matter where the cut is made in the transmitted symbol interval, one full period of data will have been transmitted.

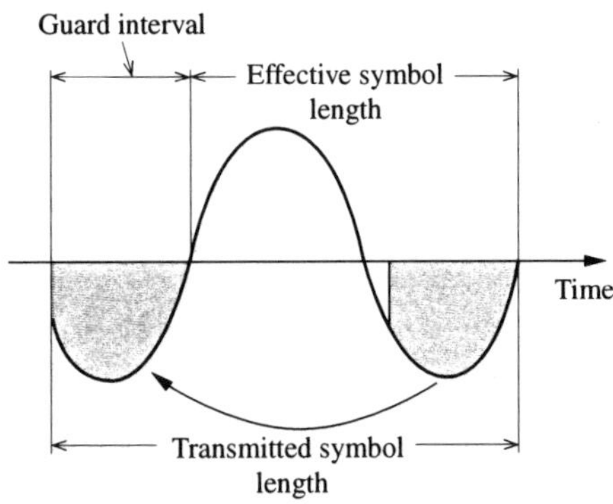

Fig. 4.15   Guard Interval

Carrier Holes

In the OFDM system, modulation, output power and other parameters can be freely selected for the individual carrier waves. Moreover, specific carrier frequencies can be selected for non-use. The selection of a contiguous group of subcarriers for non-use results in opening a gap in the spectrum, which is referred to as a carrier hole. Since the broadcast spectrum must be shared with the current system of analog broadcast television, carrier holes facilitate the joint use of the frequency band. Figure 4.16 shows an example of the current analog television frequency spectrum. The primary peaks, starting from the lower frequencies, are the video carrier (1.25 MHz), the color subcarrier (1.25 + 3.58 MHz), and the audio carrier (1.25 + 4.5 MHz). Signal power is held to low levels except in the immediate vicinity of the peaks. Therefore, the only signals produced by digital broadcast television that could be picked up as interference by the current analog system are the ones near the three peaks listed above. As illustrated in Fig. 4.17, carrier holes have therefore been placed accordingly in the frequency spectrum.

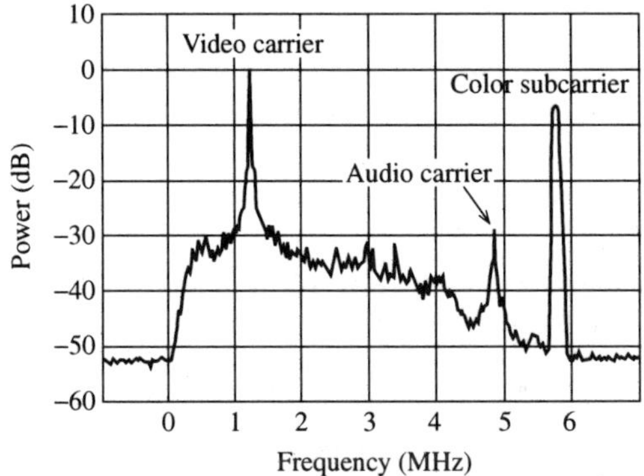

Fig. 4.16   Analog Television Broadcast Frequency Spectrum

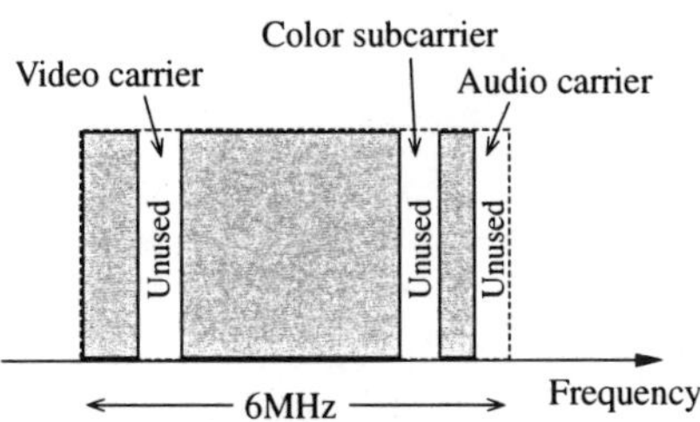

Fig. 4.17   Carrier Holes

The results of simulation experiments show that by establishing carrier holes, the interference to digital signals by analog signals can be reduced by approximately 20 dB. On the other hand, carrier holes have no influence on interference by a digital signal on an analog signal. This is due to the fact that the analog system treats the digital signal as white noise.

Information Hierarchy and Graceful Degradation
One of the notable characteristics of digital radio communication systems is an abrupt drop in the quality of reception accompanying degradation in channel conditions. (Refer to the discussion on the cliff effect in Sec. 2.3.1.) With current (conventional) broadcast television, for example, as the signal grows weaker, there is a corresponding drop in picture quality, and the viewer simply receives a picture that gradually grows poorer. However, with digital broadcasts, depending on FEC and other system technologies, picture reception remains at a high quality level, even under conditions that would be impossible for an analog TV system to handle. Then at a given signal power threshold, reception vanishes abruptly. Although it sounds extreme, certain conditions may develop where in the same channel, one household has good picture reception, while the home next door has no reception at all. The situation described above has inspired the concept of graceeful degradation, where the quality of the signal degrades more gradually. The graceful degradation concept is implemented by hierarchical layering of the source information according to its relative importance. Important information is assigned layers of a higher priority, while less essential information is assigned a lower order layer. With hierarchical information transemission, the system response to worsening channel conditions is to strengthen the FEC in the higher priority layer, while reducing or eliminating coding of the lower layer information. This action reduces the burden on the receiver systems. Hierarchical layering also makes modulating the signal easier, and increases the signal's resistance to noise (Fig. 4.18).

In general, higher *M*-ary modulation systems are more sensitive to poor transmission channel conditions, and are not very immune to noise. In OFDM, since each carrier wave is assigned its own independent modulator, information uniquely selected for tolerance to given channel conditions can be transmitted simultaneously. For example, absolutely essential information can be modulated by QPSK, while the other information can be modulated by 16-QAM or 64-QAM systems. When noise is present in the channel, even if fading becomes a problem,

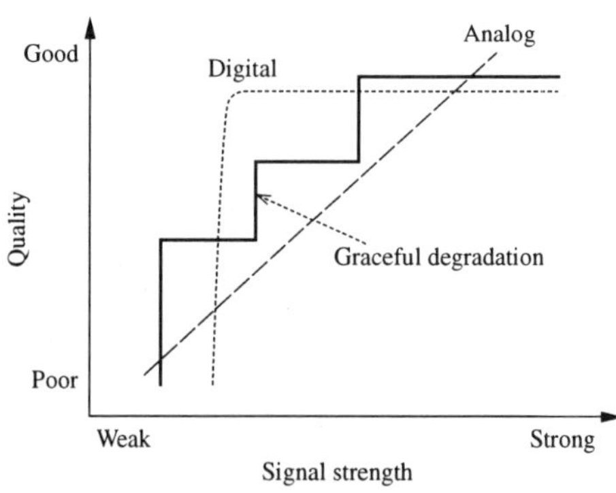

Fig. 4.18   Graceful Degradation

there is still a high probability of the QPSK signal being received error-free, and even though fine image details may not be visible, the quality of picture reception remains acceptable.

4.2.5 OFDM Problem Areas

As has been noted in the discussions above, OFDM is a multicarrier system that features good immunity to delayed wave interference and good spectrum efficiency. However, the multicarrier configuration does have certain drawbacks, and we shall discuss the problems with the OFDM system in this section.

The Effects of Channel Nonlinearities on OFDM

The problems of performance-degrading intermodulation due to nonlinear channels has previously been documented [Nagatsuka, 1994]. Channel nonlinearities cause adjacent subchannel signals to modulate each other, which creates unwanted harmonics in the signals. Here, we express amplifier input/output characteristics as

$$y = ax + bx^3,\tag{4.7}$$

where x is the input signal, and y is the amplifier output signal. Where two sine waves of frequency f_1 and f_2 are the inputs to the amplifier, its output signal is expressed as

$$
\begin{aligned}
y &= a(\sin f_1 + \sin f_2) + b(\sin f_1 + \sin f_2)^3 \\
&= a(\sin f_1 + \sin f_2) + b\sin^3 f_1 + 3b\sin^2 f_1 \sin f_2 + 3b\sin f_1 \sin^2 f_2 \\
&\quad + b\sin^3 f_2 \\
&= \left(a + \frac{9}{4}b\right)(\sin f_1 + \sin f_2) + \frac{3}{4}b\cdot\sin(2f_1 - f_2) + \frac{3}{4}b\cdot(2f_2 - f_1)\tag{4.8}
\end{aligned}
$$

+ out-of-band components.

The above equation describes the third-order intermodulation product, which appears as the distortion components below frequency f_1 and above frequency f_2 in Fig. 4.19. Since the OFDM system uses from several hundred to several thousand uniformly spaced carrier waves, the presence of these distortion components creates a separate carrier, which becomes a major source of interference. The results of a simulation that demonstrates the effects of this interference on bit error rate (BER) are shown in Fig. 4.20 [Nagatsuka, 1994]. Also, the limiting value on instantaneous power, normalized by pre-limiting average power, is also shown. The simulation used a 16-QAM modulator, and a 512-wave multiplex. The results indicate that the

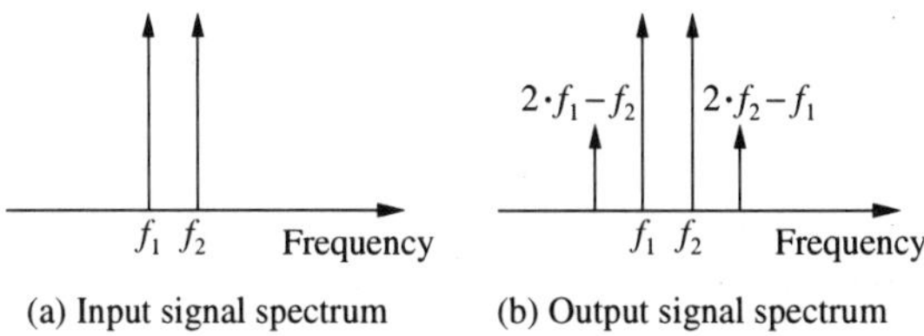

Fig. 4.19   Third-Order Intermodulation Products

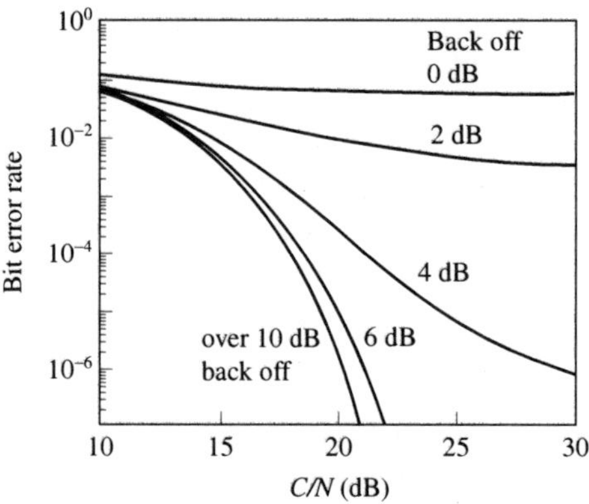

Fig. 4.20   Improving BER by Operating the Amplifier Backed Off

amplifier must be operated backed off by approximately 10 dB (back off: to operate at a lower level than the specified operating level.). These results point out that in a broadcast with an average output of 100 W, an amplifier capable of producing a peak output of 1 kW will be required.

Regenerating Symbols and Carriers

The synchronous (in-phase) regeneration of signals at the receiver is a troublesome process in any digital communication system, but becomes a particularly difficult problem in OFDM.

As was noted previously, it is very important that the effective symbol interval be accurately partitioned at the receiver in order to preserve the orthogonality of the multiple carriers of the OFDM system. Establishing guard intervals is one of the countermeasures to delayed waves, but it also provides a margin in the accuracy at which symbols are partitioned. As was illustrated in Fig. 4.13, if symbol timing is offset, the adjacent symbol improperly enters the Fourier transform, and the mixing results in intersymbol interference (ISI).

If the length of the effective symbol interval is accurately determined, then carrier spacing will also be accurate, and that meets the rigorous precision criterion required for regenerating the carrier frequencies in OFDM. These conditions become even more critical as carrier frequency spacing becomes closer.

As a means of detecting accurate carrier regeneration and symbol timing, the use of a null symbol (a symbol with a perfect zero amplitude) has been proposed. The chirp signal (Fig. 4.21) inserted at intervals of every several 10s of symbols is also being studied.

Guard Intervals and the Number of Carriers

In order to derive maximum benefit from OFDM's features, for a given volume of information transfer, the effective symbol length should be as long as possible, carrier frequencies spaced as closely as possible, and a maximum number of carriers should be used. In order to obtain such an optimum design, the system hardware

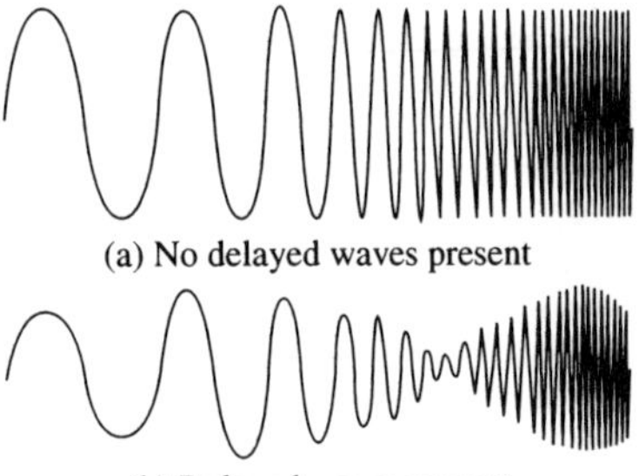

(a) No delayed waves present

(b) Delayed waves present

Note: Illustrates envelope amplitude-frequency
characteristics in channel.

Fig. 4.21   Chirp Signal

must feature extremely good frequency stability characteristics. A factor one must consider in employing a large number of frequency carriers is the increased load this places on the Fourier transform block. Also, greater scales of semiconductor circuit integration, along with higher operating speed will be required for successful implementation. On the other hand, as the number of carriers is increased, and with longer symbol lengths comes longer guard intervals (as the guard interval occupies a fixed proportion of the symbol), and this results in increased system immunity to the effects of delayed waves. The positives and negatives of increasing the number of carriers are summarized in Table 4.1. A system which employs 8,000 carriers is being investigated in the hope of realizing the single-frequency network (SFN) mentioned in Sec. 4.1.2. Work is also progressing on an LSI device containing an 8,000-stage fast Fourier transform (FFT) block.

Table 4.1 Number of Carriers (Positives/Negatives)

No. of carriers	Few	Many
Symbol length	Short	Long
Delayed wave immunity	Poor	Good
Carrier spacing	Wide	Close
Carrier replication	Easy	Difficult
FFT scale	Small	Large

Waveform Equalization Requirements

The occurrence of delayed waves in the transmission channel results in frequency-selective fading, which, in turn, changes amplitude and phase of each of the carrier waves. This produces distortion in the carriers, as is illustrated in the signal space diagram (Fig. 4.22). (A severely delayed wave results in a large change in phase and amplitude of the received signal. In the signal space diagram, a shift in phase is represented by a rotation in the symbol point, and changes in amplitude are shown

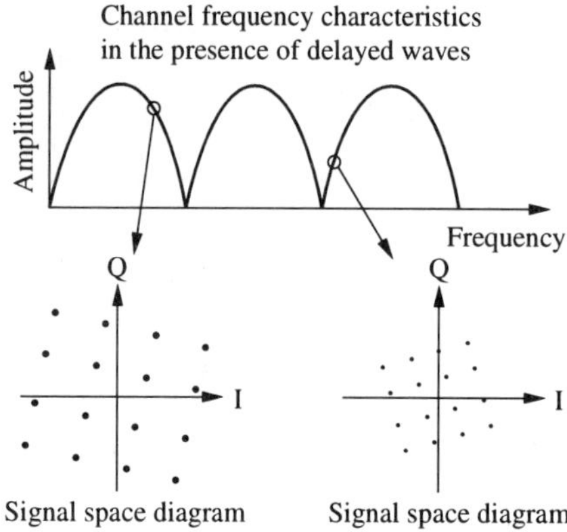

Fig. 4.22   The Effects of Frequency-Selective Fading

as a difference in distance between the received symbol point and its point at origin.) When QPSK is used as the carrier modulator, only phase information is carried on the resultant signal, but differential decoding of that phase information means that absolute phase information is not required. Consequently, the need for waveform equalization is eliminated. In fact, in the European DAB system, since differential coding is used with the QPSK-OFDM system, differential decoding is used in the receiver, so demodulation is possible without waveform equalization.

Note however, that in improving spectrum efficiency by using narrower bands to facilitate the transmission of higher volumes of information, then QAM signaling, which carries both amplitude and phase information is the choice. Thus, the additional amplitude information, which QAM requires for proper operation, necessitates phase and amplitude compensation for each carrier.

Fortunately, in the OFDM demodulator, the Fourier transform converts the received signal to frequency-domain data, waveform equalization is therefore relatively simple. The waveform equalization block for frequency-domain signals is shown in Fig. 4.23. The purpose of the pilot signal in waveform equalization is to act as a reference signal in determining the nature of the distortion that may be present in the phase and amplitude of the signal. The chirp signal discussed

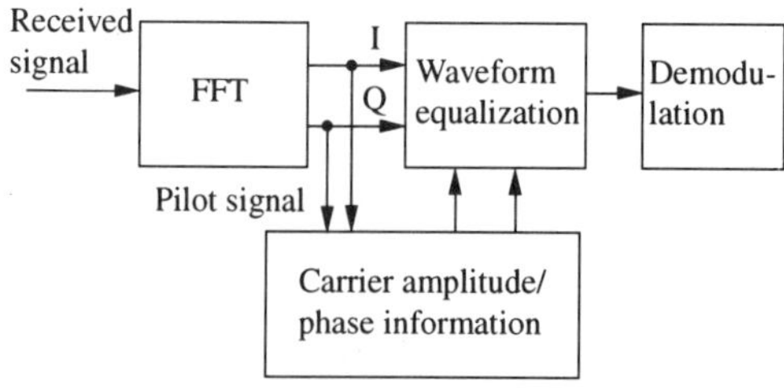

Fig. 4.23   Waveform Equalization in the Frequency-Domain

previously is used as the pilot signal in the European DAB system.

4.2.6 OFDM-related Questions & Answers

This section is designed to reinforce the reader's understanding of the basic concepts of the OFDM system. We shall discuss 11 items related to OFDM operation in a question and answer format.

Q1: What does the orthogonal part of OFDM mean?

A1: The term "orthogonal" refers to the perpendicularity of the spatial relationship of the frequencies of the carriers. In order to prevent the occurrence of interchannel interference, the orthogonality of the signals must satisfy the conditions of eq. (4.3). Theoretically, frequency spacing must be as close as possible.

Q2: Is overlapping of the spectrum side lobes permissible?

A2: This point is a difficult concept to understand, primarily because of how the spectrum is pictured in most papers related to OFDM. Generally the OFDM spectrum is illustrated as in Fig. 4.24(a). While this depiction is not incorrect, it is not representative of what we observe in the demodulation of one symbol length of data, but rather is the waveform observed over the time required to demodulate from several ten to several 100 symbols. In the spectrum pictured here, what comprises the $\sin (x)/x$ function is the inclusion of the symbol switching points. On the other hand, one symbol carefully cut from the received signal appears like the spectra we see in Fig. 4.13, which has no overlapping areas. If the side lobes did overlap, we would have interference between the carriers, and this would make it impossible to properly transmit the data sequence.

Q3: How does adding guard intervals affect orthogonality?

A3: As noted previously, when all of the carriers of the subchannels are orthogonally oriented, the frequency spacing distance between the carriers is the inverse value of symbol length. However, when guard intervals are added, this extends the length of the symbol, but does not change frequency spacing distance. What happens in this case, as illustrated in the spectrum of Fig. 4.24(b), is that the zero points between adjacent carriers become separated. If the time window is properly established in the receiver end, however, as noted in the answer to question Q2, the data sequence will be received correctly.

Q4: What is orthogonal modulation, and what type of spectrum does it produce?

A4: Orthogonal modulation is carried out by sending the complex valued data from the inverse Fourier transform through the parallel/serial converter (Fig. 4.8) and applying the real part to the in-phase component input, and the imaginary part to the quadrature (orthogonal) component input. The spectrum, if we consider it as complex valued data prior to input to the inverse Fourier transform, is aligned with the frequency axis, with real and imaginary parts distinguished only by a shift in the carrier frequency.

Q5: Is a roll-off filter really not required?

A5: The purpose of a roll-off filter is to cut the higher harmonics of the signal and

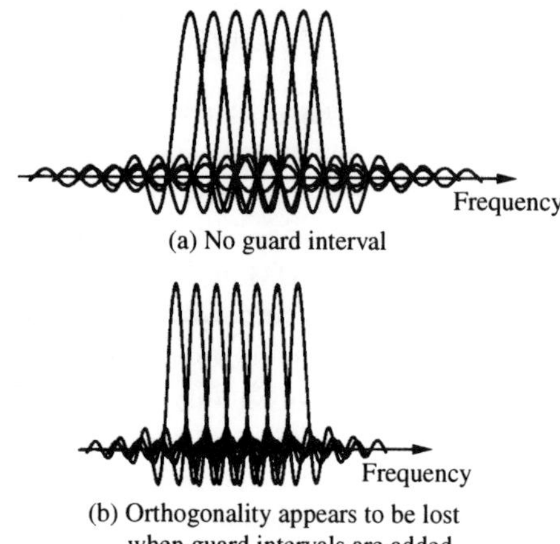

(a) No guard interval

(b) Orthogonality appears to be lost
when guard intervals are added

Fig. 4.24   OFDM Spectrum

to shape the signal so as to prevent intersymbol interference (called satisfying the Nyquist criterion). As was noted in the answer to Q2, for OFDM, the window created in the receiver excludes the switching point of the symbol, which eliminates the need for a roll-off filter.

Q6: Is OFDM suited for implementation in mobile receivers?

A6: OFDM's immunity to the effects of delayed waves gives it the characteristics desired in mobile receivers, and OFDM is used as the modulation system in the European DAB system. However, when M-ary QAM is used as the carrier modulation system, or when the number of carriers reaches into the thousands (4,000 ~ 8,000), the carrier spacing distance becomes so small that carrier regeneration becomes difficult, and in such cases, OFDM is not suitable.

Q7: Is it possible to use the differential coherent detection system commonly used in other mobile communications applications?

A7: OFDM propagates its multiple carrier waves as a "bundle" (because the inverse Fourier transform batch-modulates them), making it impossible to use differential coherent detection. Use of coherent detection is essential for OFDM. Differential decoding, however, is possible, and where this is the case, the frequencies can also be offset during demodulation time to the extent that carrier orthogonality remains preserved. Note that differential decoding at the receiver must be carried out downstream of the Fourier transform (following the division of the carriers). Differential decoding is generally only possible when it is in conjunction with modulation that does not modify amplitude (e.g., DBPSK, DQPSK, etc.). However, by changing the signal pattern, it may also be possible to use differential coherent detection with signals carrying amplitude information. (Refer to Sec. 4.4.2.)

Q8: Does an optimum number of carriers exist with OFDM?

A8: In Europe a system using 8,000 carriers has been proposed in an attempt to realize a SFN system. However, as was noted in Sec. 4.2.5, there are losses associated with carrier configuration, which indicates that there is likely a specific number of carriers that will provide optimum system performance. In Japan, systems using about 1,400 carriers are being studied for mobile receiver applications, while systems employing around 5,600 carriers are under consideration for fixed receivers.

Q9: How does a higher number of delayed waves influence BER?

A9: As one may expect, delayed waves impressed on the desired signal has the effect of making BER worse, but when the time delay of the delayed wave is shorter than the guard interval, its effects are small. On the other hand, when the power of the delayed wave is added to the power of the desired wave, we see an improvement in equivalent carrier-to-noise ratio (CNR). If the power of the desired wave and the noise are held constant, then an additional two or three delayed waves actually improves BER more than only one delayed wave added to the desired signal. Note that when many delayed waves are present in the received signal, envelope amplitude of the signal takes on a Rayleigh distribution. (Refer to Sec. 4.3.3.)

Q10: What shape does the Fourier transform window take?

A10: In current systems, the receiver's Fourier transform window is rectangular, and must always be rectangular in order to preserve orthogonality of the carrier waves. (Using a window shape other than rectangular would invite interchannel interference, and the data could not be correctly transmitted.)

Q11: How is the electric field strength of the OFDM signal measured?

A11: In analog broadcast television, signal strength is determined by measuring the voltage of the coherent part of the signal, and calibrating this measurement with antenna gain and the effective height of the antenna. However, since OFDM is a multicarrier system with each carrier modulated independently, as mentioned previously, the received signal waveform has the appearance of noise, which makes it difficult to measure the electric field strength using conventional methods.

The OFDM signal is measured by using a spectrum analyzer to measure power within the transmission bandwidth, and then converting that value to electric field strength. One cautionary note here regarding the OFDM signal is that the signal level observed on the spectrum analyzer screen varies depending on the resolution bandwidth setting of the test equipment. For example, the OFDM spectrum shown in Fig. 4.9 was measured at a resolution bandwidth setting of 30 kHz. Should this setting be changed to 3 kHz, the level would drop by 10 dB, which is precisely what we see in noise measurements. Note that in observing the spectrum of the analog TV signal (Fig. 4.16), changing resolution bandwidth has little effect on the levels of the peaks. Consequently, when measuring analog TV signals and OFDM signals simultaneously (e.g., when investigating crosstalk between analog and digital TV signals), one must be aware of the fact that different resolution bandwidths may give a false impression of the actual state of the crosstalk problem. As noted above, what is required is independently measuring and comparing the power within the transmission band.

4.3 Propagation of Ground-Waves and Delayed Waves

4.3.1 Propagation of Ground-Waves

Understanding how ground-waves are propagated is key to being able to properly evaluate signal quality and determine the television station service area. Broadcasts in the VHF and UHF bands are normally considered straight-line transmissions, which are defined using the free-space propagation, or the plane earth propagation formulas. In Japan, however, the terrain more typically consists of a very mixed oceanic and/or mountainous environment. Hence, modified versions of the propagation formulas that account for the reflections and refractive conditions of these environments are more useful to design engineers in Japan. This section provides an overview of the calculations applicable to those conditions.

Free-Space Propagation

Free space is defined (for propagation of radio waves) as a straight-line path in a vacuum or ideal atmosphere in which there are no objects to disturb propagation of the waves. Where the transmitter antenna gain is G (half-wave dipole ratio), the transmitter power is P (W), and the distance between the transmitter antenna and receiver is d (m), the electric-field strength E_0 (V/m) at the receiver is given by the following equation [Mushiake, 1961]:

$$E_0 = \frac{7\sqrt{GP}}{d} \quad [\text{V/m}].$$
$$(4.9)$$

Note that the above equation is an approximation of electrical-field strength for line-of-sight transmissions, and is applicable for tall antennas at both transmitter and receiver locations.

Plane-Earth Propagation

In terrestrial broadcasts, even though buildings and other objects which can reflect the waves may not be present, the interference which results from direct waves and waves reflected from the earth's surface changes the strength of the signal received at the receiver. Electric-field strength varies following a certain pattern in relation to receiver antenna height. This is shown by the height pattern in Fig. 4.25. Given this height pattern, electric-field strength is calculated by

$$E = E_0 \cdot 2 \cdot \left| \sin\left(\frac{2\pi h_1 h_2}{\lambda d} \right) \right| \quad [\text{V/m}],$$
$$(4.10)$$

where E_0 is the free-space electric-field strength (from eq. (4.9)), h_1 is transmitter antenna height, h_2 is the receiver antenna height, and λ is the signal wavelength.

In eq. (4.10), where the sine function term in parentheses results in an angle of less than 0.5 rad (due either to antenna height or distance between transmitter/receiver), an approximation of the trigonometric function yields

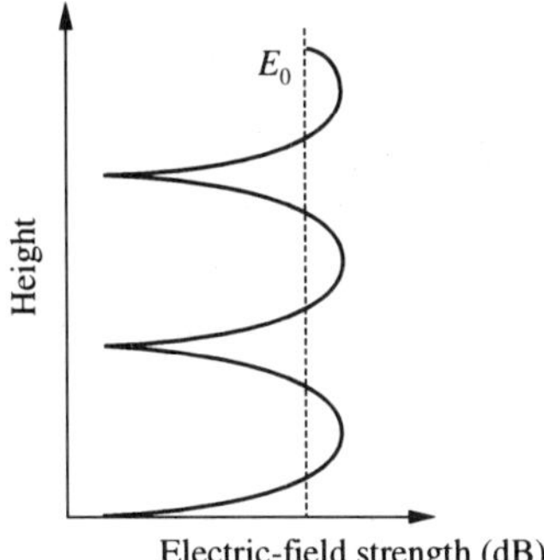

Fig. 4.25   Height Pattern

$$E = E_0 \cdot \frac{4\pi h_1 h_2}{\lambda d}$$

$$= \frac{28\pi h_1 h_2 \sqrt{GP}}{\lambda d^2} \ [\text{V}/\text{m}]. \tag{4.11}$$

Thus, as transmitter/receiver distance increases, electric-field strength diminishes in inverse proportion to the square of the distance.

As transmitter/receiver distance continues to increase, at some point the radius of the earth becomes a factor that cannot be ignored (Fig. 4.26). The effective radius of the earth takes into account the refractive index of the atmosphere, and has been determined to be approximately 4/3 the actual radius of the earth. This consideration has also been included in the standard variation of electric-field strength graph, which was published with ITU-R recommendation 370 (Fig. 4.27). When the propagation distance exceeds the effective height of the transmitter antenna, the rounding of the earth makes the transmission no longer straight-line. The transmitter and receiver are now out-of-sight, and electric-field strength drops off rapidly. Those readers desiring a more comprehensive treatment of radio wave propagation physics are invited to refer to the [Mushiake, 1961] reference.

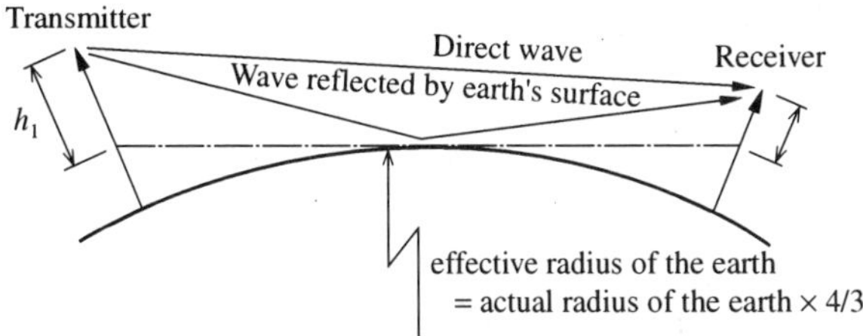

Fig. 4.26 Plane-Earth (Ground-Wave) Propagation

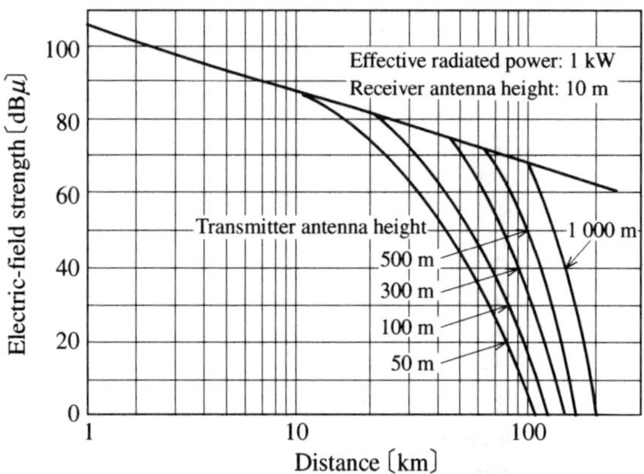

Fig. 4.27   Propagation Characteristics [ITU, 1990]

Diffraction by Mountainous Terrain

Let us now consider a straight line connecting the point of the transmitter and the point of the receiver, obstructed by a hill or building which diffracts and diminishes the power of the transmitted radio signal. If we make the assumption that the obstacle is very narrow (in thickness), and given the positional relationship between the transmitter, receiver, and the diffracting obstacle as shown in Fig. 4.28, electric-field strength is expressed as

$$E = E_0 \cdot \frac{1}{\sqrt{2}} \cdot \left[\left\{ \frac{1}{2} + C(\upsilon) \right\}^2 + \left\{ \frac{1}{2} + S(\upsilon) \right\}^2 \right]^{1/2} \quad (\text{V} / \text{m}), \qquad (4.12)$$

where $C(\upsilon)$ and $S(\upsilon)$ are the Fresnel integrals. υ is called the Fresnel parameter, and is given by

$$\upsilon = H \sqrt{\frac{2}{\lambda} \left(\frac{1}{d_1} + \frac{1}{d_2} \right)} \qquad (4.13)$$

The height of the obstacle is shown as H in Fig. 4.28, and represents the amount by which the obstacle extends above the straight line drawn between the transmitter and receiver. Figure 4.29 illustrates the diffraction lo*ss* as a function of υ, and the graph shows the relationship E/E_0. (Diffraction loss is the reduction in power of a radio wave whose propagation direction has been changed by diffraction caused by an obstacle located in its original path.)

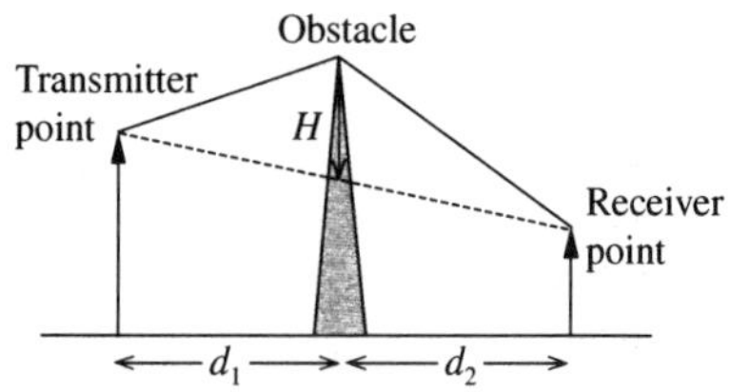

Fig. 4.28   Positional Relationship Between Transmitter, Receiver, and Diffracting Obstacle

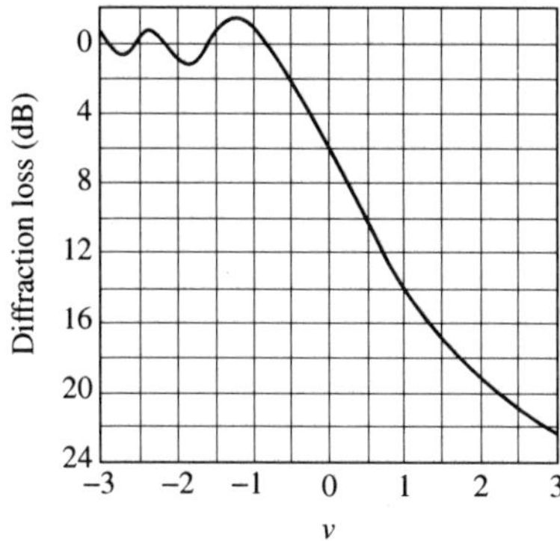

Fig. 4.29   Diffraction Loss

Over-Water Propagation

Radio wave propagation over water falls under the category of a straight-line transmission, and as such, conforms to the plane-earth propagation conditions discussed above under that heading. Note however, that radio waves reflected off a water surface remain strong at the receiver. The standard electric-field strength data for over-water propagation are also provided in ITU-R rec. 370, which was mentioned above. Where k_a is the effective earth radius (as defined above), electric-field strength for an out-of-sight transmission is expressed as follows:

$$E = 56 \cdot 2^{1/4} \cdot k_a^{5/4} \cdot (GP)^{1/2} \cdot \lambda^{-1/2} \cdot d^{-4} \cdot h_1^{9/8} \cdot h_2^{9/8} \quad [\text{V/m}]. \qquad (4.14)$$

4.3.2 What are Delayed Waves (Ghosts)?

Under "ideal" conditions, the radio waves radiated by transmitter antenna travel in a straight line to be intercepted by the receiver antenna. However, when landforms or buildings lying in the propagation path reflect the propagating waves, those that are reflected follow a slightly different path, and arrive at the receiver delayed in time from the direct waves. When this is the case in analog broadcasts, the picture image is offset into two or more partially overlapping images. This phenomenon is called ghost image interference.

In digital broadcast television, forward error correction (FEC) and other digital signal processing techniques reduce the effects of delayed waves on picture quality. Unlike the offset images we see in analog TV, delayed wave interference in digital TV expresses itself as changes in color or luminance only in limited areas (in pixel or block units) of the picture. This is due to the use of the MPEG-2 image code compression technique (introduced in Ch. 2), in which each bit carries the information for one pixel or block unit of the picture. The term "ghost image" consequently does not properly express the same result of interference for digital TV as it does for analog TV. In this text, we shall therefore use the term delayed wave interference when discussing this phenomenon.

In terrestrial digital broadcasts using the OFDM signaling system, as was discussed in the previous section, extending the symbol length of the parallel signals result in better immunity to delayed wave interference. Adding a guard interval that encompasses the delay time of the delayed waves also completely eliminates ISI, making the system extremely robust to the effects of delayed waves.

Shaking Buildings, Dancing Ghosts

Ghost images are a type of interference primarily caused by the television signal being reflected off the sides of high-rise buildings. In conventional analog broadcast television, the appearance of these "ghosts" change depending on the difference in arrival times at the receiver between the delayed waves and the direct waves, which has been a subject of considerable research. Delay times also vary depending on signal propagation distance, and again, propagation distance can be determined by very accurately measuring delay times. Wind-induced oscillations (shaking) of the antenna or the structure supporting the transmitter/receiver equipment can cause changes in propagation distance, and the resulting difference shows up as variations in delay times. The graph shown below is a frequency analysis of the data derived from measurements of structure oscillations caused by the strong winds accompanying a typhoon moving through the Kanto region (near Tokyo), and provides a clear picture of the oscillations of Tokyo Tower and other high buildings under such conditions. The plot can also be used to confirm the designed value for the resonance frequency of Tokyo Tower and other structures.

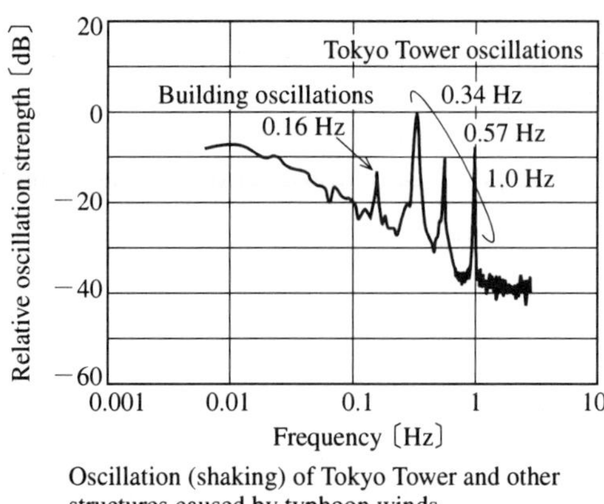

Oscillation (shaking) of Tokyo Tower and other structures caused by typhoon winds.

4.3.3 The Effects of Delayed Waves on Bit Error Probability

While OFDM signaling systems have been recognized for their high immunity to delayed wave interference, in this section we shall discuss the theoretical degradation of bit error probability performance caused by the effects of delayed waves.

Compared with single-carrier modulation systems, OFDM systems feature extended length symbols, which imparts very good resistance to the effects of delayed wave interference. However, depending on the power of the delayed waves and the amount of time delay involved, delayed waves can result in a degradation of the bit error rate (BER). As we will discuss later, BER characteristics related to delayed waves is an important parameter in the design of the circuit, and so should be treated in some detail.

In the headings that follow, we shall first consider the case of a single additive delayed wave, whose delay time is shorter than the guard interval (no ISI), and investigate the theoretical probability of bit error occurrence as derived from the received signal's amplitude probability density. Second, we will consider the case of multiple delayed waves, and in the third and final heading, we take up the case of delayed waves having a delay time greater than the guard interval (ISI is present in this case). In this study, we assume the use of coherent detection, and also assume that any variations in the amplitude of the received signal which are caused by the delayed waves are fully correctable by the demodulator, as are any phase

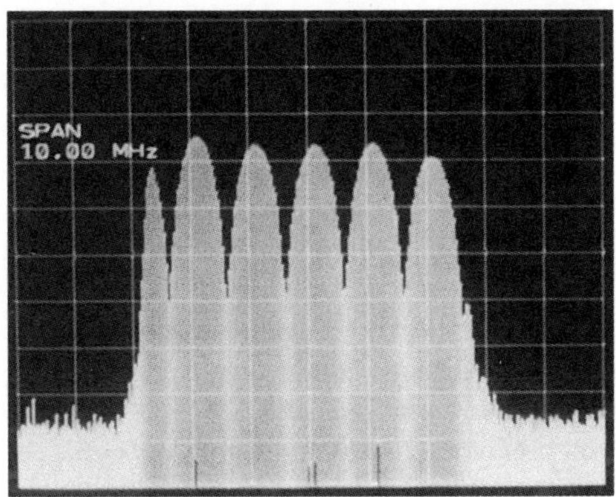

Fig. 4.30   Received Signal Spectrum (one additive delayed wave)

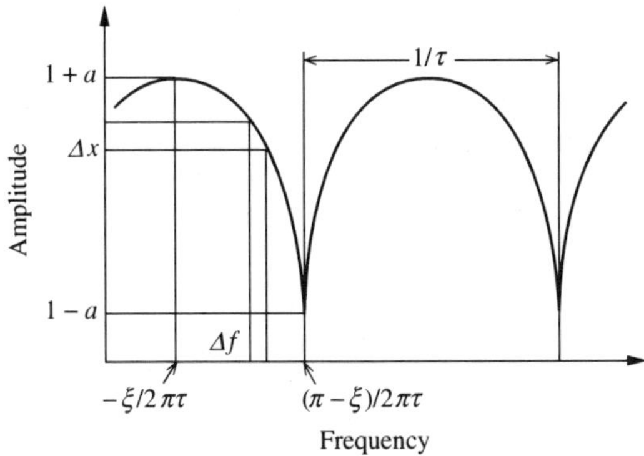

Fig. 4.31   Received Signal Amplitude Characteristics

offsets in the carrier. In this section, the discussion only applies to fixed receiver applications. (Mobile receiver applications will be covered in the next section.)

The Single Delayed Wave Case

We shall begin by determining error probability in the case where only one delayed wave is present. Where the delayed wave is additive to the desired wave, amplitude and frequency characteristics of the received signal are as shown in Fig. 4.30. Here, the additive delayed wave has a delay time of 1 μs, and a D/U (desired signal power/undesired signal power) of 0 dB. As the D/U ratio grows smaller (from an increased power of the delayed wave), the amplitude trough grows deeper. Also, as delay time increases, the period of the signal grows shorter.

Where amplitude of the received signal is represented by x, and amplitude of the delayed wave, normalized to the amplitude of the desired wave is α, delay time is τ, and initial phase is ζ, amplitude-frequency characteristics of the received signal are expressed by

$$x = \sqrt{\alpha^2 + 2\alpha \cdot \cos\gamma + 1} \tag{4.15}$$

$$\gamma = 2\pi f \tau + \zeta. \tag{4.16}$$

A diagram of this relationship is illustrated in Fig. 4.31, where the change in frequency, Δf, is related to a small increment (Δx) of the amplitude probability density of the received signal. This value can be obtained by differentiating the inverse function of eq. (4.15) with x. Here, x is the periodic function of the period $1/\tau$, and its distribution varies with delay time τ and initial phase ζ. When delay time τ is small, it is possible for its period to extend wider than the signal bandwidth. However, since initial phase ζ has a uniform distribution between 0 and 2π, the amplitude probability density obtained over a sufficiently long observation should be obtained within the interval $-\zeta/2\pi\tau \leq f \leq (\pi - \zeta)/2\pi\tau$.

The inverse function of eq. (4.15) is given by

$$f = \frac{1}{2\pi\tau} \cos^{-1}\left(\frac{x^2 - \alpha^2 - 1}{2\alpha} \right) - \zeta. \tag{4.17}$$

Differentiating over the interval $1 - \alpha < x < 1 + \alpha$ yields

$$\frac{df}{dx} = \frac{x}{2\pi\alpha\tau \cdot \sqrt{1 - \left(\dfrac{x^2 - 1 - \alpha^2}{2\alpha} \right)^2}}. \tag{4.18}$$

When the amplitude of the delayed wave is α, taking as a condition

$$\int Q(x,\alpha)dx = 1, \tag{4.19}$$

then amplitude probability density $Q(x, \alpha)$ becomes

$$Q(x,\alpha) = \frac{x}{\pi\alpha \cdot \sqrt{1 - \left(\dfrac{x^2 - 1 - \alpha^2}{2\alpha} \right)^2}}. \tag{4.20}$$

We must also take into consideration amplitude probability density $E(x, \alpha)$, normalized against average power of the delayed wave. Since average power consists of the sum of desired wave power (1) and delayed wave power (α^2), amplitude probability density can be expressed as

$$E(x,\alpha) = \sqrt{1+\alpha^2} \cdot Q\left(\sqrt{1+\alpha^2} \cdot x, \alpha\right). \tag{4.21}$$

Amplitude probability density, while varying D/U, is shown in Fig. 4.32 [Tsuzuku, 1997]. Notice here that the amplitude probability densities are shown as average power (the sum of desired and delayed wave power), normalized to 1 using eq. (4.21). Using amplitude probability density found above, BER with D/U as a parameter is expressed by

$$BER(C/N) = \int_0^\infty E(x,\alpha) \cdot A_1 \cdot \text{erfc}\left(A_2 \cdot \sqrt{C/N} \cdot x\right)dx, \tag{4.22}$$

where $\text{erfc}(x)$ is the complementary error function, and A_1 and A_2 are modulation

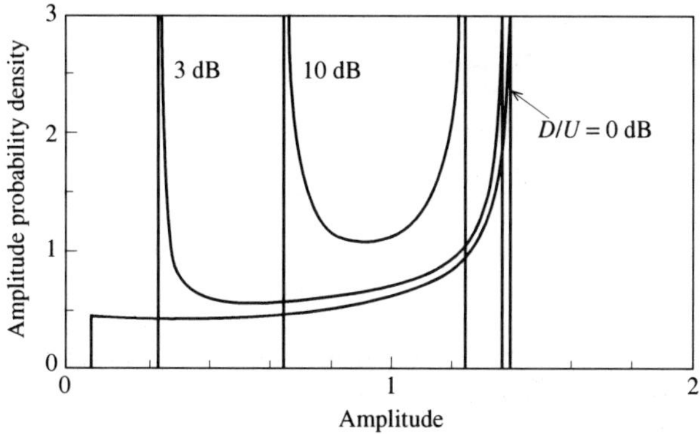

Fig. 4.32   Amplitude Probability Density Function

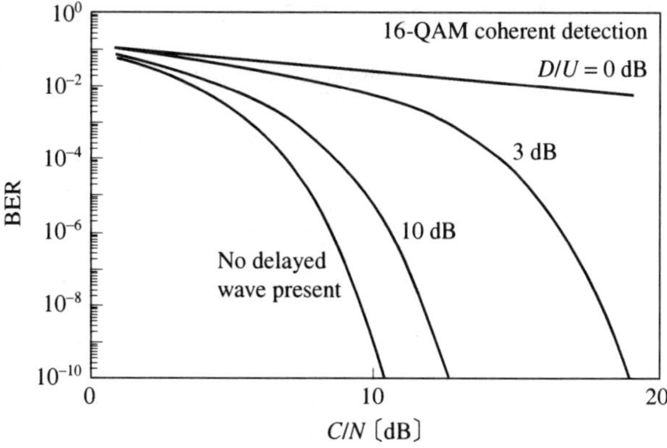

Fig. 4.33   Delayed Wave Effects on Bit Error Rate

system constants. Figure 4.33 shows an example of BER characteristics derived using eq. (4.22), where 16-QAM-OFDM was employed, and $A_1 = 3/8$, $A_2 = 1/\sqrt{10}$. Here, the C part of C/N refers to the sum of the powers of the desired and delayed waves.

The Multiple Delayed Wave Case

Bit error probability for multiple delayed waves, as in the single delayed wave case, can be found from amplitude probability density. Here, we shall only discuss the results of the computations [Tsuzuku, 1997].

Where n - 1 delayed waves result in a signal with an amplitude probability density of $P_{n-1}(x)$, if an additional delayed wave having amplitude v is added to the signal, the resultant amplitude probability density will be expressed as

$$P_n(x) = \int_{|x-v|}^{x+v} P_{n-1}(u) \cdot Q(x,u,v)du \ . \tag{4.23}$$

In the above equation, $Q(x, u, v)$ is the amplitude probability density when a delayed wave of amplitude v is added to a desired wave of amplitude u. The generalized form of eq. (4.20) is expressed by

$$Q(x,u,v) = \frac{x}{\pi uv \cdot \sqrt{1 - \left(\dfrac{x^2 - u^2 - v^2}{2uv}\right)^2}} \ . \tag{4.24}$$

Substituting amplitude probability density yielded by eq. (4.24) into eq. (4.22) gives the bit error probability. As a special case, Fig. 4.34 shows BER for multiple additive delayed waves having D/U values of 0 dB. BER is at a minimum for two delayed waves, and as the number of delayed waves increases, amplitude probability density approaches a Rayleigh distribution.

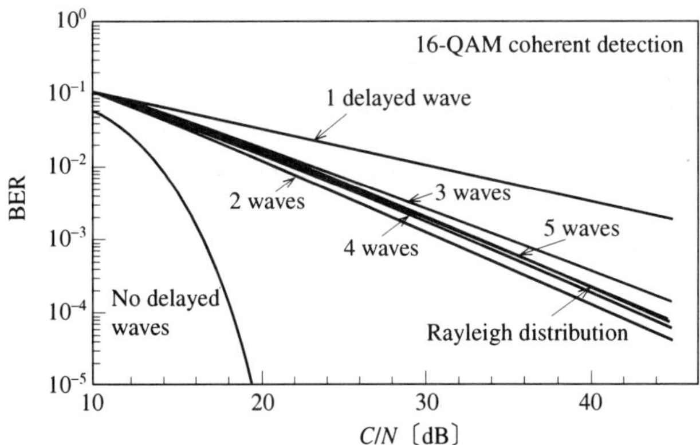

Fig. 4.34   BER in the Presence of Multiple Delayed Waves

The Extended Delay Time Delayed Wave Case

While we have already learned that OFDM is more robust than single-carrier modulation systems with regards to delayed wave interference, when the delay time of the delayed waves extends beyond the guard intervals, the resulting intersymbol interference (ISI) causes a rapid degradation in BER performance. We will describe this situation for a single delayed wave case.

When delay time of the delayed wave is longer than the guard interval, the previous symbol encroaches into the reference symbol interval, resulting in interference. This is equivalent to additive noise, in that it degrades *C/N* (for which the notation *C/N′* is used in the equation below). In such cases, eqs. (4.25) and (4.26) are used to express bit error probability [Tsuzuku, 1997]. (Note that this expression applies to a coherent detection system.)

$$C/N' = C \cdot \left[1 + \left\{ \cos\theta \, \frac{1}{D/U}(1-T) \right\}^{1/2} \right]^2 \Big/ \left(N + C\frac{2}{D/U} \right) T \quad (4.25)$$

$$BER(C/N) = A_1 \cdot \mathrm{erfc}\left(A_2 \cdot \sqrt{C/N'} \right). \quad (4.26)$$

In eq. (4.25), θ represents the delayed wave phase with respect to the carriers, and T is the proportion of time, relative to the length of the effective symbol, that the delayed wave has exceeded the guard interval. As with eq. (4.22), A_1 and A_2 are constants whose values depend on the modulation system in use.

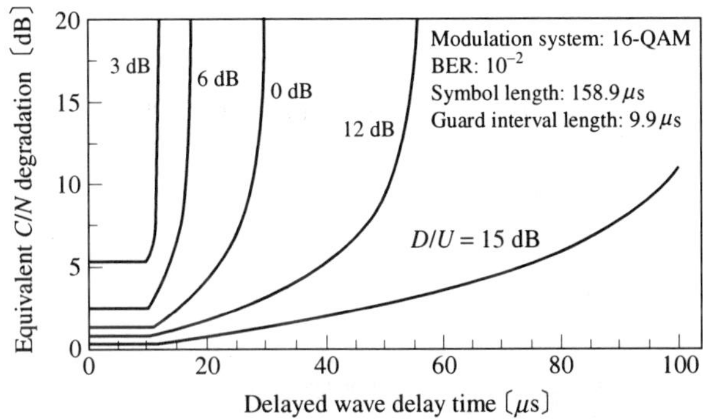

Fig. 4.35   Delayed Wave Delay Time and Equivalent *C/N* Degradation

Interchannel interference is the interference produced by signals from other channels intruding into the reference channel, but since the interchannel interference problem is negligible compared to ISI, it will be disregarded here.

Figure 4.35 illustrates the case at hand. The equivalent *C/N* degradation is shown on the vertical axis, and it represents the amount of *C/N* degradation at a specified BER (here, 10^{-2}) compared to a system free of delayed waves. Note that when delay time of the delayed wave is shorter than the guard interval, degradation is slight even for strong delayed waves. On the other hand, where delay time is longer than the guard interval, degradation becomes substantial, even though the delayed wave may be weak (high *D/U*).

4.3.4 The Link Budget

To assist in specifying the service area and the level (or grade) of service, the digital television ground-wave broadcasting system uses a link budget formulated for the specific modulation system it utilizes. A link budget refers to a numerical evaluation whose purpose is to ascertain that all the equipment in the communications or broadcasting system meet the performance criteria required of the system. We will illustrate the formulation of the link budget for a fixed receiver system using the items listed in Table 4.2.

Table 4.2 Link Budget Table (typical) (fixed television receiver)

Item	Analog	Digital (OFDM)
(1) Frequency (MHz)	700	770
(2) Modulation system	VSB	64QAM
(3) Transmission bandwidth (MHz)	4.2	5.6
(4) Spectrum efficiency (bit/s/Hz)	–	3.28
(5) Information transfer rate (Mbit/s)	–	18.3
(6) Pre-FEC bit error rate	–	$1*10^{-2}$
(7) Required C/N (dB)	–	19
(8) Equipment margin	–	1
(9) Required C/N for receiver (dB)	37	20
(10) Receiver noise figure NF (dB)	14	5
(11) Receiver noise power N (dBm)	−93.6	−101.3
(12) External source noise power (dBm)	−95.4	−103.1
(13) Total receiver noise power (dBm)	−91.4	−99.1
(14) Minimum input voltage for receiver (dBμ)	60.3	35.6
(15) Receiver antenna gain G (dB)	10	10
(16) Antenna height loss (dB) (based on 4 m standard)	0	0
(17) Effective antenna length (dB)	−18.1	−18.1
(18) Feeder loss L_f and insertion loss L_m (dB)	3	3
(19) Required electric-field strength (no variation margin) dB (μV/m)	71.5	46.7
(20) Reception time percentage (dB) (corrected from 50% to 99%)	–	4
(21) Reception site percentage (dB) (corrected from 50% to 90%)	–	8.2
(22) Required electric-field strength (μV/m)	71.5	58.9
(23) Transmitter power ratio (based on analog TV = 1)	1	0.08

Note in the link budget table that the required electric-field strength for digital television is about 11 dB lower than the 70 dBμ value specified for analog television [MPT order, 1950], and also how small the transmitter power requirements are for digital TV.

Items in the link budget table are briefly described below.

(1) Frequency: The frequency applicable here is 770 MHz (equivalent to 62 television channels). This value reflects the highest frequency in the band allocated for broadcast television.

(2) Modulation system: For fixed receiver applications, 64-QAM-OFDM was selected based on its favorable spectrum efficiency.

(3) Transmission frequency bandwidth: In the interest of preventing interchannel interference, a 6 MHz bandwidth was utilized. Guard bands placed every 200

kHz in one sideband result in an effective bandwidth of 5.6 MHz.

(4) Spectrum efficiency: A coding rate 3/4 convolutional code (constraint length of 7) and a Reed-Solomon RS(204, 188) code capable of correcting eight symbols was used for FEC. Guard intervals 1/8th the length of the symbol were inserted, and the addition of pilot and control signals reduce the original efficiency by 1/9th. The following calculation is for spectrum efficiency without including guard bands.

(4) = (number of bits transmitted per symbol) × (coding rate of convolutional code) × (coding rate of RS code) × (efficiency loss due to guard intervals) × (efficiency loss due to pilot/control signals) = $6 \times 3/4 \times 188/204 \times 8/9 \times 8/9$

(5) Information transfer rate: Specifies the number of bits transmitted per unit time. Obtained by multiplying spectrum efficiency (4) and transmission bandwidth (3).

(6) Pre-FEC bit error rate: When using FEC as specified in (4), if pre-FEC BER is less than 1×10^{-2}, post-FEC is improved to 10^{-11}, which is essentially error-free.

(7) Required *C/N* (dB): Refers to the minimum *C/N* which will satisfy the BER conditions described in (6).

(8) Equipment margin: Refers to the value which expresses the permissible degradation of actual receiver performance from theoretical performance. Causes of degradation include internal noise (self-noise) of the receiver (but excludes the noise figure portion), carrier phase offset during demodulation time, and amplifier intermodulation.

(9) Required *C/N* for the receiver (dB): Specified as the sum of required *C/N* and equipment margin. (9) = (7) + (8)

(10) Receiver noise figure *NF* (dB): In early broadcast television link budgets, specified as 14 dB, which is far too low considering the performance standards of more recent receivers. Here we use a value of 5 dB.

(11) Receiver noise power N_r (dBm): Computed as

$$N_r = k \cdot T \cdot B \cdot NF,$$

where k is the Boltzman constant, $k = 1.38 \times 10^{-23}$ (J/K), T is the temperature, $T = 300$ (K), B is the noise bandwidth, and NF is the noise figure. The total computation in units of decibels is as follows:

$$(11) = -228.6 + 24.8 + 10 \log ((3)) + (10) + 30 \ \text{(dBm)}.$$

(12) External source noise power (dBm): Specifies the man-made noise (urban noise) entering the receiver. Noise power is generally quantified in inverse proportion to the 2nd or 3rd power of noise frequency [ITU, 1982]. The urban area noise temperature at 170 MHz is approximately 11,000 K, at 470 MHz, ~ 700 K, and at 770 MHz, ~ 200 K. (Refer to Fig. 4.36.)

$$(12) = -228.6 + 10 \log (T) + 10 \log ((3)) + (10) + 30 \ \text{(dBm)}.$$

(13) Total receiver noise power *N* (dBm): Total receiver noise power is computed as the sum of the receiver noise power and the external source noise power.

$$(13) = 10 \log (10^{((11)/10)} + 10^{((12)/10)}).$$

(14) Minimum input voltage for receiver (dBμ): In specifying the minimum input voltage to the receiver, the ratio of noise power and signal power must satisfy required *C/N*. Thus,

Minimum input power = (9) + (13).

The power obtained above is converted to a voltage at the receiver antenna (open voltage) as follows:

Minimum input power = (voltage at the receiver input terminal)^2/4R

Here, R is the impedance of the receiver antenna (normally 75-ohms).

(14) = 114.7 + ((9) + (13))

(15) Receiver antenna gain G (dB): Refers to the antenna gain for a fixed receiver application. In this case, a 14-element Yagi-Uda antenna was used.

(16) Antenna height loss: Refers to the antenna loss (or gain) referenced against a 4-m antenna.

(17) Effective antenna length: Effective antenna length is computed as follows [Mushiake, 1961]:

(17) = 20 · log(λ/π) = 20 · log(300/(1)/π), where λ is wavelength.

(18) Feeder loss L_f and insertion loss: Refers to loss in the feeder cable and insertion loss (the loss due to impedance mismatch).

(19) Required electric-field strength: (19) = (14) − (15) + (16) − (17) + (18).

(20) Reception time percentage (dB) (50% rate): Refers to the correction of possible reception time (service time) from 50% for analog broadcasts to 99% for digital broadcasts [ITU, 1990]. This correction is required to account for the cliff effect, which is peculiar to the digital system.

(21) Reception site percentage (dB) (50% rate): Refers to the correction of possible reception at any given site within the service area from 50% for analog broadcasts to 90% for digital broadcasts [Okumura, 1986]. This correction is required to account for the cliff effect, which is peculiar to the digital system.

(22) Required electric-field strength dB (μV/m): (22) = (19) + (20) + (21).

(23) Transmitter power ratio: Compares digital TV transmitter power with analog TV transmitter power.

(23) = $10^{(((22) - 70)/10)}$.

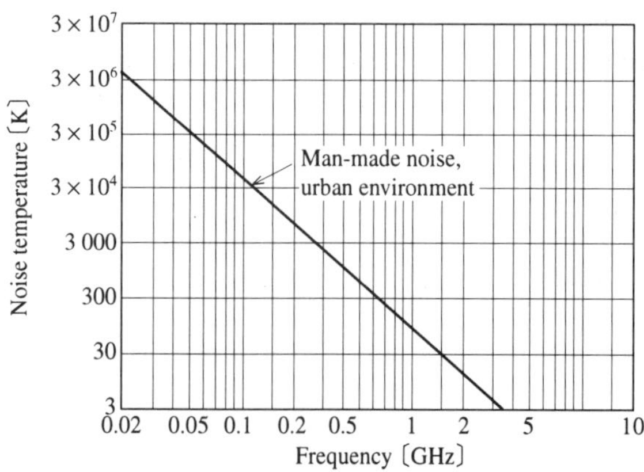

Fig. 4.36   Noise Temperature (ITU, 1982)

4.4 Digital Broadcast Television Utilizing Mobile Receivers

4.4.1 Radio Wave Propagation

Reception by Mobile Receivers

Antennas used on the vehicles that normally serve as the installation for mobile receivers are typically much shorter than the "standard" of 4-meters. Therefore, in addition to the fact that we have motion between the transmitter and receiver, a mobile television broadcast using the ground-wave medium must also contend with the effects of obstacles such as landforms and buildings periodically intruding into the propagation path. Thus, except for certain exceptions, the plane-earth propagation formula (covered previously) cannot be used without modifications.

In order to determine the nature of the effects of landforms and other natural formations on radio wave propagation, experiments and statistical analyses are required to characterize the problem at hand. We are fortunate in this respect, as the numerous studies carried out in developing mobile/portable voice communication systems should also suffice for television purposes [Okumura, 1967].

Fluctuations in Electric-Field Strength

As illustrated in Fig. 4.37, in the mobile environment, propagation characteristics can be placed in two categories. We first have the variance which depends on the distance between the transmitter and receiver (center-value of large-scale intervals). Second, there are the gradual changes that occur in the average values over an interval of several tens of meters (center-value of small-scale intervals), and third, we have abrupt changes occurring over intervals of several tens of meters (instantaneous variations).

In large-scale intervals, the center-value tends to diminish along with decreasing strength of the electric-field as distance increases between the receiver and transmitter points. Okumura's curves are commonly used to statistically quantify diminishing signal strength as a function of distance, and are described below.

The center-values of small-scale intervals are affected by landforms and other natural features. In this case the average value of electric-field strength fluctuates gradually, and statistically has a log-normal distribution (which means that the decibel value of the logarithm of electric-field strength has a normal distribution).

Instantaneous value fluctuations are due to interference between the numerous waves in a multiplexed carrier propagation path, which gives the (amplitude) envelope of the received signal a Rayleigh distribution, and is commonly referred to as Rayleigh fading.

Okumura's Curves

Y. Okumura and his colleagues conducted numerous experiments in the field of terrestrial voice communications, which were instrumental in clarifying the relationships between distance and signal attenuation, in addition to receiver

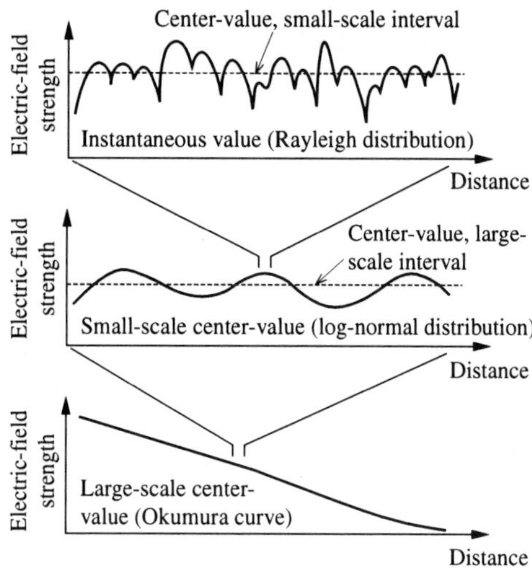

Fig. 4.37   Electric-Field Fluctuations in Mobile Environment

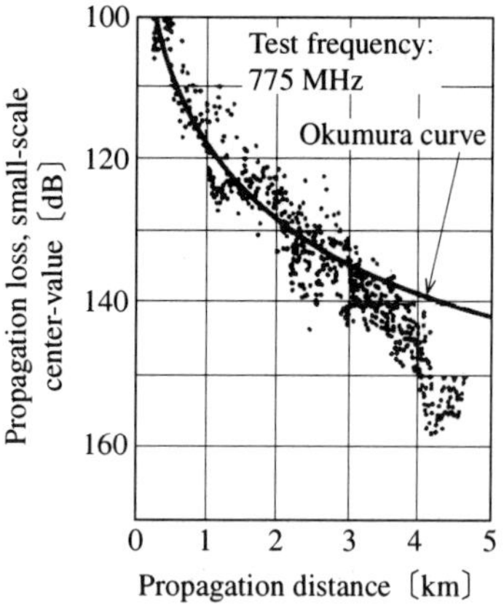

Fig. 4.38   Okumura's Curve and Measured Data Points [Moriyama, 1989]

antenna height and frequency in various urban and other terrestrial environments. The results of these experiments were statistically processed and compiled in a set of graphs called Okumura's curves [Okumura, 1967].

A center-value is used to estimate the amount of attenuation on the Okumura curve. Usually the actual attenuation will vary somewhat from the center-value. This is illustrated in Fig. 4.38, where the scatter of data points obtained experimentally generally follow the Okumura curve.

4.4.2 Problems in Mobile Receivers

As noted above, the extreme variations in the strength of the received signal make it very difficult to properly demodulate the signal in the mobile environment. As will be discussed in the following, many problems still need to be worked out, including how to handle Doppler frequency shifts and low received signal power. In the link budget for mobile television (Sec. 4.4.5), the requirement for a large electric-field strength margin becomes evident. These problems should also be thoroughly addressed in the design of the system.

Fading (in the Mobile Environment)

The fluctuations that occur in the amplitude of the received signal are called fading. Fading normally takes two different forms. Flat fading is defined by a "flat" profile in the frequency characteristics within the transmission channel. Frequency-selective fading is characterized by more severe instantaneous amplitude fluctuations. Fading is a result of the intermix (and consequent interference) of the delayed waves present in a radio signal. Flat fading is produced when the differences in delay time of the delayed waves are shorter than the inverse value of the bandwidth. On the other hand, when delay times are longer, the degree of attenuation differs depending on frequency, and in that case, we have frequency-selective fading. The 3-space map in Fig. 4.39 was constructed from a frequency-selective observation. The spectrum of the received signal shows flat characteristics within the 3 MHz band. The cause of variations in the power of the received signal is due to the phase relationships of the additive delayed waves in the band. Areas where the delayed waves are added in-phase are amplified, while the areas where waves with opposite phases are added are attenuated. Broadcast television occupies a bandwidth of 6 MHz, so a delay of only a few hundred nanoseconds will result in frequency-selective fading.

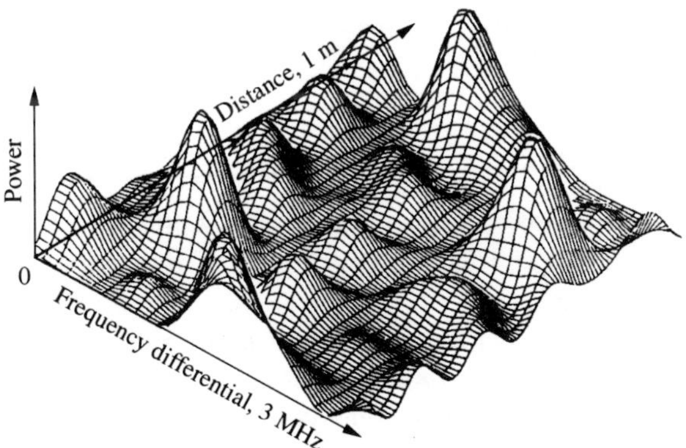

Fig. 4.39   Frequency-Selective Fading

Most research in the theory of fading has been conducted in the field of mobile communications, and the reader can find more details in specialized works on the subject [Okumura, 1986]. Only a brief outline of the subject is presented in this text.

However, what has been learned from the studies in mobile communications is that the amplitude of the signal envelope received by a mobile receiver has a Rayleigh distribution when the propagation path between the transmitter and receiver is indirect (out-of-sight). Where the electric-field of the received signal is x, the amplitude probability density, $E(x)$, for an envelope with a Rayleigh distribution is given by

$$E(x) = \frac{x}{b_0} \exp\left(-\frac{x^2}{2b_0}\right),\qquad(4.27)$$

where b_0 is the average received power.

When the propagation path between the transmitter and receiver is straight-line (the received signal is direct), the amplitude probability density of the received signal takes on a Nakagami-Rice distribution, and is expressed as

$$E(x) = \frac{x}{E_{rms}^2} \exp\left(-\frac{x^2 + E_s^2}{2E_{rms}^2}\right) \cdot I_0\left(\frac{xE_s}{E_{rms}^2}\right).\qquad(4.28)$$

In the above equation, x is the electric-field of the received signal, E_s is the electric-field strength of a normal (direct) wave, E_{rms} is the effective (rms) value of the electric-field strength of an irregular (reflected or diffracted) wave, and I_0 is a modified Bessel function.

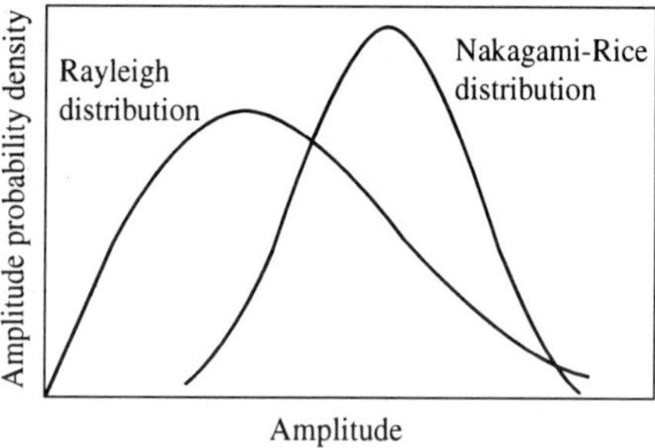

Fig. 4.40   Rayleigh and Nakagami-Rice Distribution Curves

Doppler Frequency Shifts

The Doppler frequency shift, for our purposes, refers to a change in the observed frequency of a wave due to the relative motion of the transmitter and mobile receiver. When the mobile receiver receives only a direct signal, the resulting Doppler frequency shift is corrected by the receiver's automatic frequency control (AFC), which consists of a local oscillator circuit. However, if the signal is made up of many delayed waves arriving at the receiver from many different directions, the changing Doppler frequency shifts result in a condition whereby the receiver is receiving multiple waves with different frequencies. Where this is the case, the receiver normally tunes to the frequency of the strongest signal, and because the other signals are not a part of the received signal, bit error rate suffers.

Low Received Signal Power

In general, mobile receivers employ short antennas. As noted in Sec. 4.3.1, electric-field strength depends also on antenna height, and lower antennas result in a reduced power of the received signal. The relationship between antenna gain and mobile antenna height is illustrated in Fig. 4.41 [Okumura, 1986]. The graph in Fig. 4.41 demonstrates a difference in attenuation by city size. This difference can be attributed to the height and number of large buildings in larger cities, which greatly influences the propagation of radio signals. For television broadcasts over ground-waves, comparing reception with roof-top antennas at heights of about 10 meters with mobile (automobile-mounted) antennas having heights of at most 1.5 meters, we see a reduction in received signal power of approximately 9 dB. The shape of the antenna also has a bearing on its ability to pick up radio signals. Most automobile-mounted antennas are of the whip or dipole design, while roof-top antennas are typically multiple-element Yagi-Uda antennas. A drop of 5 ~ 10 dB can be expected with the mobile antenna, due to its shape.

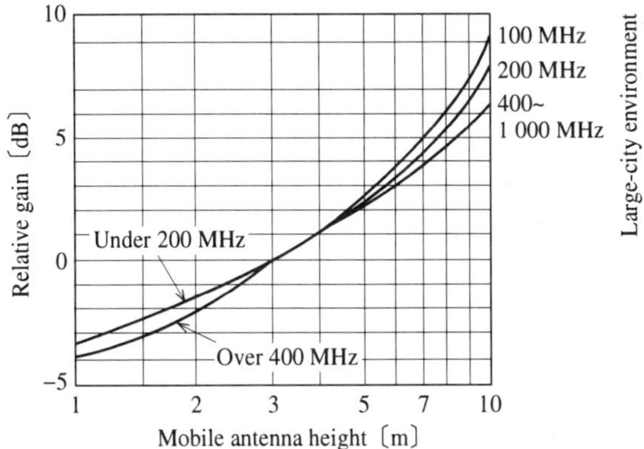

Fig. 4.41   Receiver Antenna Height versus Relative Gain [Okumura, 1986]

Coherent Detection and Differential Decoding

The process of recovering the modulating wave from the modulated carrier is termed demodulation or detection. Two basic methods are used to demodulate digital signals: coherent detection and noncoherent detection. In coherent detection, the received signal, following passage through a band-pass filter, is multiplied with a sine wave having an identical frequency as the carrier used by the modulator. The resultant signal is then passed through a low-pass filter to extract the original signal. In the modulation process, the signal was shifted on its frequency base by an amount f. In order to restore the present spectrum to its original state, the generated carrier that the signal is mixed with must have precisely the same frequency and phase as the one used by the modulator. To synchronize these signals, coherent detection requires a carrier regeneration circuit whose function is to create a reference carrier meeting the above criteria, and this adds to the complexity of the demodulation circuit. On the other hand, the reference carrier generated by the coherent detection system is free from external noise, so the system is less prone to coding errors than is the noncoherent detection system. Also, many ingenious carrier regeneration circuits are available, making the task relatively simple. Therefore, except in certain mobile communications applications where channel conditions are so severe as to make regenerating the carrier difficult, coherent detection is the favored method.

Noncoherent detection does not require a reference carrier to demodulate the signal. While the absence of the carrier regeneration circuit makes the demodulation system simpler, BER performance is not as good as in coherent detection. In differential coherent detection, which is one of the noncoherent detection methods, one symbol previous to the current symbol becomes the reference, and demodulation is accomplished by multiplying that reference with the current received signal (a reference carrier is not required). (See fig. 4.42.) The simplicity of the receiver circuit, and because noncoherent detection does not require a reference carrier, the differential coherent detection system is widely used in mobile communications applications where severe fluctuations in the channel would normally make regenerating the carrier difficult. Note however, that since both the received signal and the previous symbol used as a reference may be noisy, it is more prone to coding errors than is coherent detection.

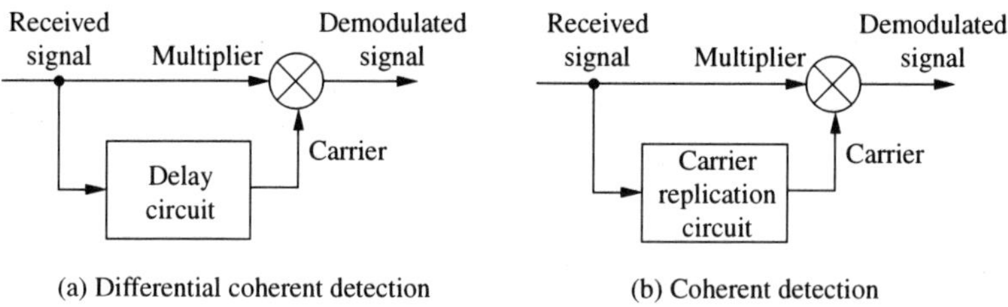

(a) Differential coherent detection   (b) Coherent detection

Fig. 4.42   Differential Coherent Detection and Coherent Detection

The OFDM modulation system, through the Fourier transform, operates on the batch modulation principle, and this precludes use of differential coherent detection. However, it can use differential decoding. If the signal is differentially encoded at the transmitter end, even if absolute phase of the symbol is an unknown, the signal can still be demodulated by taking the difference between the current symbol and the previous symbol. Thus, decoding is possible even with some amount of frequency offset in the OFDM demodulation block.

DQPSK (Differential Quadrature PSK) is a widely-used differential coding system, but 16-DAPSK (16-Differential Amplitude and PSK) and 64-DAPSK, which also include amplitude information in the encoding process have been proposed as methods of increasing transfer rates [Rohling, 1995]. The signal space diagram for 64-DAPSK is shown in Fig. 4.43. 64-DAPSK combines four sets of 16-PSK signals, each set having different amplitudes, and phase information for each

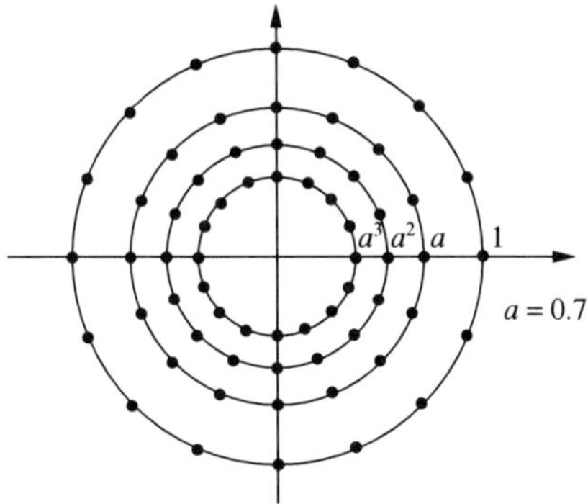

Fig. 4.43   64-DAPSK Signal Space Diagram

symbol the same as for differential 16-PSK. However, with regards to amplitude, the ratio between the amplitudes of the previous and current symbols is used for information purposes. Thus, even though we are talking about *M*-ary modulation, because the system operates on the phase difference and the amplitude ratio between symbols, utilization in differential decoding is possible.

4.4.3 Theoretical Analysis of Bit Error Rate Characteristics

In this section we shall discuss the subject of bit error rates (BER) in digital broadcast television in the mobile environment (Fig. 4.44). The first case we consider is flat fading conditions. (Flat fading is characterized by the delay time of the delayed waves being shorter than the inverse value of the bandwidth.) In the second case, we discuss digital broadcasts using ground-wave propagation in the single-frequency network (SFN), a system that is unique to digital broadcast TV. In the SFN case, we consider a mobile located in straight-line paths between two stations, which receives only direct wave signals from each of the stations. (In this

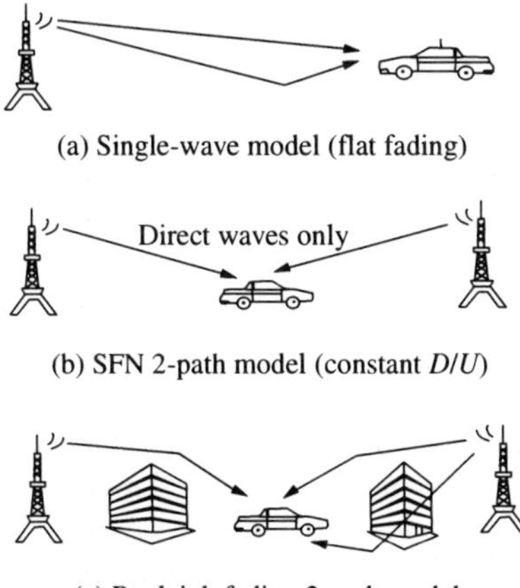

(a) Single-wave model (flat fading)

(b) SFN 2-path model (constant *D/U*)

(c) Rayleigh fading 2-path model

Fig. 4.44   Propagation Path Models

case, frequency-selective fading is possible, but waves amplitudes do not vary.) We have chosen to call this a "SFN 2-path model." In the third case, we consider an indirect (out-of-sight) transmission between the transmitter and receiver. Here, we expect reflected and diffracted waves, with the resulting conditions producing rapid changes in *D/U* and frequency-selective fading. We refer to this as a "Rayleigh fading 2-path model" [Tsuzuku, 1997].

Flat Fading (Mobile Receiver)

Fading is caused by the large variations in received signal power due to reflected and diffracted waves in an indirect (out-of-sight) path between the transmitter and receiver. As long as the delay time of the delayed waves remains small, there is little effect on the frequency characteristics of the signal envelope, and we have flat fading in this case. As was noted in Sec. 4.3.2, where $E(x)$ is the amplitude probability density of the received signal, BER is expressed as

$$BER(C/N) = \int_0^\infty E(x) \cdot A_1 \cdot \mathrm{erfc}\left(A_2 \cdot \sqrt{C/N} \cdot x\right) dx \qquad (4.29)$$

Since $E(x)$ is the Rayleigh distribution from eq. (4.27), substituting eq. (4.27) into the above equation and integrating yields the bit error rate. The results are shown in Fig. 4.45. Note that compared to an absence of delayed waves, the improvement in BER is limited, even with higher *C/N* values.

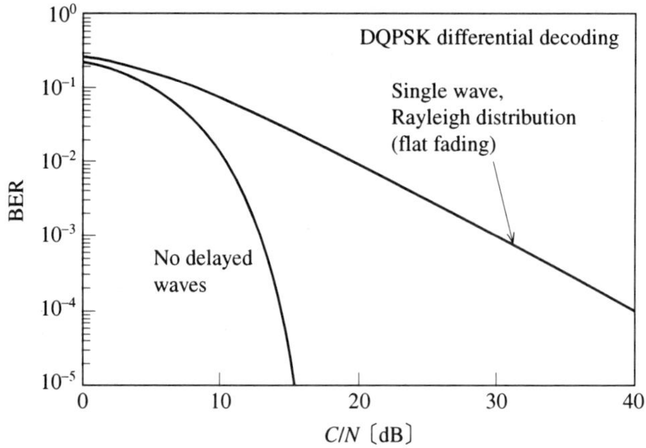

Fig. 4.45   Bit Error Rate (Flat fading)

SFN 2-Path Model

When delayed waves are present, BER performance is degraded regardless of whether or not the waves are fluctuating. Fluctuations in the delayed waves does cause further degradation, however. Figure 4.46 shows only the results of the delayed wave effects on BER. For a more comprehensive treatment of the effects of delayed waves on BER, refer to the [Tsuzuku, 1997] reference.

The Effects of Frequency-Selective Fading

Since mobile receiver antennas are short, a straight-line transmission between the transmitter and receiver occurs infrequently, and most often, the electric-field strength of the received signal has a Rayleigh distribution. In the SFN case, since the radio signals transmitted by both stations have a Rayleigh distribution, D/U goes through frequent changes. BER in this case is illustrated in Fig. 4.47, where Doppler frequency shift is taken as a parameter (Tsuzuku, 1997]. As is evident from the graph, when the Doppler frequency shift is large, BER fails to improve, even in the area where C/N values are high.

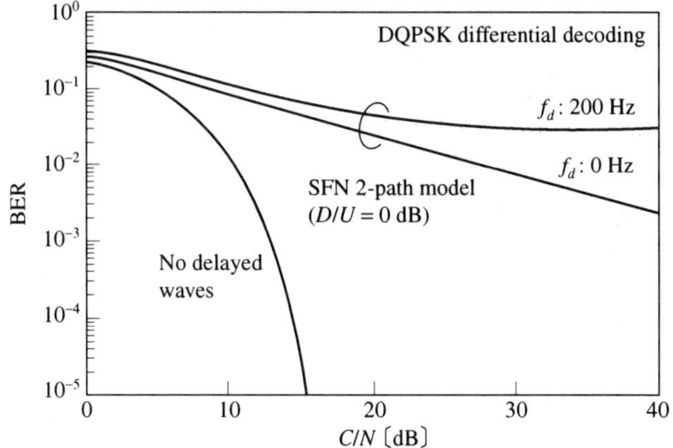

Fig. 4.46 Bit Error Rate (SFN 2-path model)

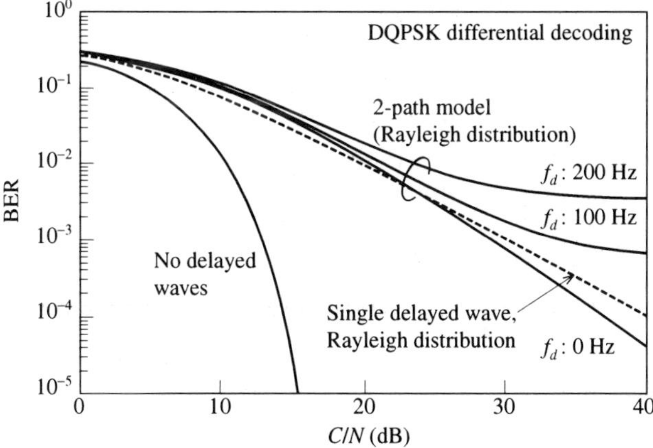

Fig. 4.47 Bit Error Rate (2-path model)

4.4.4 Link Budget (Mobile Receiver)

The purpose of the link budget was discussed in Sec. 4.3.4. Here we will briefly describe the items listed in Table 4.3 that are applicable to a mobile system. (For the

Table 4.3   Link Budget Table (Mobile Receiver)

Item	Digital (OFDM)		
Receiver type	Fixed	Mobil	Mobil
(1) Frequency (MHz)	770	770	770
(2) Modulation system	64QAM	16DAPSK	DQPSK
(3) Transmission bandwidth (MHz)	5.6	5.6	5.6
(4) Spectrum efficiency (bit/s/Hz)	3.28	2.18	1.09
(5) Information transfer rate (Mbit/s)	18.3	12.2	6.1
(6) Pre-FEC bit error rate	$1*10^{-2}$	$1*10^{-2}$	$1*10^{-2}$
(7) Required C/N (dB)	19	20	9
(8) Equipment performance degradation margin	1	1	1
(9) Required C/N for receiver (dB)	20	21	10
(10) Receiver noise figure NF (dB)	5	5	5
(11) Receiver noise power N (dBm)	−101.3	−101.3	−101.3
(12) External source noise power (dBm)	−103.1	−103.1	−103.1
(13) Total receiver noise power (dBm)	−99.1	−99.1	−99.1
(14) Minimum input voltage for receiver (dBμ)	35.6	36.6	25.6
(15) Receiver antenna gain G (dB)	10	0	0
(16) Antenna height loss (dB) (based on 4 m standard)	0	4	4
(17) Effective antenna length (dB)	−18.1	−18.1	−18.1
(18) Feeder loss L_f and insertion loss L_m (dB)	3	2	2
(19) Required electric-field strength (no variation margin) dB (μV/m)	46.7	60.7	49.7
(20) Reception time percentage (dB) (corrected from 50% to 99%)	4	4	4
(21) Reception site percentage (dB) (for mobiles, corrected from 50% to 99%)	8.2	15	15
(22) Electric-field fluctuation margin, portable/mobile receivers (dB)	–	9	9
(23) Required electric-field strength (μV/m)	58.9	88.7	77.7
(24) Transmitter power ratio (based on analog TV = 1)	0.08	74.4	5.9

description of the other items, refer back to Sec. 4.3.4.)

   The numbered descriptions below apply to the mobile receiver case.

(2) Modulation system: 64-QAM-OFDM modulation is used in fixed receiver applications, but considering the low required C/N of mobile receivers, and the utilization of differential decoding, 16-DAPSK-OFDM and DQPSK-OFDM sytems are used in mobile receivers.

(4) Spectrum efficiency: Refers to the number of bits transmitted per symbol. Four bits per symbol are transmitted with 16-DAPSK, and two bits with DQPSK. Comparing the spectrum efficiency of these two systems with 64-QAM, it is 2/3 and 1/3, respectively.

(7) Required C/N (dB): Refers to the minimum C/N which will satisfy the bit error rate in item (6).

(15) Receiver antenna gain G (dB): A whip antenna is used with the mobile receiver.

(16) Antenna height loss (based on 4-m): The height of the mobile receiver antenna was 1.5 meters. Electric-field strength reduced by 4 dB compared to 4-m antenna (Fig. 4.41).

(20) Reception time percentage (dB) (50% rate): 50% time service of analog TV corrected to 99% for digital mobile TV. Correction required to account for the

cliff effect [ITU, 1990].

(21) Reception site percentage (dB) (50% rate): 50% site service of analog TV corrected to 99% for digital mobile TV. As for time service, this correction is required to account for the cliff effect. Also, unlike fixed or portable sets, the mobile receiver cannot be located for "best reception," which influenced the reception site rate selection of 99%. Reception site percentage provides the margin with respect to fluctuations of the center-value for fading (established for urban use) [Okumura, 1986].

(22) Electric-field fluctuation margin, portable/mobile receivers: Refers to the margin with respect to instantaneous fluctuations due to fading (Rayleigh) in the mobile receiver [Okumura, 1986].

(23) Required electric-field strength dB (μV/m): (23) = (19) + (20) + (21) + (22).

To compensate for the performance degradation caused by fading, low antenna gains, and other factors, the mobile system requires a much higher transmitter power value in order to provide reception coverage of a service area equal in size to a conventional analog TV service area. In the above link budget example, the required transmitter power is 5.9-times greater for DQPSK, and 74-times greater for 16-DAPSK. If the same transmitter power served both fixed receivers and mobile receivers, the mobile service area would be reduced to an unacceptably small area.

4.5 Developments in Digital Television

4.5.1 The International Scene

Developmental efforts in high-picture quality television using the terrestrial (ground-wave) broadcast medium and next-generation television designed for interoperability with computers and other telecommunications infrastructure are primarily centered in the United States and Europe. The next-generation ATV (Advanced TV) under development in the U.S. uses a single-carrier transmission system, while in Europe, OFDM is the favored system.

Developing International Standards

The development of international standards for terrestrial digital television, as was the case for DAB, is the responsibility of the radio communications sector of the ITU (ITU-R). Particularly with regards to an internationally acceptable transmission system standard, the setting of standards for terrestrial digital television has become an area of contentious debate. Nevertheless, driven by the schedule set for ATV in the U.S., there has been rapid progress made in the area of report production. The SG 11 task group (TG 11/3) produced new recommendations covering the elemental technologies of terrestrial digital TV (e.g., video coding, modulation systems, audio coding, conditional access, service multiplexing, etc.), but in November of 1996, TG 11/3 disbanded, passing on its responsibilities to a related working group. The transmission parameters recommended by ITU-R for terrestrial digital TV are listed in Table 4.4.

Table 4.4   System Recommended by ITU-R

	North America	Japan	Europe	
Parameter	6 MHz single-carrier	6 MHz multi-carrier	7 MHz multi-carrier	8 MHz multi-carrier
Bandwidth	5.38 MHz	5.57 MHz	6.66 MHz	7.61 MHz
No. of carriers	1	1,405, 2,809, 5,617	1,705 and 6,817	1,705 and 6,817
Modulation system	8 - VSB	QPSK ~ 64 QAM (OFDM)	QPSK ~ 64 QAM (OFDM)	QPSK ~ 64 QAM (OFDM)
Symbol length	92.9 ns	$252 \sim 1{,}008\ \mu s$	$256 \sim 1{,}024\ \mu s$	$224 \sim 896\ \mu s$
FEC system	Trellis + RS	Convolutional + RS	Convolutional + RS	Convolutional + RS
Transfer rate	19.39 Mbit/s	Max. 23.2 Mbit/s	4.35 ~ 27.71 Mbit/s	4.98 ~ 31.67 Mbit/s
Required C/N	15.19 dB	3.1 ~ 20.1 dB	3.1 ~ 20.1 dB	3.1 ~ 20.1 dB

The European Scene

The results of a study on the future of digital television in Europe was released in the Reimers report. This report included a general indication of the transmission rates required for fixed, portable, and mobile receiver applications. The ongoing EP-DVB (European Project - Digital Video Broadcasting) project is also an attempt to come up with a universal European system standard.

In Great Britain, terrestrial digital TV broadcasting is scheduled to start in the latter part of 1998, and the necessary digital equipment is being procured. The British government has already allocated six channels for use by terrestrial digital TV. What is noteworthy here is the fact that the digital TV channels have been allocated in adjacent channels to existing analog broadcast channels. This means that existing transmitter/receiver antennas can be used for both digital and analog broadcasts, which minimizes the burden on the broadcast service providers, and should smooth the transition to all-digital service.

The North American Scene

The investigation into ATV was started in November, 1987 with the establishment of ACATS (Advisory Committee on Advanced Television Service) by the FCC (Federal Communications Commission in the U.S.), whose function was to make recommendations on an ATV system. In its initial stages, the committee studied a high-detail image broadcasting system, but switched to investigating a digital broadcasting system partway through the project. In November, 1995, ACATS submitted to FCC four of the best proposals as the ATSC-DTV (Advanced TV System Committee - Digital TV) standard. In this report, an 8-ary VSB modulation system was recommended for terrestrial broadcasts, and 16-ary VSB was recommended for CATV. Both recommendations were based on the Zenith Corporation proposal, and with the use of a good VSB filter, excellent results with regards to interference immunity have been reported.

The FCC has sent a schedule to television broadcasters, specifying dates by which digital TV service must be implemented (Table 4.5). Also, analog TV broadcasting licenses expire in 2006, ending analog TV services.

Table 4.5　Digital TV Implementation Schedule for the U.S.

Service region	Date
The four major networks in the top-ten major markets.	by 5/1/1999
The four major networks in the 11 ~ 30 markets.	by 11/1/1999
Remaining commercial TV stations.	by 5/1/2002
Non-commercial TV stations.	by 5/1/2003

4.5.2 The Domestic (Japan) Scene

In Japan, research and development on the OFDM system and CDM (Code-Division Multiplexing) has been primarily conducted by NHK's Science and Technical Research Laboratories. The Communications Research Laboratory (CRL), under the Ministry of Posts and Telecommunications has also been engaged in OFDM device development because of their interest in the spectral efficiency this system offers. In 1996, NHK, together with the researchers from private broadcasters and the various electrical appliance manufacturers formed a private company and set about developing the next-generation digital television system. With regards to an OFDM transmission system, research and development in Japan began a number of years ago [Hirosaki, 1980], but as for utilization of OFDM in a digital television transmission system, efforts in Japan have lagged behind European research. Japan's efforts in this field, ranging from its research organizations to its equipment manufacturers and private broadcasters, have ground to make up in this area.

An Overview of Japan's Television Advisory Structure
The Telecommunication Technology Council has been tasked with conducting deliberations regarding the transmission system for terrestrial digital television. (A report is scheduled for release in the Spring of 1999.) The general parameters of the transmission system under consideration are shown in Table 4.6, and the organizational structure of the deliberative process is outlined in Fig. 4.48. An experimental system is now being fabricated by the Ministry of Posts and Telecommunications based on provisional proposals, and field testing for verification will be conducted upon completion. An approximate schedule is shown in Fig. 4.49.

Table 4.6   Provisional Transmission System (Japan)

Parameter	6 MHz OFDM
Bandwidth	$6 \times 13/14$ MHz (approx. 5.57 MHz)
No. of carriers	1,405 (Mode 1) 2,809 (Mode 2) 5,617 (Mode 3)
Modulation system	QPSK, DQPSK, 16QAM, 64QAM
Symbol length	252 μs 504 μs 1,008 μs
Guard interval length	1/4, 1/8, 1/16, 1/32 of effective symbol length.
FEC system	Convolutional code r = 1/2, 2/3, 3/4, 5/6, 7/8 − RS(204,188)
Transmission rate	23.23 Mbit/s max.

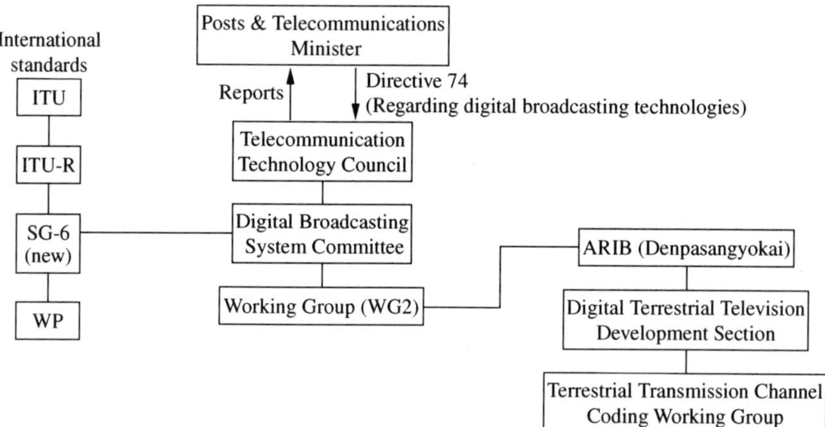

Fig. 4.48   Deliberative Structure for Terrestrial Digital Television

Fig. 4.49   Digital TV Schedule

The Features of Japan's Transmission System

As noted in Table 4.7, the transmission system selected for Japan's terrestrial digital television has a number of advantageous features. These features are briefly described below.

(a) BST and hierarchical layering: With BST (Band-Segmented Transmission), the 6 MHz band is divided into multiple groups so that each group can be modulated by the most appropriate modulation system. In the provisional system, one unit (or segment) consists of 6/14 MHz, and there are a maximum of 13 segments (approx. 5.57 MHz). (See Fig. 4.50.)

Table 4.7   Features of Japan's Transmission System

1. BST and hierarchical layering.
2. Modulation selectable for mobile reception.
3. Pilot signal system.
4. Reinforced interleaving.
5. Transmission rate suitable for HDTV.

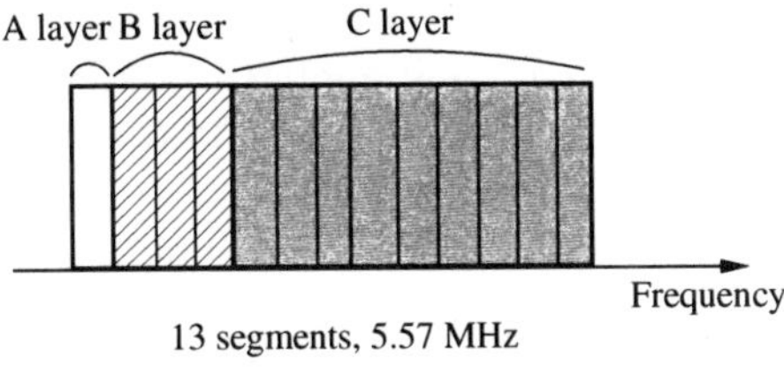

Fig. 4.50   BST and Hierarchical Layering

(b) Modulation selectable for mobile reception: The transmission parameters are as noted in Table 4.6, but one of the system's most advantageous features is the inclusion of a selectable DQPSK modulation system suitable for mobile systems. There are also three operational modes: Mode 1 for mobile reception, Mode 3 for fixed receivers, and Mode 2 for intermediate reception conditions.

(c) Pilot signal system: The two types of pilot signals are the SP (Scattered Pilot) and the CP (Continual Pilot). The pilots are inserted as TMCC (Transmission and Multiplexing Configuration Control) information specifying transmission mode.

(d) Reinforced interleaving: The provisional system uses four levels of interleaving. The interleaving structure is as follows:

1) Interleaving between the inner and outer code is in byte units (byte interleaving).

2) Mapping to each symbol is in bit units (bit interleaving).

3) Delay time can be set up to a maximum of 1 second (time interleaving).

4) Carriers can be switched to the frequency-domain (frequency interleaving).

(e) Transmission rate can accommodate HDTV: Transmission rate is a maximum of 23.23 Mbit/s. Since HDTV broadcasts require a bit-rate of approximately 18 Mbit/s, the system is capable of handling one channel of HDTV programming and one channel of standard TV programming.

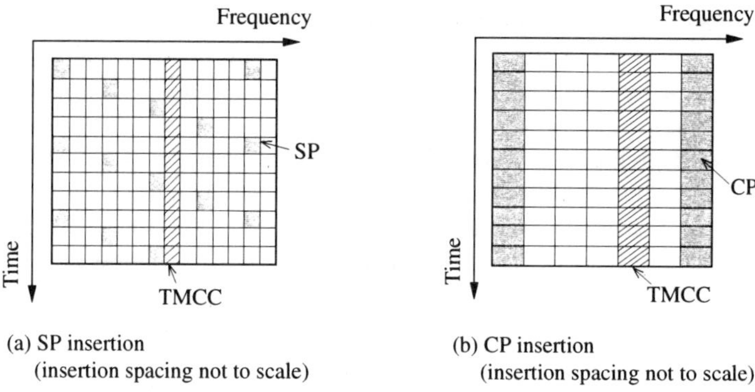

(a) SP insertion
(insertion spacing not to scale)

(b) CP insertion
(insertion spacing not to scale)

Fig. 4.51   Pilot Signal Configuration

4.6 Conclusion

The OFDM transmission system, as it applies to terrestrial digital television broadcasting, was discussed in this chapter. The effects of delayed waves, which is the most troublesome problem in terrestrial broadcasts, was covered, as was mobile receiver applications. Since this is the first implementation of an OFDM transmission system in terrestrial digital TV, unfortunately there are very few works treating this subject specifically. However, for those readers desiring a comprehensive treatment of radio wave propagation and mobile receiver systems, the [Mushiake, 1961] and [Okumura, 1986] references are recommended.

Chapter Questions

Q1: List the problem areas of the OFDM system.

Q2: What features of the OFDM system make it resistant to the effects of delayed waves?

Q3: List the problem areas of the single-frequency network (SFN).

Q4: Why must the signal power for mobile receivers be greater than for fixed receivers?

Chapter References

[Alard, 1987] Alard M. and Lassalle R. : "Principles of modulation and channel coding for digital broadcasting for mobile receivers," EBU Review-Technical, No. 224, pp. 168-190 (Aug. 1987)

[BTA. 1988] Housou Gijutsu Kaihatsu Kyougikai (BTA): "Terebi Juushinki Kaizen Iinkai Chousa Kenkyuu Houkokusho" (1988.11) (in Japanese)

[Chang, 1966] Chang R. W. : "Synthesis of band-limited orthogonal signals for multichannel data transmission," BSTJ, Vol. 45, pp. 1775-1796 (1996)

[Chang, 1968] Chang R. W. and Gibby R. A..: "A theoretical study of performance of an orthogonal multiplexing data transmission scheme," IEEE Trans. Comm., Vol. COM-33, pp. 529-540 (1968)

[Cimini, 1985] Cimini L. J. Jr. : "Analysis and simulation of a digital mobile channel using orthogonal frequency division multiplexing," IEEE Tans. Comm., Vol.COM- 33, pp.665-675 (1985)

[Hirosaki, 1980] Hirosaki B., "An analysis of automatic equalizers for orthogonally multiplexed QAM systems," IEEE Trans. Comm., Vol. COM-28, pp. 73-83 (1980)

[ITU, 1982] ITU-R Report : "Systems for the broadcasting Satellite service," Rep. 215-5 (1982)

[ITU, 1990] ITU-R Recommendation : "VHF and UHF Propagation Curves for the Frequency Range from 30 MHz to 1000 MHz," Rec. 370-5 (1990)

[ITU, 1996] ITU-R Recommendation : "Criteria required for planning digital terrestrial television service in the VHF/UHF television band," Rec. BT-655 (1997)

[Moriyama, 1989] Moriyama, Saruwatari, Mizuno, Iwama, Ryuko, Sekizawa: "Jun-maikuro Hatai Rikujou Idou Denpan Jikken Kekka", Tsuushin Sougou Kenkyuujo Dai 76 kai Kenkyuu Happyoukai (1989.5) (in Japanese)

[MPT order, 1950] Yuuseishourei: "Housoukyoku no Kaisetsu no Komponteki Kijun", Denpa Kanri Iinkai Kisoku Dai 21 gou 2 jou 11 kou (1950) (in Japanese)

[Mushiake, 1961] Y. Mushiake: "Antena Denpa Denpan", Koronasha (1961) (in Japanese)

[Nagatsuka, 1994] Nagatsuka, Tsuzuku: "OFDM ni yoru Chijou Deijitaru Housou - Hisenkei Kaisen de no Tokusei Kentou -", 1994-nen Denshi Jouhou Tsuushin Gakkai Shunki Taikai, B-381 (1994.3) (in Japanese)

[Okumura, 1967] Okumura, Ohmori, Kawano, Fukuda: "Rikujou Idou Musen ni okeru Denpantokusei no Jikkenteki Kenkyuu", Kenkyuu Jitsuyouka Houkoku, Vol. 16, No. 9, pp. 1705-1764 (1967) (in Japanese)

[Okumura, 1986] Okumura, Y. and Shinji, M., eds.: "Basic land mobile communications," IEICE, pp. 24-60 (10.1986) (in Japanese)

[Rau, 1990] Rau M. C., Claudy L. D., Salek S. : "Terrestrial Coverage Considerations for Digital Audio Broadcasting Systems," IEEE Trans. Broadcast. 364 pp. 275-283 (Dec.1990)

[Rault, 1989] Rault J. C., Castelain D., and Le Flock B. : "The coded orthogonal frequency division multiplexing (OFDM) technique, and its application to digital radio broadcasting towards mobile receivers," IEEE GLOBECOM 89, No. 12. 3, pp. 428-432 (Nov. 1989)

[Rohling, 1995] Rohling H. and Engels V. : "Multilevel differential modulation technique with 64-DAPSK," Deutsche Telekom DAPSK workshop (1995)

[Sistanizadeh, 1993] Weinstein S. B. and Ebert P. M. : "Data transmission by frequency division multiplexing using the discrete Fourier transform," IEEE Trans. Comm., Vol. COM-19, pp. 628-634 (1971)

[Tsuzuku, 1995] Tsuzuku, Fukuchi: "OFDM ni yoru Chijou Deijitaru Housou - Nishuuhasuu Housouha Chuukei (DFN) no Kentou -", B-917, p. 363 (1995.3) (in Japanese)

[Tsuzuku, 1997] Tsuzuku, Ohta, Nakamura: "Chienpa ga OFDM Densou ni ataeru Eikyou", Eizou Jouhou Medeia Gakkaishi, Vol. 51, No. 10, pp. 1493-1503 (1997) (in Japanese)

Chapter 5
Satellite Broadcast Television

This chapter describes the elements of satellite broadcasting, an area in which the development of digital techniques is progressing rapidly. We shall discuss the features of the satellite broadcasting system and the frequencies that it utilizes, and explain the effects digitalization has had in this area. The chapter starts with an overview of the current analog satellite broadcasting system, and then as a CS-band digital satellite broadcasting service, a description of the PerfecTV system broadcast over the JCSAT (Japan Communications Satellite) will be intertwined with the discussion on satellite transponders, the earth stations, and the satellite broadcasting link budget. We shall also introduce the subject of rain attenuation prediction and describe how rain affects radio wave propagation, and consequently, the system's link budget. The chapter will conclude with a discussion on future BS-band digital broadcasting technology, and the future use of the 21-GHz satellite band.

5.1 An Overview of Digital Satellite Broadcasting

5.1.1 Introduction

Early Predictions

The primary event providing the impetus to the development of modern satellites was, no doubt, the launching of Sputnik by the Soviet Union in 1957. The subsequent competition between the United States and the former Soviet Union resulted in tremendous progress in development of the communications, broadcasting, and the observation technologies deployed in today's satellites. As a testament to the rapid advances made in the various fields of technology, only 12 years after the launching of Sputnik, whether driven by purpose, or from a desire to defeat the Russians, the US managed to land a crew of astronauts on the surface of the moon and safely return them to earth.

The first proposals regarding the concept of man-made satellites carrying communications and broadcasting equipment were made in 1945, 12 years prior to the launching of Sputnik. At the time, Arthur C. Clark was serving in the British Royal Air Force, and it was he who envisioned the concept of satellites lifted into a geostationary orbit, capable of facilitating communications between distant countries. (Refer to Interlude 5.1.) Clark's proposal clearly indicated the feasibility of using satellites for television signal broadcasts. Clark also indicated in his writings that a satellite-mounted antenna of about 1 meter, and an antenna length of 30 centimeters at the earth-based receiver should provide acceptable picture quality. Clark's early predictions were quite prophetic, as today we not only have the ability to communicate world-wide using a geostationary orbit satellite system, but satellites provide useful services in a wide range of applications. With regards to television, the modern broadcast-type network is a system covering a wide area that has many of the features that Clark discussed in his early proposals.

Clark's Early Predictions of a Global Communications Network using Satellites – The Father Satellite of Communications, John Pierce [Clark, 1993]

Arthur C. Clark was a Radar Technician serving in the Royal Air Force during World War II. In early 1945, buoyed by the spirit and a newly-found sense of optimism felt with the winding down of the war, Clark allowed his mind to wander into thoughts of space travel and new ideas in communications. The fact that Clark was also a member of the British Space Travel Club, and was deeply interested in radio communications by virtue of his work with radar probably had a large influence on his ideas in this area. In his article entitled "V2 for Ionosphere Research?" that appeared in the February 1945 edition of Wireless World, Clark proposed that three satellites positioned 120-degrees apart and in 24-hour orbits around the earth could provide television signals and microwave communications service to all areas of the globe. In May of the same year, Clark distributed among his associates a memorandum of 4-5 pages entitled "The Space Station: Its Radio Applications," in which he described the concept of a geostationary orbit type relay satellite. A compilation of his writings were published in the October, 1945 edition of Wireless World under the heading of "Extra-Terrestrial Relays – Can Rocket Stations Give World-Wide Radio Coverage? --." In this article, Clark discussed the first astronautical velocity (the initial velocity required for a rocket to break free of earth's gravity field), satellite-mounted antennas, and antenna specifications in general, and further elaborated on the future technologies that would be required to achieve satellite communications. Clark's writings clearly indicated a strong premonition regarding the future deployment of geostationary communications satellites.

On another front, in the May, 1955 edition of Jet Propulsion, John Pierce of Bell Laboratories (US) wrote an article entitled "Orbital Radio Relays," without benefit of ever having read Clark's writings. In this article, Pierce reported the results of a study on a communication technique which used an active repeater (*1) in a geostationary orbit satellite carrying a communications reflector. Subsequently, the honor of putting the first man-made satellite into orbit went to the Soviet Union. However, in 1960, Pierce and his colleagues at Bell Laboratories launched the Echo experimental satellite with a passive repeater (*2), and with the successful 1962 launch of the Telstar carrying an active repeater, the first television signals were relayed across the Atlantic.

Although Arthur Clark is a well-known science-fiction writer, it seems that in his youth, John Pierce, writing under a pen name, also had some science and technology articles published in the Astounding Stories magazine. In the May, 1945 edition of the magazine, an article appeared in which the author indicates that a light ray having a single wavelength will be required for space communications. It does appear that a strong imagination is also an essential element of technological creativity.

*1 Active repeater: A repeater that amplifies and modifies the frequency of the received signal before relaying it on.

*2 Passive repeater: A repeater that relays the received signal without modification.

Features of Satellite Broadcasting

Satellite broadcast television generally implies the broadcasting of signals transmitted by a satellite or other space vehicle, which are received as a direct signal on earth. As will be noted when we discuss the administrative functions of ITU, broadcast television satellite service was initially defined by the 1963 ITU Administrative Radio Conference (EARC) in a special session conducted for the purpose of allocating satellite telecommunications frequencies, and to review the rules for general telecommunications. Further amendments were made at the WARC of 1971. According to the latest definition, satellite broadcasting service is defined as the provision to the general public direct signals or retransmitted signals via a space vehicle. Direct signal reception is divided into individual reception and community reception, according to the way the subscriber receives the signal. In individual reception, a subscriber (usually household) receives the signal using a small antenna, whereas with community reception, the signal is received by a well-equipped facility served by a large antenna. The facility then distributes the signal to multiple subscribers (the general public) within a prescribed service area.

The features of satellite broadcasting generally parallel those of general satellite communications, namely:

(1) Commonality of information: Provides a simple means of sending the same information (e.g., program) to many subscribers.

(2) Wide area of coverage: Radio signals transmitted by satellite cover a large area of the earth, which is an economical and efficient means of providing service country-wide. On the other hand, special satellite antennas and the use of higher broadcasting frequencies are required for broadcasts covering only a designated area (such as the service area covered by a current terrestrial broadcast.)

(3) Disaster resistant: Satellite service is uninterrupted during times of natural disasters on earth (earthquakes, floods, etc.) which can damage land-based

broadcasting facilities.

(4) Uniform coverage over a large geographical area: The steep angle-of-arrival of the signal at the receiver provides uniformly good reception in normally difficult areas such as extremely mountainous terrain or isolated islands. The high arrival angle of the signal also largely eliminates the ghost image problem typically caused by large buildings in urban areas; a common problem in terrestrial broadcasting systems.

(5) Good mobility: Satellite communications links are easily established in the field. Mobile transmitter stations can be used to provide on-scene relay coverage of news events occurring anywhere in the world.

An additional feature of satellite broadcasting is the numerous channels it makes available to subscribers. Also, with new frequency band allocations, broadcasting over a wider channel bandwidth provides a higher quality picture than terrestrial broadcast television.

Frequency Allocations for Satellite Broadcasts [Endoh, 1988, in Japanese]

The successful transmission of television signals between North America and Europe relayed by the Telestar satellite, and completion of a number of other preliminary trials in the field of satellite communications brought about the need for frequencies that could be used in space communications. These frequencies were allocated at the 1963 ITU Special Session ARC (EARC), facilitating the initiation of satellite services such as provided by Intelsat and various weather satellites. The 1963 EARC was also the first conference where the problem of defining satellite broadcasting services was taken up for debate. The WARC-ST (World ARC-Space Telecommunications) opened in 1971 to consider matters related to space telecommunications. This was also the first conference which clearly defined what should constitute a broadcasting satellite. This conference allocated six frequency bands for use in satellite broadcasting: 700-MHz, 2.6-GHz, 12-GHz, 22-GHz, 42-GHz, and 84-GHz.

The 1977 WARC-BS conference was convened to consider the commercial use of satellites broadcasting in the 12-GHz band, and satellite broadcast channels were allocated for the ITU first-region (Europe,Russia, Africa) and the ITU third-region (Asia, Oceania), taking into consideration the current analog systems in use at the time. A few adjustments and additions were made to this allocation in the 1979 WARC. Channel allocations in the 12 GHz satellite broadcast band for the ITU-defined second-region (North America, South America) were made in the Second Regional Administrative Conference (RARC-SAT-83) held in 1983. With regards to 12-GHz band broadcasts, Japan pioneered in this field, and was already operating a smoothly functioning system. On the international scene, as the performance and price of receiving facilities became more favorable, and with the advent of practical digital technologies, developments in this area became even more productive. As new digital technologies appeared, ITU was obliged to reconsider the WARC-77 12 GHz channel plan, which was the primary purpose of WRC-97. (WRC, World Radio Conference, is the new title replacing WARC.)

Thus, as a broadcast television medium, the 12-GHz satellite band is maturing, both in a technological and a practical sense. As satellite broadcasting takes on a bigger role, however, and as anticipated improvements in performance and quality are realized, the 12-GHz band will most likely not be able to accommodate future needs. Research and development on a higher frequency satellite broadcasting systems are therefore being actively pursued, both in Japan and abroad. Thus, with regards to the frequencies assigned for wideband HDTV satellite broadcasting, whose original purpose was to provide the means of transmitting higher volumes of information than 12-GHz band satellites, the WARC-ORB-1 (1985) conference decided to abandon the 22.5 ~ 23 GHz assignments made previously to the second- and third-regions, and instead consider a universal radio frequency band plan. The decision regarding this was continued to the 1988 WARC-ORB-2 conference, however. The wideband HDTV satellite broadcasting matter was discussed at WARC-ORB-2, but all decisions were put off for a future WARC meeting. International frequency allocations for wideband HDTV satellite TV were finally made at WARC-92. World-wide frequency band allocations were also made for CD-grade audio for mobile (automotive) satellite digital sound broadcasting at WARC-92. Frequencies allocated for satellite broadcasting at WARC-92 are listed in Table 5.1.

Table 5.1　Satellite Broadcasting Frequencies (WARC-92)

	Satellite sound broadcasting (mobile service assumed)			Wideband HDTV satellite broadcasting	
Frequency	1.5-GHz band (1.452~1.492 GHz)	2.3-GHz band (2.310~2.360 GHz)	2.5-GHz band (2.535~2.655 GHz)	21.4 ~ 22 GHz	17.3 ~ 17.8 GHz
Regions	World-wide, except USA	USA, India	Japan, Korea, China, Singapore, Thailand, India, Pakistan, Sri Lanka, Bangladesh, Russia, Ukraine, Belarus	ITU 1st region (Europe, Russia, Africa) ITU 3rd region (Asia, Oceania)	ITU 2nd region (North, South America)

As Japan is included in the ITU third-region, its satellite broadcasting frequencies are listed in Table 5.2. The course in reaching the decision to use the 21 GHz band as the satellite broadcasting frequency for next-generation television has had many twists and turns, but the band finally settled on is close to the one proposed by the Japanese delegation in the 1971 WARC-ST [Kuwata, 1982].

5.1.2 Analog Satellite Broadcasting Systems

The Analog Satellite TV Scene

From inception, communication satellites have filled the role of providing television broadcast signals that cover very large service areas. As mentioned previously, the

Table 5.2   Japan's Satellite Broadcast Frequency Band Allocation

Frequency	Range	Notes
700-MHz band	620 ~ 790 MHz	Conditional, not used in Japan.
2.5-GHz band	2.5 ~ 2.69 GHz * 2.535 ~ 2.655 GHz	*Satellite sound broadcasting, allocated at WARC-92.
12-GHz band	11.7 ~ 12.2 GHz * 12.5 ~ 12.75 GHz	In service - satellite TV. *Community reception (power restrictions).
21-GHz band	21.4 ~ 22 GHz	Wideband HDTV satellite TV, allocated at WARC-92.
40-GHz band	40.5 ~ 42.5 GHz	
80-GHz band	84 ~ 86 GHz	

1971 WARC-ST conference defined satellite broadcasting service and established the frequency bands that the services may utilize, which paved the way for subsequent testing using various broadcast satellites (ATS-6 (USA), CTS (Canada), BS (Japan), OTS (Europe)). Since the 1977 WARC-BS, a number of experimental and in-service satellites operating in the 12 GHz broadcast channel plan have also been placed into orbits. Examples of these operational systems include Japan's BS satellites, and the European TDF, TV-SAT, and BSB satellites. Broadcast television services utilizing communication satellites are provided by Luxembourg's ASTRA, Canada's ANIK-C, and Japan's JSAT satellites.

With regards to analog satellite TV broadcasts, the TV signal is multiplexed with the audio signal and is broadcast as an FM signal. The 12-GHz satellite broadcast channel plan was based on the assumption that frequency modulation would be used, but as long as technical parameters such as the interference protection ratio are satisfactory, digital, or any other suitable modulation system can also be utilized. As the channel plan in Table 5.3 indicates, the channels allocated to Japan include the eight odd-numbered channels from 1 to 15, which are transmitted as right-hand circularly polarized waves emitted from broadcast satellites whose geostationary orbits place them at 110-degrees east longitude. The satellites in this

Table 5.3   Japan's Channel Allocation and Program Provider Service
in the 12-GHz Satellite Band (as of 9/1997)

Channel	Center-frequency (GHz)	Program
1	11.72748	
3	11.76584	
5	11.80420	JSB (WOWOW)
7	11.84256	NHK satellite No. 1
9	11.88092	Hi Vision (HDTV)
11	11.91928	NHK satellite No. 2
13	11.95764	
15	11.99600	

orbit also serve North and South Korea. The frequency bandwidth of each channel is 27 MHz [Endoh, 1988].

The Analog Satellite TV Scene in Japan
With the successful completion of trial broadcasts conducted using the geostationary orbit BS satellite (launched 4/1978), Japan followed this in May of 1984 with the BS-2a satellite, which ushered in service in the 12-GHz band (channel bandwidth: 27 MHz). The next two launch attempts to place additional satellites into orbit ended in failure, but by March of 1997, the number of subscribers receiving satellite service in Japan had reached 11,240,000. With the 1990 launching of BS-3a, NHK started service over two satellite channels, and JSB (Japan Satellite Broadcasting Corp.) and SDAB (Satellite Digital Audio Broadcasting) initiated pay television service and PCM audio multiplexed broadcasting, respectively. The broadcasting of HDTV started in 1989 with a trial period of service for one hour per day over the BS-2b satellite. This system employed MUSE (Multiple Sub-Nyquist Sampling Encoding), and in November, 1991, trial service was expanded to eight hours per day over BS-3b. This service period was gradually expanded, until by 10/97, operational service (still experimental) was for 17 hours per day.

Analog TV satellite broadcasts using a communication satellite started in 1992 with J-SAT (Japan Satellite Systems, Inc.) and SCC (Space Communications Corporation) providing satellite audio broadcasts and satellite broadcast television services over a common receiver type satellite operating in the 12.5-GHz band. These broadcast signals differ depending on the orbital positions of the satellites. Bandwidth for JC-SAT is 27 MHz, and for the Super Bird is 36 MHz. The radio waves from these satellites are linearly polarized, either vertically or horizontally.

5.1.3 Digital Satellite Broadcasting Systems
Digital Satellite Broadcasting Features
In addition to the normal functions found in satellite broadcasting systems, digitalization of the system produces the unique features listed below:
a) Multiple channel capability: Comparing equal bandwidths, the recent advances made in digital information compression techniques (i.e., MPEG) permit the transmission of information over more channels than analog systems. This feature is particularly advantageous to commercial broadcasters, as it greatly reduces the per-channel costs of using the satellite transponder. Also, considering the fact that the spectrum is a limited resource, reducing the bandwidth a given channel occupies makes good sense.
b) High-quality service, good functionality: By employing digital modulation systems with strong immunity to noise, in addition to reinforced FEC methods, a uniformly high level of service can be provided over any area where the electric-field strength at the receiver is above a given level (which defines the service area). Also, because digital systems are compatible with computers and other signal processing equipment, extra functions, such as the addition of program-

related information into the signal feed, are also available. In future systems, we can also look forward to download type broadcasts in which program material and the software with which the program can be decoded and customized can be accessed by the subscriber.

c) Media interoperability: With the rapid developments in operating speeds and with the increased availability of functions, the computer technologies that provide the means for personalization and the ability to network are quickly creating the realization of a new form of media in which communications and television are integrated.

The Digital Satellite Broadcasting Scene

Digital broadcasting by satellite has generally moved into the implementation phase, both in Japan and abroad, and the number of program providers is increasing rapidly. In the United States, DirecTV (DBS) started offering 175 channels of satellite broadcast service (HS 601) in 1994, and by 8/1997 had proved its popularity with 2,790,000 subscribers (including USSB). Also in 8/97, Primestar began offering about 160 channels of digital programming over the GE-2 satellite, and has signed 1,940,000 subscribers. At the same time, Echo Star was providing approximately 140 channels, with 640,000 subscribers.

In the European region, eight broadcast satellites, or communication satellites with an assigned broadcast mission, are serving France, Spain, Germany, and other Northern European countries with digital broadcasting services. Due to a low number of subscribers, none of these ventures have proven financially successful. However, digital satellite broadcasting service over the CS-band to Great Britain, France, Germany, and Italy utilizing the mid-range output ASTRA and EUTELSAT is increasing [ASC, 1996], [MPT, 1996]. Also in Europe, the EP-DVB (European Project on Digital Video Broadcasting) has been established to set digital broadcasting standards that will provide compatibility between the satellite, terrestrial, and cable media.

In the Asian region, Hong Kong has commenced broadcasting Star TV over Asiasat, and Korea has a digital satellite broadcast using Koreasat.

The Digital Satellite Broadcasting Scene in Japan

Broadcasting of digital satellite television in Japan was initially via communication satellites. Digital multichannel broadcasting started in October, 1996 with Perfec TV, which utilized the JCSAT-3. Service increased in 1997 and 1998 with the addition of digital satellite broadcasting by JskyB and DirecTV Japan.

Perfec TV is broadcast by multplexing several channels through one transponder. Currently each transponder (video/audio bit-rate of 24 Mbit/s) is capable of broadcasting four to seven channels. The per-channel information transfer rate for a four-channel broadcast is 6 Mbit/s, and for six channels is 4 Mbit/s. Programs containing a high percentage of still pictures can be transmitted at a lower rate (~2 Mbit/s), regular moving pictures at 3 Mbit/s, and sports and other action-type programs require a 6 Mbit/s rate. Using 20 transponders allows approximately 100 channels of television programming, and over 100 channels of CD-grade sound broadcasting. The services provided by the broadcaster are essentially picture and sound in a digital multichannel feed. By using an IC card that can be inserted into the subscriber's receiver, services can be billed according to the number of channels received, or even by the individual programs the subscriber views. The latest programming schedule is also continuously transmitted over a designated channel, which is an important service in a multichannel system.

JSkyB commenced service in the Summer of 1997, broadcasting six channels of PerfecTV, and in 1998, increased its digital television service offerings to about 70 channels over JCSAT-4. A receiver is being developed that will provide compatibility between PerfecTV and JSkyB. Also, the longitudinal difference in the orbits of JCSAT-3 and JCSAT-4 is 4 degrees, and an antenna has been developed that is capable of receiving broadcasts from both satellites. Consequently, subscribers to PerfecTV and JSkyB can get reception from both the above satellites.

With the 1997 launching of the SCC-C (Super Bird) satellite, DirecTV Japan began a period of trial broadcasts, and in 12/1997, commenced digital TV service. However, with regards to antenna compatibility, since the orbits of the SCC-C and the JCSAT-3/4 satellites are separated by a considerable distance, it is difficult to develop a regular parabolic dish antenna with the capability of receiving DirecTV, JSkyB, and PerfecTV using the same dish. In order to realize such an antenna, special measures are needed to obtain the proper directivity characteristics. Table 5.4 gives a summary of the digital satellite broadcasting services using

Table 5.4   Digital Satellite Broadcasting Services using Communication Satellites (Japan)

	JSky B	PerfecTV	DirecTV
Provider	JSky B	Japan Digital Broadcasting Service, Inc.	DirecTV Japan, Inc.
Affiliations	News, Corporations, Software banks, Sony, Fuji Television	Japan Satellite Systems, Sumitomo Corp., C. Itoh & Co., Mitsui & Co., Nissho Iwai Corp., TBS	Space Communications Corp., Hughes Space & Communications, Culture Convenience Club, Dai Nippon Printing Co., Matsushita Electric Industrial Co.
No. of channels	1. 6(TV) 2. 70 (TV)	100 (TV) 105 (audio)	100 (TV)
Satellite	1. JCSAT-3 2. JCSAT-4	JCSAT-3	Super Bird C
Service start	1. 1997 (by Perfec TV) 2. 1998.4	1996.10	1997.12
Service provided	Digital video/audio/data Multichannel TV Subscriber (by channel or program) fee service	Same as column to left.	Same as column to left.

communication satellites in Japan. Note that the providers of JSkyB and PerfecTV merged as SkyPerfecTV in 1999, and were joined in 2000 by DirecTV. (All are now SkyPerfecTV.)

The status in Japan of digital broadcasting over BS-band satellites (BS digital broadcasts) was investigated by the Ministry of Posts and Telecommunications (MPT), which as a result, issued the directives summarized below. In view of these directives, a study regarding technical standards was undertaken by the Telecommunications Technology Council (TTC), which in 1998 determined that a governing body should be established prior to the official start of broadcasting, and official commencement of digital satellite television services was set for the year 2000 (Refer to Sec. 5.5.1.) A summary of the MPT directive follows:

(1) Satellite television is to be broadcast digitally, with primary emphasis on HDTV (High-Definition TV).

(2) Analog broadcasts, including Hi-Vision, are to be phased into the digital system. Therefore, the BS-4b satellite must secure digital transition channels for broadcasting the same programming as provided by BS-4a analog services.

5.2 Wave Propagation in the Satellite Link

In formulating the link budget for a satellite broadcasting system, a clear understanding of how perturbers such as rain and obstacles in the transmission channel affect propagation characteristics is essential to being able to properly quantify the system parameters. Fortunately, the propagation path of the satellite broadcasting consists mostly of empty space. That, coupled with the fact that the steep angle-of-arrival of the signal virtually eliminates the problem of ghost interference (common in terrestrial broadcasts) permits the channel to be characterized using the most basic theories of wave propagation. In particular, for satellites that broadcast over the 12 GHz and higher frequency bands, the plane-wave theory of wave propagation in a vacuum usually covered in elementary communications textbooks should suffice for our purposes. However, even though the atmospheric portion of the path between the satellite and the earth may appear insignificant, any rain encountered in the atmosphere produces attenuation and scattering of the radio waves, resulting in a general degradation in the quality of the signal. Satellite broadcasting for mobile applications, while affected less by local environmental conditions than are terrestrial mobile broadcasts, nevertheless must contend with reflection-induced multipath (delayed waves), Doppler frequency shifts, and trees and other obstacles that tend to perturb the signal. These factors, in addition to the normal perturbers affecting non-mobile systems result in varying degrees of signal degradation.

In the discussion that follows, we shall examine rain attenuation as it affects the signal quality of fixed-receiver satellite broadcasting systems, particularly at frequencies above 10 GHz, and will examine some of the countermeasures to this problem. For more details on the propagation of radio waves in mobile satellite communications, the reader is invited to refer to the Satellite Communications reference [Iida, 2000], which is one of the texts in the Wave Summit Course series published by Ohmsha/IOS Press.

5.2.1 Rain Attenuation Characteristics

Radio Wave Propagation Characteristics of Interest

The major factor influencing satellite broadcast signals in frequency bands of 10 GHz and higher is rain. Rain encountered by the signal in the transmission channel results in signal attenuation, depolarization, increased noise due to rain absorption, and interference due to rain scatter. The general concept of this subject is pictured in Fig. 5.1.

Rain attenuation occurs due to the absorption of radio waves by the rain drops, and to the scattering effect, both resulting in diminished received signal power. The rain attenuation-induced difference in the received signal power between a clear day signal and a rainy day signal is normally expressed in decibels. Depolarization of radio waves is due to the shape of raindrops. Raindrops, which we may think of as spherical shapes, are actually formed by the forces of the atmosphere into disk-like shapes. Because of their shape, the amount of rain attenuation of linearly polarized waves varies depending on whether the wave is horizontally- or vertically-polarized.

Depolarization degradation is a product of polarization interference, due to the fact that the raindrops are non-spherical. Thus, as illustrated in Fig. 5.1, because these horizontally- or vertically-polarized waves pass through the aspherical raindrops, a portion of the power originally transmitted is received depolarized. When circular polarization is used, these conditions can switch right-hand polarization to left-hand, and vice versa. These phenomena have an effect on the dual polarization communication system, which is a system designed to efficiently utilize frequency resources by transmitting different information on two orthogonally-oriented polarized waves (vertical and horizontal). Depolarization degradation results in interference between the two polarized signals, reducing the signal transmission quality of the communications link. Moreover, link quality of digital satellite broadcasts at higher than 10 GHz are particularly susceptible to rain-induced depolarization degradation, and rain attenuation can become severe in these systems. However, some studies indicate that if the rain attenuation problem can be overcome, dual polarization digital satellite broadcasting may become a practical reality [Ohta, 1998].

The increase in noise temperature due to rain-induced absorption attenuation is caused by noise being re-radiated from the raindrops acting as absorbers of the radio waves. The relationship between noise temperature, T_s (K), and absorption attenuation, A (dB), is given in eq. (5.1). As rain-related factors which degrade signal quality, note here that we must consider increases in noise power due to absorption attenuation, in addition to the reduction in signal power accompanying rain attenuation.

$$T_s = T_m (1-10^{-A/10}). \tag{5.1}$$

In the above equation, T_m (K) is the equivalent temperature of the absorbing

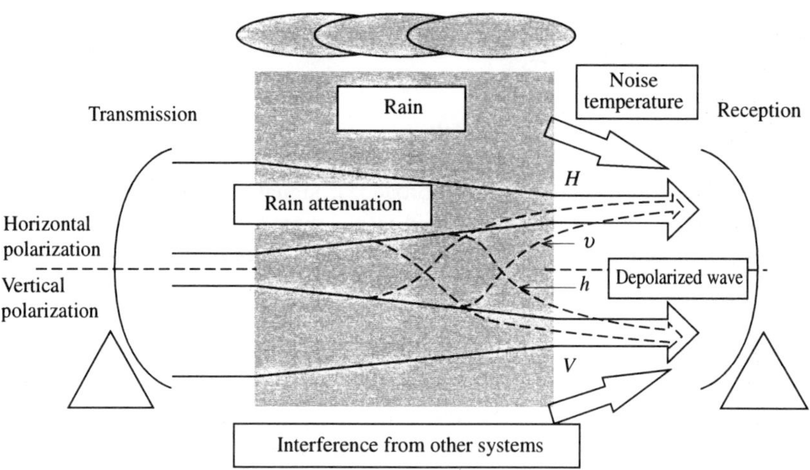

Fig. 5.1   Propagation Path Through a Rain Cell

medium, which, in the case of rain, ITU recommendations specify as 260 K [ITU, 1994 a].

Rain scatter occurs when the transmitting antenna beam passes through a rain cell, and the raindrops scatter a portion of the radio waves. The resulting scattering acts as interference to other communication systems operating in the same frequency band. It is therefore particularly important in satellite broadcasting to take measures that will assure link quality exceeds a prescribed level at the satellite transmitter, and when rain is falling at the earth-based transmitter, transmission power control (TPC, designed for compensating rain attenuation) is often used to maintain power at the satellite receiver end at a constant level. However, in the process of increasing transmitter power to overcome the effects of rain scatter, this method also has the possibility of increasing interference to other systems. The procedure for estimating the level of interference when TPC is used is provided in ITU recommendations [ITU, 1994 b]. Determining the strength of the interference at receivers, and locating the stations operating in the same frequency band for the various communication systems in a given area is referred to as establishing coordination distances.

Of the various factors affecting radio wave propagation, rain attenuation has significant influence on signal quality. In the section below, we shall therefore discuss how rain attenuation fits in the system's link budget.

Rain Attenuation in the Link Budget

The purpose of the link budget is to assure that transmission quality standards are maintained at a level which will provide a specified level of broadcasting service. In analog satellite broadcasting systems, this standard is defined as a signal-to-noise ratio (SNR), and in digital satellite systems, by the bit error rate (BER) of the system. The time parameters over which this standard of service is provided is referred to as the service availability of the system. If, for example, service availability is prescribed as 99.9% of the time on an annual basis, the satellite system must be designed to provide a signal with a BER better than the threshold level specified for at least 99.9% of the time. This means that the system designer must know quantitatively the probability of the signal not meeting the BER threshold for 0.1% of the time (annually). Of the degradation factors we are concerned with, the effects of rain attenuation can be determined by its cumulative distribution, which is illustrated in Fig. 5.2. The data graphs shown in Fig. 5.2 were constructed from measurements taken in the satellite links of Japan's CS- and BS-band experimental geostationary satellites. The measurements were taken at Kashima, Ibaraki Prefecture using the following band and elevation angle information, and applicable dates: 11.7-GHz, 37-deg. elevation angle (1978~1981); 19.5-GHz, 48-deg. elevation angle (1978~1986) [Fukuchi, 1983]. The vertical axis in Fig. 5.2 shows the annual time percentage in which the the horizontal axis value is exceeded. Note that this 0.1% annual time point corresponds to a rain attenuation value exceeding 6 dB for 19.5 GHz, and exceeding 2 dB for 11.7 GHz. In the link budget, in order to meet the required CNR (carrier power/noise power ratio) which will satisfy transmission link quality and modulation requirements, transmitter power must be increased to the point where the clear-weather CNR becom larger than the CNR plus rain attenuation value.

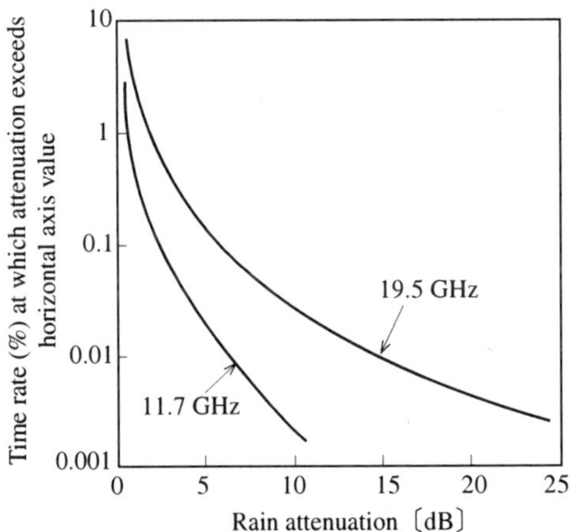

Fig. 5.2   Rain Attenuation Cumulative Distribution

5.2.2 Rain Attenuation Prediction
Estimating Rain Attenuation Cumulative Distribution

ITU recommendations provide methods of estimating the cumulative distribution of rain attenuation for use in formulating the link budgets for satellite communication systems [ITU, 1994 a]. The procedural steps are listed below, and Fig. 5.3 provides a diagram for reference.

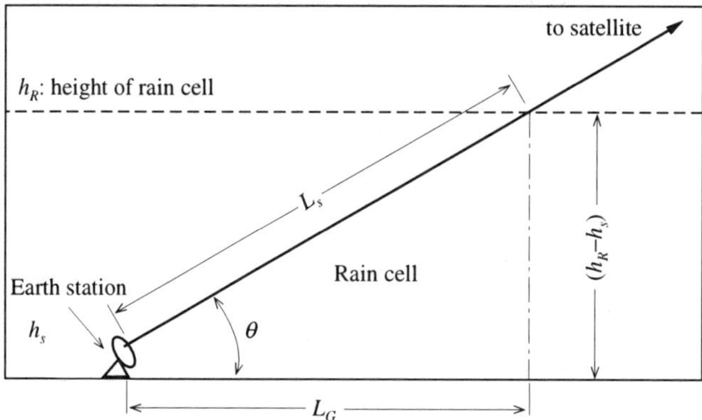

Fig. 5.3   Parameters for Estimating Rain Attenuation Cumulative Distribution

《Required Data》

$R_{0.01}$: Average annual rain intensity (integrated over a time interval of one minute) at a cumulative time rate of 0.01% for the locale in question (mm/h)

h_s : Elevation above sea-level of earth station (km)

θ : Elevation angle to satellite.

ϕ : Latitude (absolute value) of earth station (deg)

f : Frequency (up to 30 GHz) (GHz)

《Estimation Procedure》

Step 1 : Calculate the equivalent height of the rain cell, h_R, from ϕ.

$$h_R = \begin{cases} 3.0 + 0.028\phi, & 0 \leq \phi < 36° \\ 4.0 - 0.075(\phi - 36), & 36° < \phi \end{cases} \quad \text{(km)}. \tag{5.2}$$

Step 2 : When θ exceeds 5 degrees, calculate the length of the rain cell, L_s, by

$$L_s = \frac{(h_R - h_s)}{\sin\theta} \quad \text{(km)}. \tag{5.3}$$

When θ is less than 5 degrees, curvature of the Earth becomes a factor, and L_s is given by

$$L_s = \frac{(h_R - h_s)}{\left(\sin^2\theta + \dfrac{2(h_R - h_s)}{R_e}\right)^{1/2} + \sin\theta} \quad \text{(km)}, \tag{5.4}$$

where R_e is the equivalent Earth's radius, given as 8,500 km when h_s is less than 1 km.

Step 3 : Calculate the horizontal projection distance, L_G, of the rain cell length from

$$L_G = L_s \cos\theta \quad \text{(km)}. \tag{5.5}$$

Step 4 : Obtain $R_{0.01}$. If data is available for the specific locality, that data should be used. Otherwise ITU recommendation PN 837-1 [ITU, 1994 c] provides a world map with rainfall data given in zones (rainfall intensity data by cumulative time).

Note however, that this map classifies Japan into only two zones, which is a poor representation of actual conditions. If the estimation is being made for a station in Japan, using a source for $R_{0.01}$ that is more specific to the locality in question will give better results [Morita, 1978], [Karasawa, 1991].

Step 5 : Calculate the correction factor $r_{0.01}$ to account for the spatial nonuniformity of rainfall. When $R_{0.01}$ is less than 100 mm/h, use

$$\gamma_{0.01} = \frac{1}{1 + L_G / L_0},$$
(5.6)

where $L_0 = 35 \exp(-0.015 R_{0.01})$ (km). (5.7)

When $R_{0.01}$ exceeds 100 mm/h, use 100 mm/h for $R_{0.01}$ in eq. (5.7).

Step 6 : Calculate the specific attenuation (γ_R) per kilometer in a cell with uniform rain intensity using factors a and b, whose values depend on $R_{0.01}$ value and frequency. The procedures for calculating a and b are provided in reference [ITU, 1994 d], and are additionally included in Sec. 5.6 at the end of this chapter.

$$\gamma_R = a R_{0.01}{}^{b} \quad (\text{dB/km}).$$
(5.8)

Step 7 : Rain attenuation $A_{0.01}$, as it applies to cumulative time 0.01%, is estimated by the following equation:

$$A_{0.01} = \gamma_R L_s r_{0.01} \quad (\text{dB}).$$
(5.9)

Step 8 : Rain attenuation with respect to cumulative time p (from 0.001% to 1%), using $A_{0.01}$ and p as factors, is calculated using common logarithms as follows:

$$\frac{A_P}{A_{0.01}} = 0.12 \, p^{-(0.546 + 0.043 \log p)}.$$
(5.10)

Estimating Cumulative Distribution of Worst-Month Rain Attenuation

Since degradation of signal quality attributable to wave propagation factors cannot be completely eliminated in satellite broadcasting systems (or communication systems in general), one purpose of the link budget is to establish a period of time over which signal quality parameters are allowed to drop below prescribed standards. If that period is on an annual basis, then the annual cumulative distribution of rain attenuation (discussed above) can be used to determine a rain attenuation margin. Depending on the radio communication system in question, the "any month" criterium is sometimes applied. This is generally interpreted to mean the month in which the worst conditions apply, and these conditions define the worst-month rain attenuation cumulative distribution. For example, in the present 12-GHz analog satellite broadcasting service, where 1% is the time rate for the worst month, CNR is permitted to drop below 14 dB. Of the 12 monthly sets (cumulative distributions for each month of the year), the worst-month cumulative distribution has the greatest time rate with respect to the amount of rain attenuation.

For a given signal level which takes into account rain attenuation and other signal degradation factors, the applicable ITU recommendation [ITU, 1994 e] gives the approximate relationship between annual cumulative time rate P (%) and worst-month cumulative time rate P_W (%) as follows:

$$P = 0.3P_w^{1.15}. \tag{5.11}$$

If the annual cumulative distribution for rain attenuation can be estimated with an acceptable accuracy, eq. (5.11) can be used to determine the worst-month cumulative distribution. This relationship of time rates is not only applicable to estimating rain attenuation cumulative distribution, but eq. (5.11) can also be developed so as to indicate rain intensity and the degree of polarization (by type of polarization). The resulting values should also be applicable to all areas of Japan [Fukuchi, 1985].

The technology standard for current 12-GHz band analog satellite broadcasting systems calls for a CNR of over 14 dB for a worst-month time rate of 99%. In this case, rain attenuation specific to the 12-GHz band at a worst-month cumulative time rate of 1% is estimated, and satellite transmitter power is boosted, or other measures are taken to provide the attenuation margin. In Japan's analog satellite broadcasting system, the link budget in all cases specifies a rain attenuation margin of 2 dB. For analog broadcasting systems, CNR can be set slightly lower than 14 dB, which assures an acceptable level of signal quality degradation. With digital systems, however, the use of error correcting codes leaves open the possibility that should CNR drop below a certain value, the sudden increase in BER could result in a service shut-down. Thus, in order to guarantee the same level of service availability as prescribed for analog systems, the worst-month service availability time rate for digital systems must also be set at over 99%. With regards to the 21-GHz band allocated for future broadcast satellites, ITU is also proposing a worst-month service availability rate of 99.7%. In adapting this standard to meet quality-of-service requirements in Japan, and to accommodate the the higher broadcast frequency and the increase in service availability rate, note that the rain

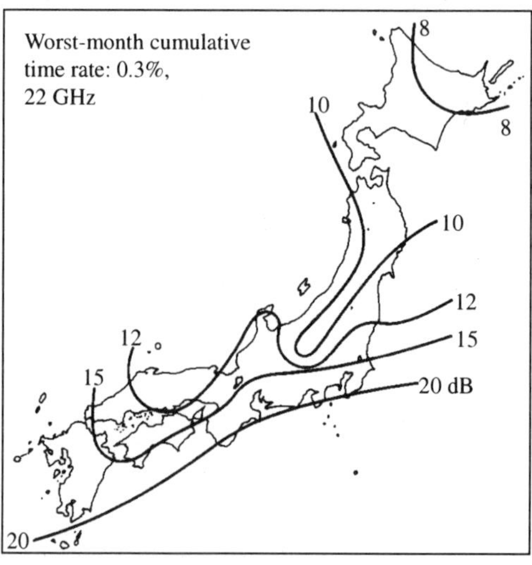

Fig. 5.4   Rain Attenuation Distribution at 22-GHz (Worst-month: 0.3%)

attenuation margin in the southwestern part of the country and along the coastal areas of southern Honshu must be 15 ~ 20 dB (Fig. 5.4).

5.2.3 Rain Attenuation Countermeasures

As was discussed above, the rain attenuation margin for digital satellite broadcasting systems must be higher than their analog system counterparts. Rain attenuation countermeasures are therefore an important component of digital system design and operation. Since we must assume that the earth-end of a satellite broadcasting system consists of a subscriber household with minimum reception facilities, the downlink CNR essentially determines the overall link quality. This is why rain attenuation countermeasures in the downlink are so important. The uplink is serviced by a broadcast station with a high-powered transmitter and large antenna, so rain attenuation in the uplink can be compensated by either increasing transmitter power to a level where the effects of rain scatter are overcome, or by switching to a station where it is not raining (site diversity).

Table 5.5 lists the various satellite broadcasting system rain attenuation countermeasures for the downlink. The broadest categories for attenuation countermeasures are static methods, adaptive (or dynamic) methods, and diversity methods. Note that these methods are not mutually exclusive; some techniques must be used in combination with others in order to gain the desired effect.

Static methods include simply employing a high-output amplifier in the satellite, or using high-gain satellite antennas to increase the effective radiated power from the satellite. Other static methods are utilizing modulation systems which are strongly immune to noise, and forward error coding (FEC) with high coding gain characteristics. Digital satellite broadcasting typically involves the use of a concatenated error correction code such as Reed-Solomon (RS) and convolutional coding combinations.

Adaptive methods normally consist of measures such as increasing satellite transmitter power commensurate with rainfall intensity, and adaptively controlling broadcast bandwidth. One of the objectives of satellite broadcasting systems is to keep subscriber facilities as simple as possible, and utilizing combinations of power-control techniques such as amplifier output control and satellite-mounted multibeam antennas (with the power of each beam independently controllable) is an effective method of accomplishing this. These methods have been the subject of studies which investigated the effects of rain attenuation countermeasures on terrestrial spatial correlation characteristics, as they are affected by rainfall intensity [Fukuchi, 1994]. In broadcast television, extending service availability time is a desired objective, even at the expense of a certain level of signal quality. In analog systems, picture quality is a function of link quality, and degrades gradually in what we refer to as graceful degradation. But with digital systems, picture quality falls off very rapidly once a given threshold in link quality is reached. One technique being studied in Japan (and elsewhere) to counteract this phenomenon is the use of hierarchical information transmission, which transmits source information according to its level of importance. The use of combinations of modulation

Table 5.5   Downlink Rain Attenuation Countermeasures for Satellite Broadcasting Systems

	Countermeasure	Brief Description	Notes
Static	Boosting effective radiated power.	On-board high-output amp. On-board high-gain antennas.	Satellite power generation limited. All rain margins not supportable.
	Mod/demod systems, error coding.	Robust modulation systems. Robust error-correcting codes.	Reduces bit-rate per unit bandwidth. Subscriber needs high-speed receiving facilities.
Adaptive	On-board common resource sharing[*1]	Satellite radiates higher power in specific directions of rainfall.	Power-combining techniques required. Doubtful in areas of heavy rainfall.
	Hierarchical encoding, hierarchical modulation.	Hierarchical information transmission, important information more immune to noise.	Graceful degradation[*2] Controls temporal- and spatial-resolution.
Diversity	Frequency diversity	Diversity in 12-GHz, 21-GHz bands (including hierarchical transmision).	Cannot be used independently in 12-GHz and 21-GHz channels.
	Orbital diversity	Provided by multiple satellites in different orbits.	Minimum effects for strong rainfall conditions near ground. Receiver antenna requires multibeam reception capability.
	Time diversity	Retransmits the signal (or part of the signal).	Subscriber needs high-capacity intelligent memory. Effective at even higher broadcast frequencies.
	Site diversity	Simultaneous reception at stations separated by distances of several 10s of km.	Ideal for Head-End of CATV, but not for individual subscribers.
Other	Improves G/T of receiving facilities. Increases data compression level without loss in signal quality.		

*1 Power and broadcast bandwidth are considered as an on-board common resource (e.g., increasing antenna beam power in a rain cell (zone)).

*2 "Graceful degradation" in signal quality is preferable to sudden outage at a given threshold. This is one of the techniques designed to achieve that goal.

systems capable of modulating these hierarchical signals assures that the most critical information is transmitted at a higher priority. This method should also be effective as a rain attenuation countermeasure.

Methods that depend on diversity consist of having available to the signaling system two or more selectable paths of transmission whose rainfall rates are not highly correlated. This allows selection of the transmission path with the least amount of rain attenuation at any given time. The various diversity modes permit the switching of frequency bands, satellite orbits, reception site, and reception times, and of course, each mode has its advantages and disadvantages.

The switching of reception sites is the preferred method in the uplink of large-scale satellite communications or broadcasting systems, but is not suited for the normal household subscriber. In the future, as more "intelligent" functions are added to household receiving facilities, time-diversity techniques should be realizable [Fukuchi, 1992]. Time-diversity utilizes the fact that rain is never continuous in one location, and works by either retransmitting the program information in a later time slot, or retransmitting high definition information components using hierarchical techniques. In order to implement time-diversity, however, the receiving facilities must have high capacity program storage capability, and the intelligent receiving facilities must have the ability to keep track of what information is missing from the program, and know how to access the missing information. If these conditions are met, subscribers can build their own personal libraries, stocked with high-quality programs.

The time-diversity effects are shown in Fig. 5.5. This figure illustrates a situation of rain attenuation in the path of a 19-GHz satellite transmission, and the graphs show the calculated results of retransmitting the signal in different time slots as measured by attenuation values. Note that when the retransmission is made several hours later, equivalent rain attenuation has dropped off to insignificant values. In addition to the 21-GHz satellite broadcast band, this method should also be applicable as a rain attenuation countermeasure in the future 40- and 80-GHz bands as well.

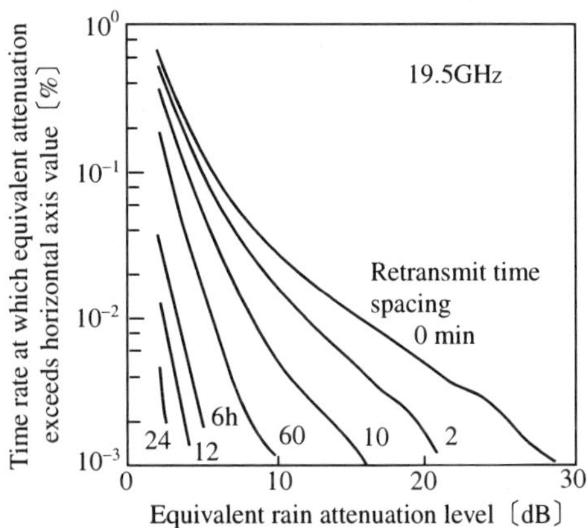

Fig. 5.5　Equivalent Rain Attenuation Cumulative Distribution when using Time-Diversity

5.3 Satellite Digital Broadcasting Systems (DBS)

Research and development projects on digital broadcasting systems (DBS) supported by sattelite are active, both in Japan and abroad. The European countries universally employ the DVB (Digital Video Broadcasting) standard. The DVB-S (DVB-Satellite) standard specifies QPSK for modulation, and a convolutional-Reed-Solomon (RS) concatenated inner-outer code for FEC. MPEG standards are also used for source coding, and TDM (Time-Division Multiplexing) for multichannel service. The current standard for Japanese digital satellite broadcasting is essentially based on DVB standards. An outline of DVB-S standards is given in Tables 5.6 and 5.7.

Table 5.6   Basic DVB-S Standards

Baseband video coding	MPEG-2 VIDEO	Channel coding and modulation	
Baseband audio coding	MPEG-2 AUDIO	Input bitstream	MPEG-2 transport stream
Data service coding	DVB standard	Energy dispersal	Dispersal by $X^{15} + X^{14} + 1$ polynomial.
Multiplex	MPEG-2 SYSTEMS	Error correction : outer-code : inner-code	Reed-Solomon code, RS(204, 188) Convolutional code (constraint length: 7) Coding rate: 1/2, 2/3, 3/4, 5/6, 7/8, 1
Service information system	MPEG-2 SYSTEMS + DVB (independent standards)	Interleaving	Convolutional interleaving at 12-level depth
Standard scrambling system	Standards in planning.	Modulation system	QPSK (Gray coded) Filter response roll-off rate: 0.35

Note: Refer to Chapter 2 and Section 6.2.1 regarding MPEG-2 standard.

Table 5.7   Information Transfer Rates and Transponder Bandwidths for the DVB-S Standard

Bandwidth (-3 dB) (MHz)	Bandwidth (-1 dB) (MHz)	R_s (BW/R_s = 1.28) (Mbaud)	R_u: (QPSK + 1/2 convolutional) (Mbit/s)	R_u: (QPSK + 2/3 convolutional) (Mbit/s)	R_u: (QPSK + 3/4 convolutional) (Mbit/s)	R_u: (QPSK + 5/6 convolutional) (Mbit/s)	R_u: (QPSK + 7/8 convolutional) (Mbit/s)
54	48.6	42.2	38.9	51.8	58.3	64.8	68.0
46	41.4	35.9	33.1	44.2	49.7	55.2	58.0
40	36.0	31.5	28.8	38.4	43.2	48.0	50.4
36	32.4	28.1	25.9	34.6	38.9	43.2	45.4
33	29.7	25.8	23.8	31.7	35.6	39.6	41.6
30	27.0	23.4	21.6	28.8	32.4	36.0	37.8
27	24.3	21.1	19.4	25.9	29.2	32.4	34.0
26	23.4	20.3	18.7	25.0	28.1	31.2	32.8

R_u: Indicates the information transfer bit-rate at the MPEG-2 multiplexer output terminal.
R_s: Indicates symbol transfer rate.
BW: Bandwidth (-3 dB)

The satellite DBS consists of the on-board high-output transponders, the earth stations which transmit the broadcast signals, and the various reception facilities. A block diagram of satellite DBS is given in Fig. 5.6. Figure 5.7 shows a picture of JCSAT-3, the DBS satellite currently (1997) being used for digital TV service in Japan.

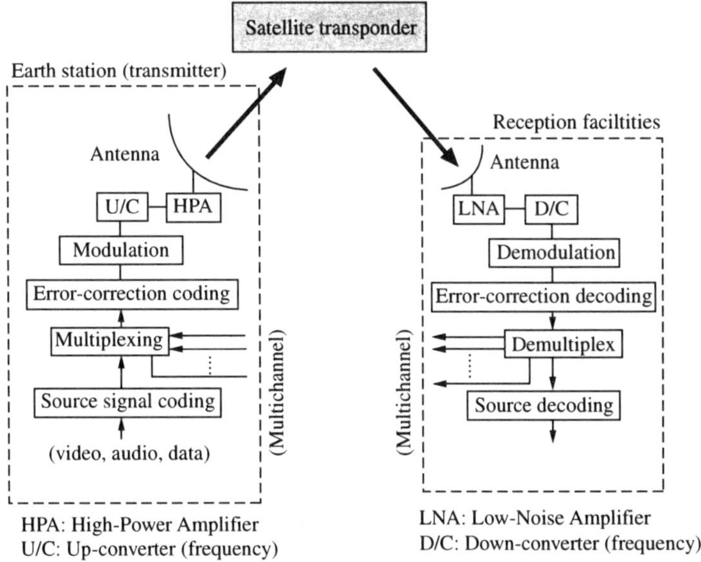

Fig. 5.6   Basic Configuration of the Satellite Digital Broadcasting System (DBS)

Fig. 5.7   JSAT-3 DBS Satellite [Photo courtesy of JSAT.]

5.3.1 Satellite Features

Transponders

Figure 5.8 shows the basic configuration of a through repeater type device, normally referred to as a transponder. The receiving antenna of this type of transponder collects the electromagnetic energy of the signal transmitted by the earth station, and the transponder amplifies the signal and converts it to a different frequency. Following filtering to obtain the proper bandwidth with respect to center-frequency, the signal is amplified by the power amplifier and radiated from the

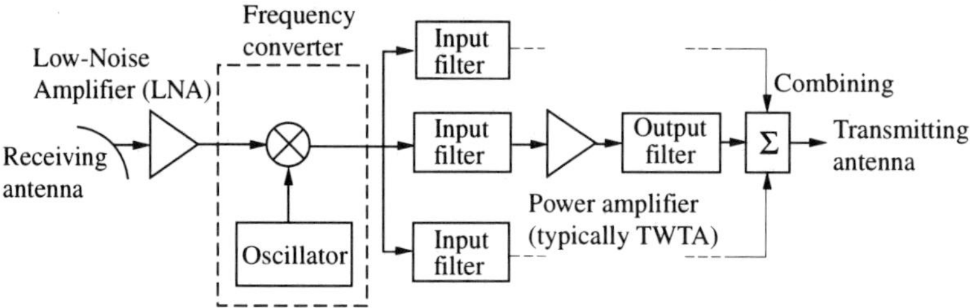

Fig. 5.8   Configuration of a Typical Satellite Transponder

transmitting antenna.

In order to assure reception of the small diameter antennas typically owned by subscriber households, the signal from the satellite transponder must be transmitted at maximum permissible power. The traveling-wave tube amplifier (TWTA) is currently the most popular type of amplifier used in transponders, but the solid-state power amplifier (SSPA) appears to be the technology of the future. In the TWTA, increasing the amplification of input power drives output to saturation, which gives it AM/AM characteristics. In analog BS-band broadcasts using conventional FM modulation systems, the amplifier operates at saturated output, and transmits at maximum power. On the other hand, communication satellite amplifiers are operated backed off by 1 dB from saturation, which reduces distortion in the transmitted signal. Operating at back-off implies operating the amplifier at some point below saturated output power, and in the case of JSAT-3 (PerfecTV), output back-off is 1 dB. The amount of back-off at which the amplifiers are operated, with respect to the link budget, is a trade-off between received C/N_0 and signal quality (lack of distortion). Increasing the input power to TWTA results in changes to the output phase (AM/PM characteristics) of the amplifier, and the resulting nonlinear phase characteristics causes signal distortion, which degrades signal quality. Again, distortion can be reduced by operating the power amplifier at a backed off output level.

Figure 5.9 (a) and (b) illustrate the input-to-output (I/O) characteristics (AM/AM) of the JCSAT-3 transponder [JSAT, 1996], and the I/O characteristics (AM/AM, AM/PM) of the TWTA used in COMETS (COMmunications and broadcast Engineering Test Satellite: refer to 5.4.2 and 5.5.2).

The output power ratings for transponders in two DBS satellites and an analog broadcast satellite (BS-3a) are listed in Table 5.8. Transmitter output of a communications satellite (CS) is lower than that of transmitters designed for satellite broadcasting, but with the recent improvements in receiving facility performance and the introduction of digital transmitting equipment, reception from the CS-band is now possible on small diameter (about 45 cm) household type disk antennas. In digital transmission systems, employing information compression techniques to reduce the volume of video information and utilizing modulator/demodulators with built-in error correcting functions allows signal power to be reduced, yet picture quality remains at the same level as analog broadcast

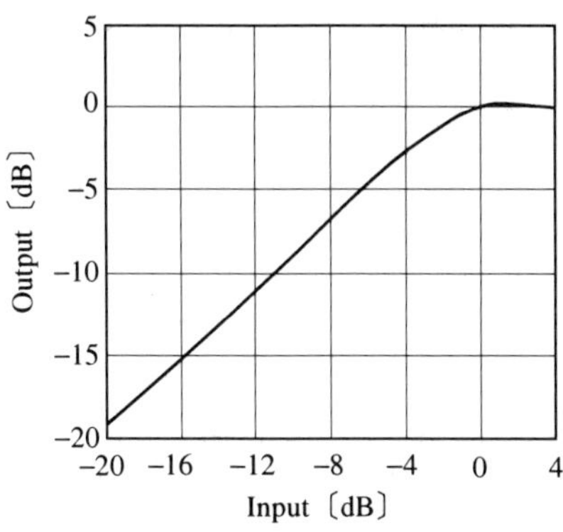

(a) JSAT-3 transponder I/O characteristics [Courtesy of JSAT]

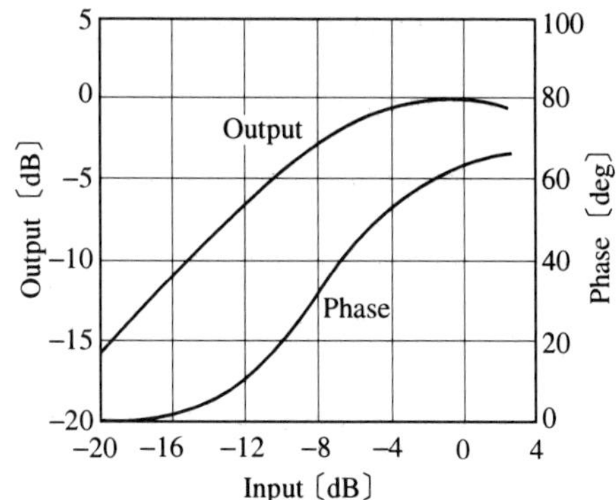

(b) I/O characteristics of the TWTA used in COMETS [Courtesy of National Space Development Agency of Japan (NASDA)]

Fig. 5.9   I/O Characteristics of the JSAT-3 Transponder and the COMETS Broadcast Satellite TWTA

Table 5.8   Satellite Output Power

	Satellite		
	JCSAT-3	COMETS	BS-3a (analog broadcast)
Output	60 W (HPA)	200 W	120 W
EIRP	55 dBW (EIRP)	68 dBW	57 dBW

television. (Refer to Sec. 1.3.)

Satellite Maintenance

The current practice is to place broadcast satellites in a geostationary orbit approximately 36,000 km above the equator. In this type of orbit, the position of the satellites remain approximately stationary with respect to a point on the surface of the earth. However, asymmetry of earth's gravitational forces, gravitational attraction of the sun and earth's moon, and the sun's radiation all work to change the satellite's trajectory. The satellite must therefore include secondary propulsion equipment in the form of thrusters, which are periodically activated to make corrections in the orbital path, and to control the satellite's attitude. These orbital corrections maintain the position of the satellite within ± 60 km of its designated location. Tasks such as these are referred to as station keeping.

When one transponder is also used for communications purposes (such as TDMA or other multiple access systems), shifts in the orbital position of the satellite can have an adverse effect on communications reliability. Since broadcast satellites generally use single-wave modulation for one of its transponders, changes in the satellite orbit will not significantly alter received signal quality. However, like frequency resources, satellite orbits are also becoming scarce, so effective utilization of orbits and satellite management functions operable from earth-stations is an important part of system design.

In order to properly control satellite operations, one must be able to accurately monitor conditions within the satellite from the ground. Information accessed through the monitoring system include the satellite's attitude, internal temperatures, power, and electrical status (voltage, current). This information is sent to earth as telemetry data. Once the earth-station has received and analyzed the telemetry data, the control center sends out appropriate commands to correct any indicated malfunctions. The operations for maintaining satellite equipment operations and internal environmental conditions are referred to as housekeeping. As the attitude of the satellite greatly affects pointing direction, and consequently, antenna directivity, and internal temperatures affect performance of the transponders, maintenance programs are critical to assuring that the subscriber receives a continuous and stable signal.

5.3.2 Earth-Station Description

By earth-station, we are referring to the studios which broadcast television programs, and the subscriber facilities that receive the programming. The following description is based on a multichannel digital broadcasting system using a communications satellite (JCSAT-3) operating in the 14/12-GHz band [TTC, 1995], [Nakagawa, 1996].

Broadcasting Station (Studio)

Figure 5.10 illustrates the basic configuration of a broadcast studio (station) operating in the satellite DBS. (The example used here is the PerfecTV system.) As the diagram indicates, redundancy in the system is provided by three geographically-separated stations to assure continual access to the uplink.

The major components of the transmission system are the baseband system (BB), encoder system (ENC), and the transmitting and receiving system, designated the radio frequency equipment (RF).

The baseband system consists of the switching terminal, VTR, video/audio monitoring system, and the automatic transmitting equipment which controls operation of the baseband system equipment. The programs offered by the service provider operating the network are sent to the encoder system in accordance with the program selection information produced by the program management system (PMS).

The encoder system is a terminal consisting of a source coding block (video, audio, data), and a multiplexing block. The unmodified information transfer rate for the digital TV video signal is 216 Mbit/s, but in order to distribute the signal to subscribers, the MPEG-2 compression standard is used to compress the video information, enabling a transfer rate of 4 ~ 6 Mbit/s. In the multiplexing system, the compressed video information is multiplexed with the audio information and other

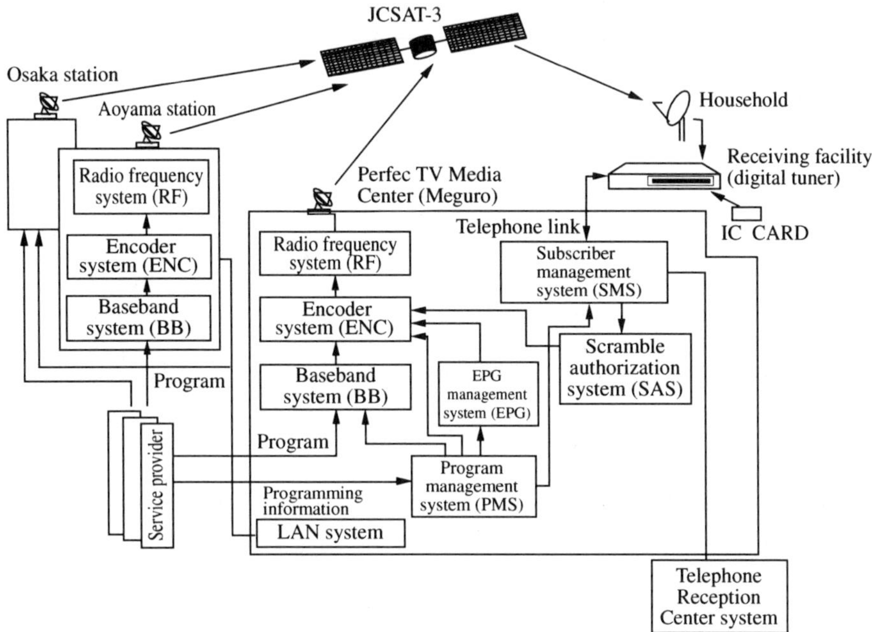

Fig. 5.10   Basic Configuration of a Broadcast Studio [Courtesy of JSAT.]

data signals, including contract management information and the Electronic Program Guide (EPG). Contract management information includes the key information as provided for in each subscriber's contract. The Subscriber Management System (SMS), based on the subscriber data base, directs the Scramble Authorization System (SAS) to compile the key information and send it to the multiplex block of ENC. When satellite DBS utilizes a communications satellite (JCSAT), for normal definition television the signals can be multiplexed so that one transponder can provide 4 ~ 8 channels of service at an information transfer rate of 24 Mbit/s.

The data sequence of the multiplexed signal is then scrambled and applied to the channel coding block for Forward-Error Correction (FEC) coding. Various codes can be used for FEC, but since the channel in a digital satellite broadcast has noise-controlled conditions, and performance of FEC is largely determined by the size of the receiving antenna, a high coding-gain error correcting code is used in most cases. Where coding-gain is the primary consideration, a concatenated inner-outer code consisting of an RS code and a convolutional code will be effective in most cases. A diagram showing the FEC and modulation blocks is given in Fig. 5.11. Between the RS(204, 188) outer code encoder and the rate-3/4 convolutional code (inner code) encoder is the interleaver which functions to randomize error bursts and improve the performance of the concatenated error correcting code.

QPSK (rate-3/4 convolutional code) modulation is typically used in satellite DBS because of its ability to function even with nonlinearities present in

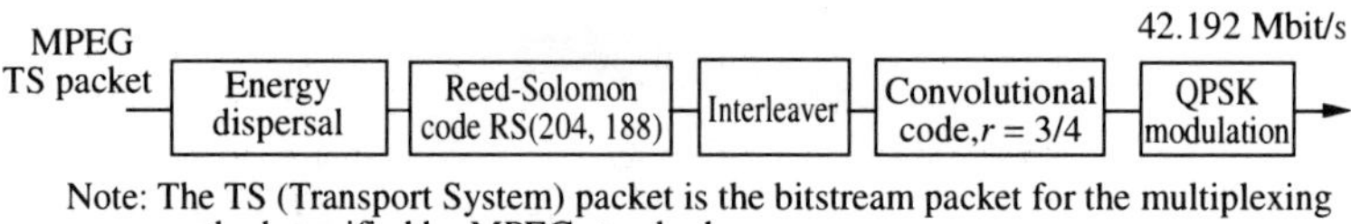

Note: The TS (Transport System) packet is the bitstream packet for the multiplexing method specified by MPEG standards.

Fig. 5.11   Error-Correcting/Modulation Blocks

transponder output. QPSK also has good BER performance even in band-limited and power-limited channel environments. Note, however, that research has recently shifted to modulators with built-in error-correcting functions such as trellis-coded 8-PSK and other high M-ary coded modulation systems.

In the transmitting and receiving system (RF equipment), the modulated signal is converted from an IF- to an RF-frequency and then amplified by the High-Power Amplifier (HPA). Assuring successful transmission in the satellite uplink is critical, so facilities capable of transmitting at a high enough power to overcome any attenuation in the signal due to unexpected changes in weather conditions are required. The main station in the CS (JSAT) satellite DBS transmits over a 7.6-meter antenna using a 300-Watt HPA. Also, by monitoring the return signal from the satellite, transmitter power can be controlled, and the transmitter has the capability of compensating for rain attenuation by up to 10 dB.

Receiving Facilities

Figure 5.12 illustrates a block diagram of the receiving facilities in the CS satellite digital broadcasting system. Since the receiving facilities typically consist of consumer electronic equipment owned by a subscriber, the components should be compact and inexpensive. The receiving antenna is normally approximately 45-cm

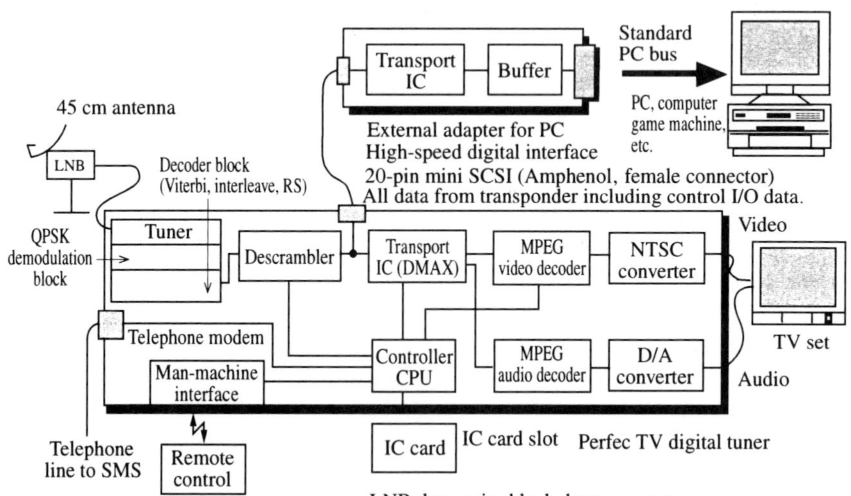

Fig. 5.12   Receiving Facilities [Courtesy of JSAT.]

in diameter. The components following the antenna are the low-noise block down-converter (LNB), Low-Noise Amplifier (LNA), and the frequency converter, all of whose purpose is to amplify the signal which has been attenuated by transmission path losses to an appropriate level. Recently, noise figures (NF) of the LNB have been improved, and devices with a NF of 0.5 dB in the 12-GHz band have been realized.

Phase noise in the frequency converter of the LNB diminishes the quality of the regenerated carrier in the receiving circuit. In a digital transmission system, phase noise in the regenerated carrier causes interference between the orthogonal signals, which degrades signal quality. This is especially a problem with M-ary modulation systems (8-PSK, etc.), where phase noise is a significant source of signal degradation. Consequently, phase noise in the LNB, in addition to amplifier noise figures, must be taken into account in the design of receiving equipment used in CS digital broadcasting systems.

The QPSK modulated signal selected by frequency with the tuner is decoded by the Viterbi decoder, passed through the deinterleaver, and then the RS code is decoded. Figure 5.13 shows the BER performance of a CS digital broadcast system (JCSAT).

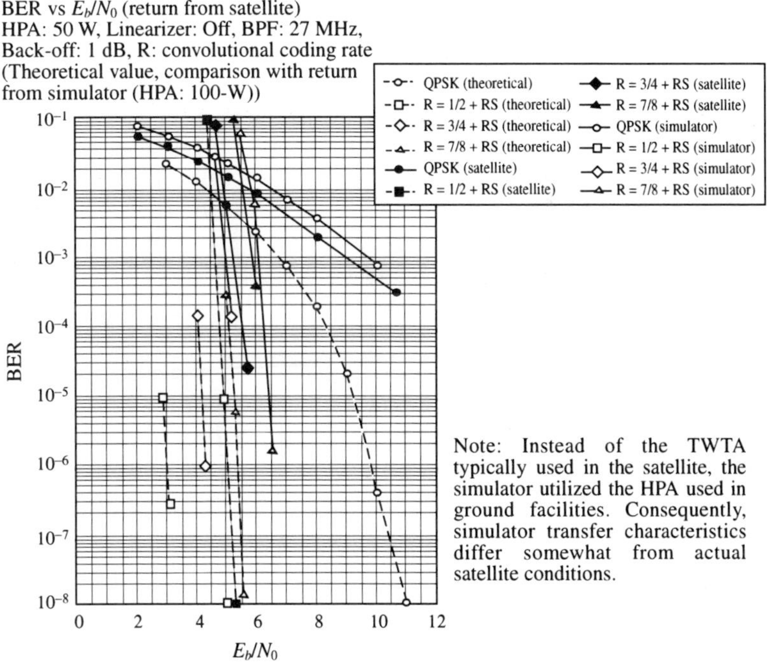

Fig. 5.13   Satellite Digital Broadcast BER Performance (JCSAT-3) [Taken from the response report to Inquiry No. 74, TTC]

Only the contracted channel packet portion of the FEC-encoded information is descrambled, and the channels are then separated by the demultiplexing equipment. This is followed by decoding of the video and audio information by their respective MPEG decoders. The video information is then converted to NTSC standards, and the audio information is passed through a D/A converter and output to a standard television receiver.

One of the advantages of digital systems is that they are compatible with computers. If the receiving equipment has a personal computer interface, it is possible to download a television broadcast into a computer. Also, with a telephone modem, PPV (pay per view) subscriber records can be sent directly to the SMS.

5.4 Satellite Digital Broadcast System Link Budget

5.4.1 Degradation Factors and Quality Standards

Factors that must be taken into account in determining signal quality are the limits on received C/N_0 (due to noise in the signal bandwidth), transponder nonlinear characteristics and bandlimiting, signal distortion due to amplitude and phase-frequency characteristics within the band, and cochannel interference from other systems operating in the same frequency band. Factors resulting in signal strength variations include rain attenuation and atmospheric absorption attenuation.

When noise in the uplink and downlink, which includes intermodulation noise ($N_{0\ im}$) and interference noise from other signals (I_0), can be interpreted as *Gaussian noise*, total received $C/N_0)_t$ is given by the following equation [Iida, 2000]:

$$C/N_0)_t = \{ C/N_0)_u^{-1} + C/N_0)_d^{-1} + C/N_0)_{im}^{-1} + C/I_0)^{-1} \}^{-1} \text{ (raw value)} \tag{5.12}$$

$$C/N_0)_u = EIRP_{ear} - L_{p,u} + G/T)_{sat} - k + L_{\gamma,\,u} \quad (\text{dB} \cdot \text{Hz}) \tag{5.13}$$

$$C/N_0)_d = EIRP_{sat} - L_{p,d} + G/T)_{ear} - k + L_{\gamma,\,d} \quad (\text{dB} \cdot \text{Hz}) \tag{5.14}$$

$$L_p = 10 \log(4\pi D^2/\lambda^2) \quad (\text{dB}). \tag{5.15}$$

The symbol meanings for the above equations are as follows:

$C/N_0)_t$: Total C/N_0.

$C/N_0)_u$: Uplink C/N_0.

$C/N_0)_d$: Downlink C/N_0.

EIRP : Effective Isotropic Radiated Power.

L_p : Free-space path loss.

L_r : Rain attenuation.

N_0 : Noise power per hertz.

G/T : Figure of merit of the receiver.

λ : Wavelength.

D : Distance between the satellite and earth.

k : Boltzmann constant, -228.6 dBW/K-Hz.

Notation suffix u : Uplink.

Notation suffix d : Downlink.

Notation suffix ear : Earth-station.

Notation suffix sat : Satellite.

The relationship between required C/N_0 (which sets the signal quality standard), E_b/N_0, and C/N is expressed in the following equation:

$$C/N_0 = E_b/N_0 + R = C/N + B_w \quad \text{(dB} \cdot \text{Hz)}, \tag{5.16}$$

where E_b is energy per information bit, R is the information transfer rate, and B_w is the equivalent noise bandwidth.

The required E_b/N_0 is determined by the specified quality level of service (information error rate), the modulation/demodulation system employed, and by the error-correction system. For example, when coherent detection BPSK or QPSK is used, BER with respect to E_b/N_0 is given by

$$BER = 1/2\,\text{erfc}\left(\sqrt{E_b/N_0}\right) \quad \text{(raw value)}, \tag{5.17}$$

where erfc ( ) is the complementary error function.

In order to satisfy the 10^{-4} error rate (no error correction) required in the link budget, the modulation system must provide a required E_b/N_0 of 8.4 dB. For more details on the various modulation/demodulation and error-correction systems, refer to Chapter 3. Also, in the link budget, the difference in total C/N_0 (eq. (5.12)) and required C/N_0 (eq. (5.16)) is called the power margin (or link margin, or simply margin).

When a single TWTA is serving multiple signals and operating as a common amplifier near saturation point, the intermodulation term in eq. (5.12) must be taken into consideration. By the central limit theorem, the higher the number of input signals, the closer the intermodulation noise resembles Gaussian noise.

Intermodulation noise is also included as a part of the transponder nonlinear characteristics, but for a single-wave input it results in amplitude- and phase-distortion, which degrades signal quality. For an example, refer to Fig. 5.14, which shows the signal distribution at the decision points for 16-QAM signals influenced by nonlinear distortions. The amount of distortion-induced degradation can be analogically inferred by computer simulation, and in the link budget should be considered as nonlinear degradation in the channel. In systems such as 16-QAM where information is carried overlapped in the amplitude direction, this degradation can be significant, but in constant envelope modulation systems such as QPSK or 8-PSK, it is not too critical. However, even with constant envelope type modulators, the muliplication effect that receiver filter bandlimiting has on nonlinear distortions

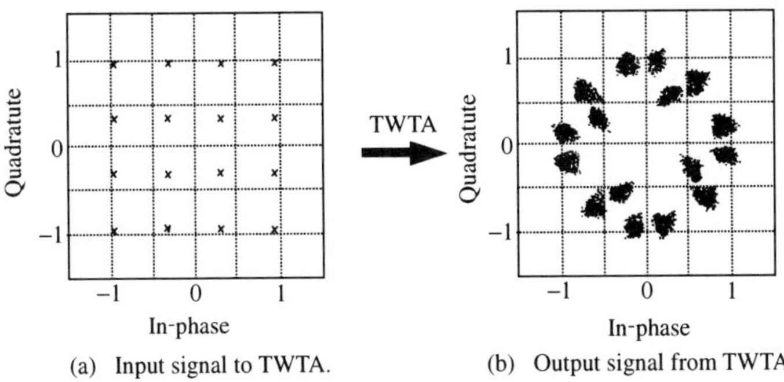

(a) Input signal to TWTA.　　　(b) Output signal from TWTA.

Fig. 5.14   Distribution of 16-QAM Signals Subject to Nonlinear Distortion

can also reduce signal quality.

The amplitude and phase frequency characteristics of the satellite and earth-station equipment can be sources of signal degradation in high bit-rate, wideband signal transmissions. As a countermeasure to this, the use of equalizers to compensate frequency characteristics in the transmitting earth-station and receiving equipment should be considered. If this countermeasure fails to correct signal degradation in the link budget to an acceptable level, then channel frequency characteristics should be considered as a possible source. In satellite DBS equipment with a signal bandwidth of about 27-MHz, it is relatively easy to obtain flat frequency characteristics.

With regards to degradations caused by channel nonlinearities and channel frequency characteristics, it is difficult to make a perfectly accurate calculation in the link budget, but these factors are not too critical. Normally these degradations are accounted for in the power margin, and when QPSK is used as the modulation system, approximately 2 dB should be added to the margin.

In the 12- and 21-GHz bands, rain attenuation in the channel is important and must be accounted for. For details regarding the subject of rain attenuation, refer to Sec. 5.2.

Satellite link service is generally segregated by service area, antenna directivity, and signal polarization. When there are other services operating in the same or adjacent channels, cochannel or adjacent-channel interference can be a problem. In such cases, the degradation caused by C/I (carrier power-to-interference power) can be calculated as part of total C/N (or C/N_0) in eq. (5.12). In the strictest sense, cochannel and adjacent channel interference (I) is not Gaussian noise, but it is close enough to be considered its equivalent.

5.4.2 Link Budget Example

Table 5.9 illustrates an example of a link budget for a CS digital broadcasting system (JCSAT) operating at 14-GHz in the uplink and 12-GHz in the downlink. This link budget takes into consideration copolarization interference between adjacent satellites, and cross-polarization interference between the reference satellite and its nearest neighbor. With regards to orbital positioning, the worst-case scenario was hypothesized as four satellites per side spaced at four degress in a geostationary orbit, for a total of eight satellites. When the interference from the third satellite away from the referenced satellite is small enough to ignore, only a satellite's two nearest neighbors are considered as interference sources. In this example, the desired wave and the interferring wave are assumed to have been transmitted with the same EIRP.

For a bandwidth of 27 MHz, an information transfer rate of 42.192 Mbit/s, and where QPSK is used, required $C/(N + I)$ is set at 8 dB. This figure was taken from a transponder simulation in which QPSK rate-3/4 convolutional + RS(204, 188) coding was used, and was increased above the C/N level of 6 dB for error-free transmission. This was based on the premise that with QPSK rate-7/8 convolutional + RS(204, 188) coding, a C/N level of 8 dB achieves error-free performance.

Table 5.9   12-GHz Digital Broadcasting Satellite Link Budget Example (JCSAT-3) (Taken from the TTC response report to Inquiry No. 74.)

(Satellite EIRP: 54 dBW, Receiving antenna: 45 $cm/29 - 25 \log \phi$, Worst-month service time rate: 99.0%)

Broadcast satellite: JCSAT-3				Transmit location: Tokyo

Earth-station specs:
Max. transmit power (W): 300
Transmission losses (dB): 0.58
Max. power density (dBW/Hz): -49.1

Receive location: Tokyo
Satellite specs:
Max. transmit power (W): 60
Max. power density (dBW/Hz): -57.4
Max. EIRP (dBW): 55.0

Item	Unit	Clear sky	Uplink rain	Downlink rain	Notes
Uplink					
Satellite PFD	dBW/m²	−92.0	-92.0	−92.0	
Satellite G/T	dB/K	9.0	9.0	9.0	
Rain attenuation (uplink)	dB	–	-12.0	–	Uplink worst-month non-availability rate: 0.068%
Earth-station EIRP	dBW	70.4	82.4	70.4	Transmit power control: 12.0 dB
Antenna gain (boresight)	dBi	58.2	58.2	58.2	Transmitting antenna dia: 7 m
Copolarization (4.4°)	dBi	12.9	12.9	12.9	= 29 - 25 log (4.4)
Cross-polarization XPDe (boresight)	dB	35.0	35.0	35.0	
Cross-polarization XPDe (4.4°)	dB	2.9	2.9	2.9	= 19 - 25 log (4.4)
Satellite antenna XPDs	dB	33.0	33.0	33.0	
Uplink C/Nu	dB	26.7	26.7	26.7	
Copolarization, 2 adjacent satellites, C/I (up)	dB	42.3	42.3	42.3	
Cross-polarization, reference satellite C/I (up)	dB	27.9	19.2	27.9	
Satellite XPDs component	dB	30.0	30.0	30.0	
Earth-station XPDs component	dB	32.0	19.5	32.0	
Cross-polarization, 2 adjacent satellites, C/I (up)	dB	52.3	52.3	52.3	
Downlink					
Satellite EIRP	dBW	54.0	54.0	54.0	
Satellite antenna XPDs	dB	33.0	33.0	33.0	
Rain attenuation (downlink)	dB	–	–	−1.3	Downlink worst-month non-availability rate: 1.000%
System noise temperature	dBK	21.3	21.3	23.2	NF 1.1.dB antenna noise temp: 50 K
Directivity error	dB	0.3	0.3	0.3	
Receiving antenna dia.	md	0.45	0.45	0.45	Receiving antenna efficiency: 70%
Antenna gain	dBi	33.9	33.9	33.9	
G/T	dB/K	12.7	12.7	12.7	
Copolarization (4.4°)	dBi	12.9	12.9	12.9	= 29 - 25 log (4.4)
Cross-polarization XPDr (boresight)	dB	25.0	25.0	25.0	
Cross-polarization XPDr (4.4°)	dBi	7.9	7.9	7.9	= 24 - 25 log (4.4)
Downlink C/Nd	dB	14.9	14.9	11.5	
C(Nu + Nd)	dB	14.6	14.6	11.4	
Copolarization, 8 adjacent satellites, C/I (down)	dB	17.0	17.0	17.0	
Cross-polarization, reference satellite, C/I (down)	dB	21.4	21.4	21.2	
Satellite XPDs component	dB	30.0	30.0	30.0	
Receiver XPDs component	dB	22.0	22.0	21.9	
Cross-polarization, 2 adjacent satellites, C/I (down)	dB	23.0	23.0	23.0	
Total copolarization C/I|Co	dB	17.0	17.0	17.0	
Total cross-polarization C/I]Ad	dB	18.6	14.6	18.4	
Total C/N + I	dB	11.9	10.7	9.9	
Required C/N + I	dB	8.0	8.0	8.0	
Required C/N + I margin	dB	3.9	2.7	1.9	

Note: PFD: Power flux-density, XPD: Cross-polarization discrimination, G/T: Figure of merit (receiver)
     (Courtesy of [Iida, 2000])

Table 5.10   21-GHz Digital Broadcasting Satellite Link Budget Example (COMETS)

Comm/broadcast satellite: COMETS				Transmit site: Kashima, Ibaraki Pref.	
				Receive site: Tokyo (Koganei)	
Transponder characteristics assured bandwidth: 120-MHz				(using the Kanto beam)	

Item	Unit	Clear sky	Uplink rain	Downlink rain	Notes
Frequency	GHz	27.8	27.8	27.8	Ibaraki Pref., Kashima transmitter
Transmit power	dBW	23.3	23.3	23.3	
	W	213.7	213.7	213.7	
Antenna dia.	m	5.0	5.0	5.0	
Feed loss	dB	4.5	4.5	4.5	
Antenna efficiency		0.55	0.55	0.55	
Transmitter antenna efficiency	dBi	60.7	60.7	60.7	
Transmitter EIRP	dBW	79.5	79.5	79.5	
Tracking loss	dB	0.3	0.3	0.3	
Free-space loss	dB	212.8	212.8	212.8	
Rain loss	dB	0.0	9.0	0.0	9 dB: Non-availability rate 1%
Atmospheric absorption loss	dB	0.7	0.7	0.7	
Transmitter antenna gain	dBi	45.7	45.7	45.7	5 dB offset/boresight
Feeder loss	dB	1.4	1.4	1.4	
Receive power	dBW	−90.0	−99.0	−90.0	
System noise temperature	K	912.0	912.0	912.0	
	dBW/Hz	−199.0	−199.0	−199.0	
G/T	dB/K	14.7	14.7	14.7	
Uplink C/N_0	dB · Hz	109.0	100.0	109.0	
Frequency	GHz	20.7	20.7	20.7	COMETS transmitter
Transponder gain	dB	113.0	122.0	113.0	
Transmit power	dBW	23.0	23.0	23.0	
	W	200.0	200.0	200.0	
Transmitter antenna gain	dBi	45.9	45.9	45.9	2 dB offset/bore-sight
Feeder loss	dB	2.6	2.6	2.6	
Transmitter EIRP	dBW	66.3	66.3	66.3	
Free-space loss	dB	210.2	210.2	210.2	
Rain loss	dB	0.0	0.0	5.4	5.4 dB: Non-availability rate 1%
Atmospheric absorption loss	dB	1.1	1.1	1.1	
Antenna dia.	cm	75.0	75.0	75.0	Tokyo, Koganei receiver
Antenna efficiency		0.66	0.66	0.66	
Transmitter antenna gain	dBi	42.4	42.4	42.4	
Tracking loss	dB	0.5	0.5	0.5	
Feeder loss	dB	0.2	0.2	0.2	
Receive power	dBW	−103.3	−103.3	−108.7	
System noise temperature	K	281	281	281	(NF: 2.5 dB, feed noise: 13.1 K,
	dBW/Hz	−204.1	−204.1	−204.1	antenna noise: 40 K)
Rain-atmospheric noise temperature	dBW/Hz	−210.7	−210.7	−205.3	
G/T	dB/K	17.7	17.7	17.7	
Downlink C/N_0	dB · Hz	99.9	99.9	93.0	
Total C/N_0	dB · Hz	99.4	97.0	92.8	
Signaling system: TC 8-PSK + RS(204, 188)				129 Mbps (bit-rate), BER: $<10^{-10}$	
Required C/N_0	dB· Hz	89.7	89.7	89.7	
Link margin	dB	9.7	7.3	3.1	

The receiving antenna diameter of 45 cm is typical of what household subscribers use, as is the receiver noise figure of 1.1 dB and antenna noise temperature of 50 K. The satellite transponder is operated at an input back-off of 1 dB. Service availability rate of the link is 99%, and rain attenuation in the downlink is 1.3 dB.

Table 5.10 lists an example link budget for the experimental COMETS satellite operating with a 27-GHz band uplink and a 21-GHz downlink (K_a band). Use of the TC 8-PSK signaling system is assumed, which, as illustrated in Fig. 5.15, is an effective system at the high information transfer rates required in a satellite link and with regards to efficient use of frequency resources. The value for required C/N_0 in the link budget is what was actually obtained (experimentally) in the return link with the equipment installed (TC 8-PSK). The link budget was formulated on the assumption of a link availability rate of 99% during periods of rainfall. The antenna diameter used in the experiment was 75-cm, although reception over smaller antennas is possible with improvements in the NF of the receiver. At a 99% time rate, the TC 8-PSK modulation system is capable of a throughput of 129 Mbit/s.

Because of the cliff effect in a digital broadcast (refer to Ch. 1), if the link C/N_0 is larger than required C/N_0, service availability remains unaffected, but if link C/N_0 is smaller than required C/N_0, the link becomes inoperable (service is interrupted).

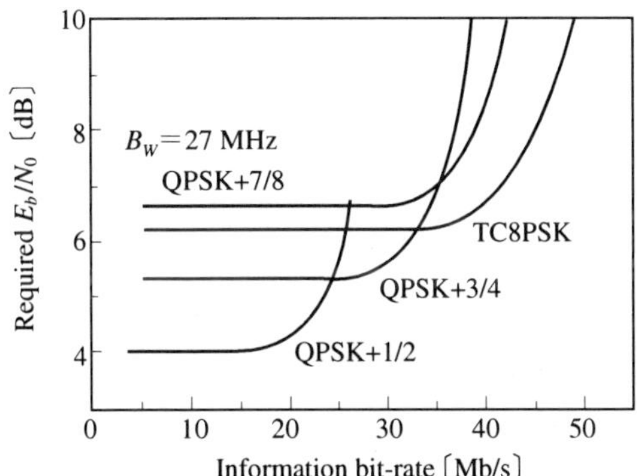

Fig. 5.15   Transfer Characteristics of TC 8-PSK and QPSK (Taken from the TTC response report to Inquiry No. 74.)

In a K_a band satellite broadcast, given the same rainfall intensity, rain attenuation is much greater than in the 12-GHz CS satellite band. Where clear sky link margins are equal, the link availability rate for the K_a band is lower, which means that non-availability time rate of the link is higher. With regards to rain attenuation countermeasures in the K_a band, investigations are being conducted on hierarchical transmitting methods, which, in response to a lower C/N_0, reduce the total amount of information transmitted while assuring a higher service availability rate.

5.5 Future Digital Satellite Broadcasting Systems

5.5.1 Digital Broadcasts over Broadcast Satellites (BS)

How the BS Digital Broadcasting System came about

Japan was one of the early leaders in implementing FM analog type satellite broadcasts, and currently has over 10 million subscriber households. As a broadcasting medium, Japan's satellite systems are therefore on a firm foundation. As digital broadcast capability over communication satellites was quickly realized both in Japan and abroad, the investigations into what form Japan's BS *digital* broadcasting system should take fell primarily to the Ministry of Posts and Telecommunications. Part of this process included considering how satellite systems would economically impact the average viewer, and how easy the system would be to operate. As transmit power would be much greater in BS than in CS, careful consideration was also given to the applicable international rules, signal polarization, and interference standards. Meanwhile, enormous progress was being made on receiver performance, and new device and digital technologies quickly filtered down to the consumer level, which all worked to blur the line between CS and dedicated BS performance. The differences between CS and BS satellites had largely disappeared. However, as was noted in Subsection 5.1.3, industry officials continued to anticipate new and advanced broadcasting media in the 21st-century which would require digital broadcast over BS, and the Ministry of Posts and Telecommunications subsequently issued directives specifying that with the introduction of HDTV (over digital BS), a transition be made from the current analog satellite system to full-digital BS systems. Based on this directive and on previous studies regarding the necessary conditions for BS digital broadcasts, the Ministry officially made the decision to go with the BS digital broadcasting system

Table 5.11 Video Formats for the BS Digital Broadcasting System

Video format	Scanning lines (effective lines)	Horizontal pixels	Scanning system	Aspect ratio	Other parameters[3]
1,125/59.94/ 2:1 system	1,125 (1,080)	1,140 1,920	Interlace	16 : 9	MP @ HL
525/59.94/ 2:1 system	525 (480)	720	Interlace	16 : 9, 4 : 3	MP @ ML
525/59.94/ 1:1 system	525 (480)	720	Sequential	16 : 9	MP @ H14L
750/59.94/ 1:1 system[1]	750 (720)	1,280	Sequential	16 : 9	MP @ HL square pixel
1125/59.94/ 1:1 system[2]	1,125 (1,080)	1,140 1,920	Sequential	16 : 9	Proof-tested standard, square pixel

*1: Conditional upon performance proof-testing (as of 2/1998).
*2 Conditional upon performance/technology standard proof-testing (as of 2/1998).
*3 Symbols (i.e., MP @ HL, etc.) specified in MPEG-2 standards (see Ch. 2).

in February, 1998.

General Details of the BS Digital Broadcasting System

The five different video formats available for use in the BS digital system are shown in Table 5.11. These formats conform to the MPEG-2 standards, and since they are for future use, note the predominance of HDTV, sequential scanning, and the 16:9 aspect ratio. The square pixel system has been through proof testing, and will likely be recognized as the video format standard. The square pixel system has also been greeted with high expectations in the computer industry, and work is now in progress to develop the computer/TV receiver as the basic interface for multimedia services. The audio signal is also encoded under MPEG-2 audio compression standards.

Table 5.12 lists the basic components of the signaling system used in the BS

Table 5.12   Signaling System Components of the BS Digital Broadcasting System

System	Description
Modulation system	Trellis-coded 8-PSK (TC 8-PSK) (Can also be combined with QPSK, BPSK, and other M-ary PSK systems.)
Error-correction system	Concatenated inner-outer code. 〔inner-code〕 Trellis-coded 8-PSK. Convolutional code (constraint length = 7) for QPSK/BPSK. Multiple (convolutional) coding rates selectable. 〔outer-code〕 Reed-Solomon code, RS(204, 188)
Bandwidth (symbol-rate)	The three proposed bandwidths are: 27 MHz (21.1 Mband) 33 MHz (27 Mband) 36 MHz (29 Mband)
Bandlimiting	Root-distributed raised cosine roll-off filter (roll-off rate: 0.35) Aperture correction on transmitter side.
Interference protection ratio	FM-to-FM, or digital-to-FM interference: Conforms to international wireless communications rules regarding BS/analog systems. (cochannel: 31 dB, adjacent channel: 15 dB) FM-to-digital, or digital-to-digital interference: Interference protection ratio not specified, but for digital systems, is determined from required C/N.
Multiplex system and controls	Based on MPEG-2 standards. Controlled by TMCC (Transmission and Multiplexing Configuration Control).
Conditional access reception	By MULTI 2 encryption algorithm.
Other	Interleaving: Block interleaving Energy dispersal: By pseudorandom binary sequence operations.

digital system, and the channel coding blocks are illustrated in Fig. 5.16. The digital BS signaling system is designed for compatibility with the current analog satellite broadcasting system as prescribed by applicable international wireless communications rules, and which protects the integrity of any existing analog satellite system, both in Japan and its neighboring countries.

On the transmitter side, as illustrated in Fig. 5.16, following multiplexing of the MPEG-standard transport stream (TS) the FEC outer-code is encoded and the energy spectrum is dispersed. This is followed by combining of the signals in TMCC (transmission and multiplexing configuration control), and frame generation. The framed signal is then interleaved, the FEC inner-code is added, and finally the signal is modulated and transmitted as a time-division multiplexed (TDM) signal.

At the receiver end, the signal is isolated and reconstructed in the reverse order. With regards to conditional access reception, the same MULTI 2 encryption algorithm as used in the CS digital broadcasting system is also employed in the digital BS system, but some of the rules and details regarding this are left to the service provider

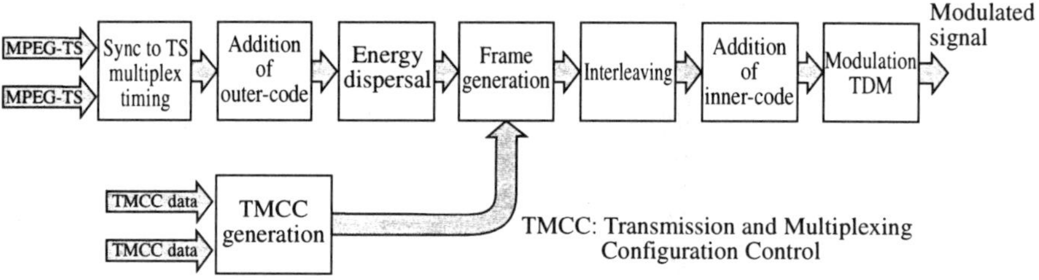

Fig. 5.16   BS Digital Broadcasting Signal Transmission Coding System

An example of the receiver configuration for this type of digital BS system is shown in Fig. 5.17. The IF signal from the BS converter is input to a BS digital tuner having conditional access capability, and the resulting video, audio, and data signals are sent to an HDTV or a standard TV receiver. The BS tuner is also connected to the telephone line for subscriber management tasks, and looking to the future, for the provision of integrated media services.

A comparison of the features of BS digital and CS digital broadcasting systems is given in Table 5.13.

Where BS digital broadcasting goes from here.

The standards for BS digital broadcasting were tentatively decided in 1997, and testing of the BS-3 satellite began that same year. Based on the results of those tests, and on the directions provided by the Telecommunications Technology Council (TTC, Japan), the form that the BS digital broadcasting system should take was established in February, 1998. The initial decisions gave service providers and electrical appliance manufacturers a framework on which to proceed in the

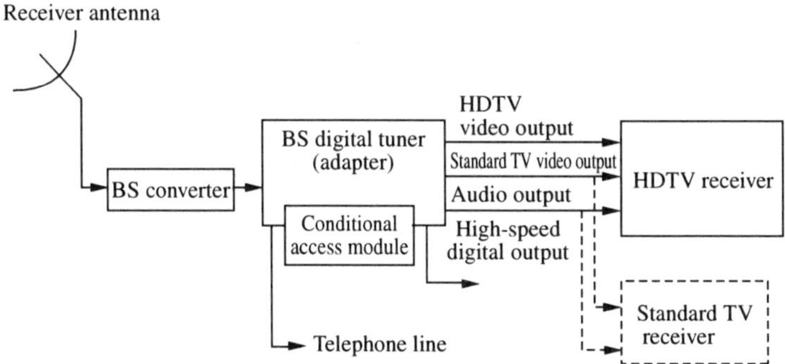

Fig. 5.17   BS Digital Receiver Configuration

Table 5.13   A Comparison of Japan's BS Digital Broadcasting and CS Digital Broadcasting Systems

Item	BS system	CS system
Channel plan (frequency, polarization, power, satellite orbits)	Conforms to the rules established by WARC-BS (including WARC-97).	Conforms to the general ITU agreements.
Basic signaling system	Trellis-coded 8-PSK	
Service initiation	Dec. 2000.	Started in 1996.
Source coding and multiplexing	Based on MPEG-2 (HDTV standard).	Based on MPEG-2 (standard TV).
Other	The square pixel system and other features included in the video format for future integration with computer services.	

development of equipment and facilities for the system. The Ministry of Posts and Telecommunications then determined the detailed parameters of the system, and in 1998 finalized the framework needed to begin broadcasting. Service is officially scheduled to begin in the year 2000.

5.5.2 21-GHz Band Satellite Broadcasting

As was noted in Sec. 5.1, the 21-GHz band was allocated as a next-generation satellite broadcasting frequency at the WARC-92 conference (1992). Originally this was intended to provide studio-grade HDTV services, but as the need for multimedia services over broadcast satellite became more apparent, the 21-GHz band also began to assume the role as medium for broadcasting large volumes of information in general.

Features of 21-GHz Satellite Broadcasting

Because of its higher frequency broadcast band, *21-GHz* band satellite broadcasting compares with 12-GHz band services as follows [Fukuchi, 1995]:

《Advantages》

(1) Wideband channels are easily realized, which makes it possible to transmit large volumes of information.

(2) System equipment and other components can be designed small and convenient.

(3) Since the 21-GHz band is a new frequency allocation, broadcasting services should have more flexibility.

(4) Small, high-gain antennas can be used. Also, a smaller spreading of the electromagnetic beam relates to the next two advantages.

(5) Multiple beams from the satellite can be radiated at a higher flux-density, which makes local satellite broadcasting possible (broadcasts aimed at a specific area).

(6) Since it is possible to focus the beam from the ground, satellites can be spaced closer in geostationary orbits and can still be distinguished. This makes better use of the available orbits.

《Disadvantages/Cautions》

(1) The atmosphere, particularly regarding the effects of rain attenuation, can be a significant problem. Interactions between the atmosphere and the propagated signal are also a problem, and achieving acceptable link availability rates is technologically difficult.

(2) The high signal power that must be transmitted from the satellite to earth in order to overcome rain attenuation increases the importance of developing frequency-sharing techniques. Also, since the 21-GHz band is near the most important band used in radio astronomy, interference countermeasures are required.

Satellite broadcasts, in the wider sense of satellite communication systems, are the one source that the individual citizen can rely on to meet the growing needs for information and communications. Given the recent trend toward developing programs containing ever greater amounts of information, and the current transition to multimedia systems, the 21-GHz satellite system, having an ability to compress large amounts of coded information into short sequences and utilizing highly efficient digital signaling systems, should be the natural vehicle to meet future demands for information services. With the digitalization of broadcast television and efforts to integrate the various information-communications media, the 21-GHz satellite system should play a complementary role with ground-based B-ISDN networks, and is expected to fill a vital function in flexible, high-speed information communications networks. It is therefore essential to develop the technologies that with regards to capacity and reliability, will place the 21-GHz satellite broadcasting system on an equal footing with ground-based networks.

21-GHz Satellite Broadcasting Services Concepts

When considering society's needs in light of the previously discussed features and related technological developments in the 21-GHz broadcasting system, we can assume that broadcasting services will be provided as an integrated service with other communication systems, and that the various broadcast service information will be combined digitally to form a high-capacity digital broadcast. Such a broadcast format is referred to as Integrated Services Digital Broadcasting (ISDB). In addition to the services provided by 12-GHz band satellite broadcasts, some examples of the types of services offered by the 21-GHz system are listed below:

(1) Direct-to-subscriber broadcasting of studio-grade HDTV (wideband HDTV).

(2) Realistic pictures and information broadcast with resolutions higher than HDTV.

(3) Multichannel, high picture quality broadcasts.

(4) Features and performance made possible by wideband signal broadcasts, examples of which are listed below:

• Program-related information and multiple scenes (from sporting events, etc.) thatthe subscriber can freely select for his or her own viewing pleasure.

• Near Video on Demand (NVoD) for multiple programs broadcast in different time slots. (See Ch. 7 for details.)

• Information in download broadcasts available by subscriber request.

Since the above service information is broadcast digitally, the future subscriber household will be able to use digital storage media and a computer to process and combine the information as desired. Also, with multiple beam antennas on-board satellites, flux-density (power) of the signal can be controlled for broadcasts covering the whole country, or for local satellite broadcasting, which offers new possibilities in satellite broadcasting services. These are only a few examples of future service concepts, and as new technologies come onto the market, it is expected that the 21-GHz satellite system will fill an appropriate role in delivering any new services that might be proposed in the future. As the 21-GHz system is integrated with ground-based networks, it is also expected to play an important part in the resulting high-capacity digital media systems.

HD-SAT Project

The RACE program (Research and development in Advanced Communication technologies for Europe) was established under the HD-SAT Project [Dosch, 1994], [Oliphant, 1995] for the purpose of conducting R & D on wideband HDTV signal technologies. The official title of the project is "Bandwidth efficient coding and modulation for wideband HDTV applications supported by satellite and cable networks," which explains the purpose of the project. A brief outline of the projects is given below, but in June of 1995 (the final year of work), a demonstration of its research results was presented at ITS '95 (the 19th International TV Symposium and Technical Exhibition) conducted at Montreux, Switzerland. The demonstration consisted of the experiments conducted using Germany's 20/30-GHz band communications satellite KOPERNIKUS, broadcasting MPEG-2 standard HDTV pictures over the satellite's 36-MHz band transponder. In this experiment, FEC error-correction, QPSK and 8-PSK signaling, and time-division multiplexing were utilized, in addition to a 2-level hierarchical information transmission system used as an effective rain attenuation countermeasure.

(1) HD-SAT Project period: 1992 ~ 1995

(2) Purpose: The transmission of studio-grade HDTV, multichannel digital sound, advanced data services to the home subscriber via satellite and CATV.

(3) Particulars: To study the technical aspects of wideband HDTV transmission, in addition to the economics, technical standards, and planning accompanying implementation. Critical consideration was also given to system compatibility with present and future multimedia networks.

(4) Satellite broadcasting service outline: The goal was to make it possible for a subscriber household to receive broadcasts at 45 Mbit/s over a 60 ~ 90 cm antenna. Worst-month link availability rate was set at 99.6%. However, since rain attenuation was dealt with in this system by using two or three types of modulation systems, time-division multiplexing, and hierarchical transmission methods, link availability was interpreted to mean the assured time rate at a minimum quality level of service.

The diagram below shows the three hierarchical levels of information transmission in the HD-SAT system studied.

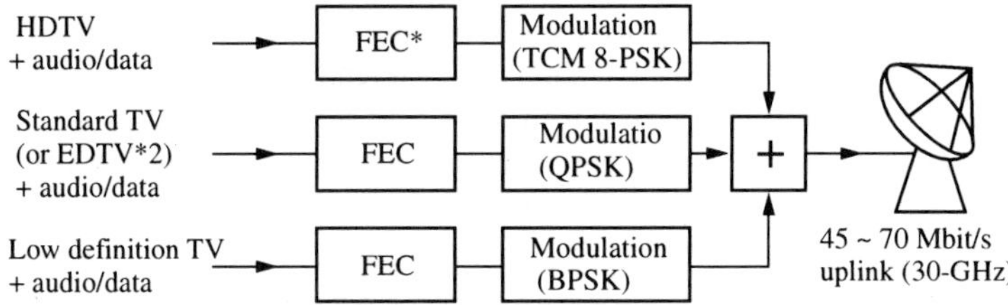

Note: *1 FEC: Forward-Error Correction
*2 EDTV: Extended definition TV 10

(a) Hierarchical transmission system

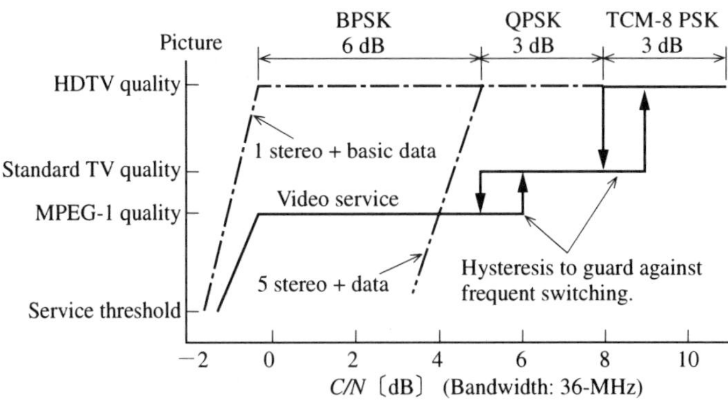

(b) Changes in picture quality vs. link quality (C/N)

Hierarchical Transmission method investigated in HD-SAT Project.

Trends on the International Scene

In Japan, as well as Germany and Italy, the immediate objective of research conducted on 21-GHz satellite broadcasts has been to realize studio-grade HDTV service. On the other hand, assisted by many recent advances in information compression techniques (i.e., MPEG standards), we are now seeing very good quality HDTV pictures produced by 12-GHz satellite systems, a system which does not have similar problems with rain attenuation. In Europe, this situation appears to have resulted in less enthusiasm for continued research on the 21-GHz system. It still remains that in the HD-SAT Project (under Europe's RACE program, which ended research in 1995), hierarchical transmission methods were demonstrated to be very successful in counteracting rain attenuation.

The COMETS Experiments in 21-GHz Band Satellite Broadcasting

This leads us to the introduction of COMETS, Japan's 21-GHz band experimental broadcasting satellite [Fukuchi, 1993]. COMETS is a 2-ton class satellite lifted into orbit in February, 1998 by an H-II rocket. Its primary mission was to provide the research platform for studying "communication techniques in K_a and millimeter wave bands, high-level broadcasting techniques in the the 21-GHz band, and inter-satellite communications techniques." Unfortunately, because of a malfunction in the launching vehicle the satellite failed to achieve its designated geostationary orbit, and subsequently, the above experiments, to the extent possible, were conducted with the satellite in a circling orbit. The on-board equipment used in this mission were developed in a joint project by NASDA (National Space Development Agency of Japan) and the Ministry of Posts and Telecommunications' Communications Research Laboratory (CRL). A diagram of the COMETS satellite is shown in Fig. 5.18.

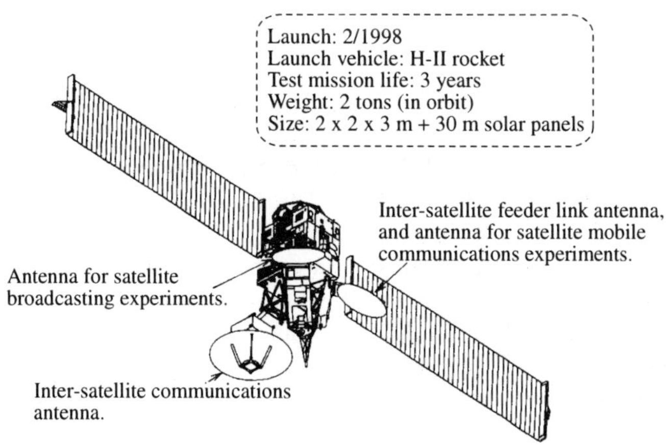

Fig. 5.18   COMETS Experimental Satellite [Courtesy of NASDA.]

Table 5.14   21-GHz Band High-Level Satellite Broadcast Transponder Specifications

Item	Specification
Frequency	Downlink 20.7 GHz Feeder link 27.3 GHz 27.8 GHz
Signal transmission bandwidth	120 MHz maximum
Polarization	Right-hand circular polarization
Transmitter output	At high output: 200 W or greater At low output: 63 W rating
On-board antennas	2.3 m multibeam (2-beam, directed to Central Honshu, Kyushu) Gain: greater than 44 dBi [*] Interbeam isolation: greater than 35 dB [*]

Note: * denotes specification at edge of service area.

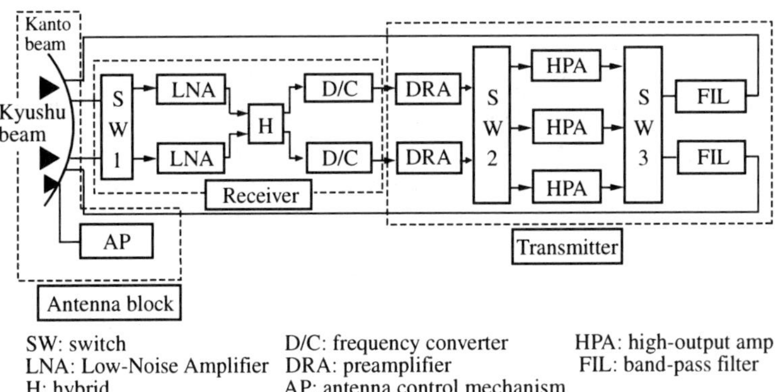

SW: switch   D/C: frequency converter   HPA: high-output amp
LNA: Low-Noise Amplifier   DRA: preamplifier   FIL: band-pass filter
H: hybrid   AP: antenna control mechanism

Fig. 5.19   Equipment Configuration for COMETS

The purpose of the COMETS program high-level satellite broadcasting mission was the world's first proof-testing of a satellite broadcast over the newly allocated 21-GHz broadcast band. The general specifications of the equipment used in the mission are listed in Table 5.14. The 120-MHz bandwidth (maximum) enabled the transmission of various types of information at various transfer rates. The COMETS transponder bandwidth is more than four-times greater than that of the current 12-GHz band system. The satellite also carries a 2-beam (multibeam) antenna aimed at Central Honshu (Kanto-Koshinetsu) and Kyushu, over which local satellite broadcasting tests were conducted. (Since this is an experimental satellite, only a 2-beam multibeam antenna is carried.)

Figure 5.19 shows a block diagram of the transponder configuration. The transponders are basic through-repeater types (relaying the signal, only modifying frequency and amplification), but in the 21-GHz band, the satellite carries three high-powered (200 W) TWTA systems (one a redundant system) for simultaneous transmission over two systems.

To put the experiments conducted with COMETS in broad categories, we have the measurement of satellite transponder charactertistics, the basic techniques of high-level satellite broadcasting, the technologies supporting the high-level broadcasting system, and signal reception techniques. The organizations involved in the experiment include CRL, NASDA, NHK, and a number of research universities.

5.5.3 Mobile Digital Satellite Sound Broadcasting

Mobile Digital Satellite Sound Broadcasting – Particulars and Services

Until recently there have been few systems to speak of whose purpose is to broadcast services to automobiles and aircraft. There are some FM broadcasts receivable by moving vehicles, but since these systems were not originally designed for mobile reception on city streets surrounded by high-rise buildings, signal quality is often degraded by multipath. The broadcasting industry in Europe and Canada, seeing a need in this area undertook the task of developing a digital sound broadcasting system capable of delivering CD-quality audio, whose purpose was to provide road information and various other types of data to moving vehicles. Where these systems are operational, broadcasting has started up using terrestrial (ground-wave) systems with OFDM (orthogonal frequency-division multiplexing) signaling systems which are quite immune to multipath. Research on these systems has now gone global, and the feasibility of implementing satellite sound broadcasting systems is being investigated in a number of countries. As was noted in Sec. 5.1, frequencies for this system were allocated at WARC-92.

The International Scene

Terrestrial mobile digital sound broadcasting initially began in the European countries, and systems are now being developed and implemented world-wide. Wide-area and global coverage, in addition to multimedia communications applications are now being studied using the European Space Agency's Archimedes satellite, an orbiting satellite in a highly inclined orbit. Realization of satellite broadcasts should bring services into areas of the world where terrestrial broadcasts are economically unprofittable. Questions regarding the economic viability of commercial ventures and other matters are now being reexamined in the European Union, but as a results of studies undertaken so far, the following plans have been developed:

(1) The utilization of six satellites in six orbits spaced 60-degrees apart, serving the northern hemisphere.

(2) A perigee ranging from 800 to 1,400 km and an apogee of 26,800 km, which enables high-angle reception even in high latitude areas of the northern hemisphere. Orbital periods of approximately 8-hours, with service possible for four hours near the apogee.

(3) Compatibility with terrestrial systems by using QPSK-based OFDM modulation. (ITU recommendations classify the system as a System A, satellite broadcast mode III, 192 carriers.)

(4) 100 channels of CD-quality sound; 200 ~ 300 channels of FM broadcast-quality
   sound.

In North America, there are a number of different plans regarding digital satellite broadcasting systems, but the form that digital sound broadcasting over geostationary satellite will take is as part of the Worldspace program. Current plans for Worldspace are to launch a series of satellites into orbit starting in 1998, with sound broadcasts in the 1.5-GHz band using a circularly polarized signal with QPSK modulation. (The scheduled satellite launch sequence is AFRISTAR (Africa), AMERISTAR (North/South America), and ASIASTAR (Asia).) The receiver terminals to be used in the Worldspace system are folding-antenna type portable devices costing about $200 (US currency), and the nature of the broadcasts are primarily intended to serve the highly-populated emerging nations.

In Japan, appliance manufacturers and radio broadcasting industry officials have announced plans to provide sound and various information services for mobile applications over a JSAT stationary orbit satellite. The system will use the 2.5-GHz band allocated to Japan, but plans regarding commercial matters are yet to be finalized.

Mobile Satellite Digital Sound Broadcast Experiments using ETS-VIII
In Japan's program to provide digital sound service and other high-level communications services in the mobile environment, the ETS-VIII (Experimental Test Satellite VIII, a geostationary satellite) is being developed for proof-testing of the system. Launch of the satellite is scheduled for the year 2002. As Fig. 5.20 illustrates, this experimental satellite carries very large (10 m) antennas. Its high effective radiated power should reduce the load on receivers. The satellite will broadcast over the 2.5-GHz band. As the satellite is still in the experimental stage, the basic transmission tests will be conducted using various modulation systems (QPSK, QPSK-OFDM, etc.). Multimedia broadcasts to mobile receivers, and various other applications tests are planned.

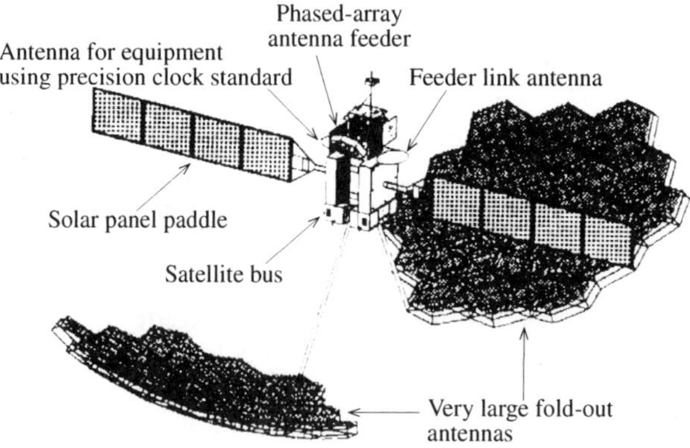

Fig. 5.20   ETS-VIII System Diagram [Courtesy of NASDA.]

5.6 Appendix: Frequency-Dependence of Rain Attenuation Coefficients

The relationship between rain intensity and rain attenuation can theoretically be determined by the forward scattering amplitude of the raindrops and the size distribution of the raindrops. These results determine the rain attenuation per unit path length (usually 1 km), which is called the specific attenuation. Specific attenuation is approximated by

$$\gamma = aR^b \quad \text{(dB/km)}, \tag{5.18}$$

where a and b are constants whose values depend on frequency, polarization, elevation angle, and drop size distribution.

Parameters for a and b are provided by ITU recommendation PN 838 [ITU, 1994 d] for vertical polarization (subscript v), and horizontal polarization (subscript h) with respect to frequency. A set of these coefficients is provided in Table 5.15. The values for a and b specific to other parameters can be approximated by the following equations:

$$a = [a_h + a_v + (a_h - a_v)\cos^2\theta \cdot \cos 2\tau]/2$$
$$b = [a_h b_h + a_v b_v + (a_h b_h - a_v b_v)\cos^2\theta \cdot \cos 2\tau]/2a, \tag{5.19}$$

Table 5.15   Rain Attenuation Coefficient Calculation [From ITU recommendation PN 838.]

Freq. (GHz)	a_h	a_v	b_h	b_v
1	0.0000387	0.0000352	0.912	0.880
2	0.000154	0.000138	0.963	0.923
4	0.000650	0.000591	1.121	1.075
6	0.00175	0.00155	1.308	1.265
7	0.00301	0.00265	1.332	1.312
8	0.00454	0.00395	1.327	1.310
10	0.0101	0.00887	1.276	1.264
12	0.0118	0.0168	1.217	1.200
15	0.0367	0.0335	1.154	1.128
20	0.0751	0.0691	1.099	1.065
25	0.124	0.113	1.061	1.030
30	0.187	0.167	1.021	1.000
35	0.263	0.233	0.979	0.963
40	0.350	0.310	0.939	0.929
45	0.442	0.393	0.903	0.897
50	0.536	0.479	0.873	0.868
60	0.707	0.642	0.826	0.824
70	0.851	0.784	0.793	0.793
80	0.975	0.906	0.769	0.769
90	1.06	0.999	0.753	0.754
100	1.12	1.06	0.743	0.744
120	1.18	1.13	0.731	0.732
150	1.31	1.27	0.710	0.711
200	1.45	1.42	0.689	0.690
300	1.36	1.35	0.688	0.689
400	1.32	1.31	0.683	0.684

where θ is elevation angle and τ is the polarization angle from the horizontal direction (45-deg. for circular polarization).

5.7 Conclusion

Digital satellite broadcast television systems have already been realized in the United States, European Union, and in a number of Asian countries, including Japan, Korea, and Hong Kong. As a result of these various programs, many of the digital satellite communications technologies have reached a high level of maturity, but one of the greatest benefits in the field of digital satellite broadcasting has been the realization of multichannel services using a single transponder. (Prior to digitalization, one transponder was required for each broadcast channel.) In the future, we should be able to extend these achievements to include broadcasts to mobile receivers, provide subscribers with digital information such as data and software, and the development of interactive media through integration with ground-based communications networks. Another likely future development is the formation of integrated digital networks, achieved by combining digital satellite broadcasting and satellite communication systems.

The material covered in this chapter gives a good overview of the basic subjects of satellite communications and broadcasting technologies. For those readers desiring additional information on these subjects, the [Iida, 2000] and [Endoh, 1988] references are recommended.

Chapter Questions

Q1: List the advantages and disadvantages of satellite broadcasting as it compares with terrestrial and CATV services.

Q2: Where uplink C/N is 20 dB, and downlink C/N is 12 dB, calculate the total C/N in the digital satellite broadcasting link budget. Also, when required C/N is 8 dB, calculate the link margin (assuming no interference).

Q3: Discuss the factors and problem areas that affect the determination of service availability rate in the K_a digital satellite broadcast band.

Q4: Estimate the worst-month rain attenuation cumulative distribution for 22-GHz satellite broadcasts (circular polarized) received in Tokyo (assume 0 km elevation), and calculate rain attenuation levels for worst-month time rates of 1% and 0.1%, respectively. For this problem, use a rain intensity ($R_{0.01}$) of 53 mm/h for a whole year time rate of 0.01%, and a satellite elevation angle of 45-degrees.

Chapter References

[ASC, 1996] Jisedai Eisei Tsuushin-Housou Shisutemu Kenkyuujo: "Sekai no Tsuushin-Housou Eisei Shisutemu no Genjou to Doukou", Heisei 7-nenban (1996.7) (in Japanese)

[Clark, 1993] Arthur C. Clark, author; Komatsu, S., translater: "Chikyuu mura (Global villege) no Kanata – Mirai kara no Dengon –", Doubunshoin (1993) (in Japanese)

[Dosch, 1994] Dosch, C., : "Inter-operability of digital HDTV satellite broadcasting (21.4-22GHz) with the existing and future media infrastructure : Status of the HD-SAT project," EBU Technical Review, pp. 51-63 (Summer 1994)

[Endoh, 1988] Endoh, Izumi: "Housoueisei no Kiso Chishiki", Kenrokkan (1988) (in Japanese)

[Fukuchi, 1983] Fukuchi, H., Kozu, T., Nakamura, K., Awaka, J., Inomata, H. and Otsu, Y. : "Centimeter wave propagation experiments using the beacon signals of CS and BSE satellites," IEEE Trans. Antennas and Propag., Vol. AP-31, No. 4, pp. 603-613 (July 1983)

[Fukuchi, 1985] Fukuchi, H., Kozu, T. and Tsuchiya, S. : Worst month statistics of attenuation and ZPD on earth-space path," IEEE Trans. Antennas and Propag., Vol. AP-33, No. 4, pp.390-396 (April 1985)

[Fukuchi, 1992] Fukuchi, H. : "Slant path attenuation analysis at 20 GHz for time-diversity reception of future satellite broadcasting," Proc. URSI-F International Open symposium, (Ravenscar, UK), No. 6. 5 (1992)

[Fukuchi, 1993] Fukuchi, Ohuchi: "Tsuushin Housou Gijutsueisei (COMETS) ni yoru 21 GHz tai Eisei Housou Jikken Keikaku", Terebijyon Gakkai Gijutsu Houkoku, Vol. 17, No. 57, pp. 1-6, ICS '93-58 (Oct. 1993) (in Japanese)

[Fukuchi, 1994] Fukuchi, H. : "Quantitative analysis on the effect of adaptive satellite power control as a rain attenuation countermeasure," Proc. IEEE Antennas and Propagation International Symposium, (Seattle, U. S. A.), pp. 1332-1335 (1994)

[Fukuchi, 1995] Fukuchi, Ohshima: "21 GHz tai Eiseihousou Shisutemu no Kenkyuu Doukou to COMETS Eiseihousou Jikken Keikaku no Gaiyou", ITU Journal, Vo. 25, No. 12, pp. 52-60 (1995) (in Japanese)

[Iida, 2000] Iida, T.: "Satellite Communications", Ohmsha (2000) (in Japanese)

[ITU, 1994 a] ITU-R Recommendations : "PN 618-3 : Propagation data and prediction methods required for the design of earth-space telecommunications systems," ITU (1994)

[ITU, 1994 b] ITU-R Recommendations : "PN 452-6 : Prediction procedure for the evaluation of microwave interference between stations on the surface of the earth at frequencies above about 0.7 GHz," ITU (1994)

[ITU, 1994 c] ITU-R Recommendations : "PN 837-1 : Characteristics of precipitation for propagation modeling," ITU (1994)

[ITU, 1994 d] ITU-R Recommendation : "PN 838 : Specific attenuation model for rain for use in prediction methods," ITU (1994)

[ITU, 1994 e] ITU-R Recommendation : "PN 841 Conversion of annual statistics to worst-month statistics," ITU (1994)

[JSAT, 1996] Nihon Sateraito Shisutemuzu Kabushikigaisha: "Eisei Tsuushin Senyou Saabisu goriyou no tame no Gijutsushiryou (JCSAT-3 gou Eiseiyou)", Dai 2-han (1996.3) (in Japanese)

[Karasawa, 1991] Karasawa, Y. and Matsudo, T., : One minute rain rate distributions in Japan derived from AMeDAS one-hour rain rate data," IEEE Trans. Geoscience and Remote Sensing, Vol. 29, No. 6, pp. 890-898 (Nov. 1991)

[Kuwata, 1982] Kuwata: "Housoueiseiyou Shuhasuu no Bunpai wa dono you ni kimerareta ka – 1971-nen no Omoide –", Housou Gijutsu, pp. 952-956 (1982.10) (in Japanese)

[MPT, 1996] Yuuseishou Tsuushin Seisakukyoku Kanshuu, Kokusai Eisei Tsuushin Kyoukai

Hakkou: "Eisei Tsuushin Nenpou", Heisei 7-nenban (1995) (in Japanese)

[Nakagawa, 1996] Nakagawa, Kamise, Hashimoto: "Nihon no CS Deijitaru Housou Jikken to Jigyoukeikaku", Terebijyon Gakkaishi, Vol. 50, No. 1, pp. 55-58 (1996) (in Japanese)

[Ohta, 1998] Ohta, H. and Fukuchi, H. : "Effect of rain-and-ice-induced depolarization on dual polarized 21 GHz-band digital satellite broadcasting system," Elect. Lett., Vol. 34, No.4, pp. 330-332. (Feb, 1998)

[Oliphant, 1995] Oliphant, A. and Combarel, L., : "Digital broadcasting demonstrations by HD-SAT and dTTb at Montreux'95," EBU Technical Review, pp. 43-49 (Summer, 1995)

[TTC, 1995] Denki Tsuushin Gijutsu Shingikai Toushin: "Shimon Dai 74 gou [Dejitaru Housou Houshiki ni Kakawaru Gijutsuteki Jouken] no uchi 12.2 ~ 12.75 GHz wo Shiyou suru Eisei Dejitaru Houshiki (27 MHz Taiikihaba wo Shiyou suru mono) no Gijutsuteki Jouken" (1995) (in Japanese)

Chapter 6
CATV (Cable Television)

In this chapter, emphasis is placed on digital cable TV technologies, and a general description of these subjects is presented. The chapter starts with a brief history of CATV and an overview of systems and channel configurations. A description of the basic concepts of digital cable television is then given, including expectations of signal and reception quality. The subject of link budgets is then discussed. The chapter concludes with a discussion of standards (present and future) for digital cable networks. Throughout this text the acronym CATV will be used to mean cable television.

6.1 CATV – History and Background

6.1.1 A Brief History

The Early Years

Early CATV began as a measure intended to improve the poor reception for the fringe areas where terrestrial broadcasts service was not adequate [Miyakawa, 1986], [Ishiguro, 1984], [Ishiguro, 1986]. This was accomplished by setting up a master antenna and using cables to distribute the signal to nearby homes. In the early days of television, priority was given to facilitating the spread of terrestrial broadcast television, and CATV was considered as a "behind-the-scenes" operation, often branded as a hindrance to the harmonious development of standard broadcast television. However, because Japan's Broadcast Law specifies that "NHK has an obligation to disseminate broadcast television over as wide an area as possible," NHK set up CATV systems to remedy the problem of poor reception in isolated areas by hills and mountains. Thus, CATV's role in early television was to supplement terrestrial broadcasting in the poor reception areas. Over time, as many high-rise buildings were constructed in large cities, many urban areas began suffering problems with multipath (ghost) interference, even though in most cases affluent signal strength were available. This resulted in many complaints by the residents, bringing about a switch in the role of CATV from supplementing terrestrial broadcasts in areas where signal strength was weak, to acting as an interference countermeasure in urban areas where signal strength was strong.

Looking at the historical background of the basic CATV technologies; as was noted in Chapter 2, the video signal utilizes a bandwidth of approximately 4 MHz, and in the early days distributing a signal of that bandwidth over a cable was not a simple matter. The immediate problem faced by CATV was finding out a cable that could transmit a signal with wide spectrum from DC to high frequency region. The possibility of a coaxial cable was first proposed in the 1930's, but the problem at that time was in supporting and isolating the center conductor of the cable. An insulating material with enough frequency characteristics was simply not available. What sufficed for a "coaxial cable" at the time was a cable referred to as a "virtual wideband cable," which was an air-gap type coaxial cable developed by the Germans prior to World War II, and whose center conductor was supported by a wrapped coil of silk string. The first practical coaxial cables capable of carrying television video signals came in the 1960's, when an economical encapsulating material having superb high frequency characteristics was constructed from polyethylene. At about the same time, transistors also appeared that had frequency characteristics suitable for use in the newly proposed negative-feedback wideband amplifiers, and this resulted in the realization of a practical wideband amplifier. The development of polyethylene-insulated coaxial cables and wideband amplifiers significantly expanded the capabilities for CATV, which subsequently grew rapidly in popularity in countries such as the US, Canada, and Japan.

The Growth Years
After recovering from the ruin at the end of World War II, the Japanese people's lives became more stable, and as a part of attaining the affluence that came to be admired by many in the rest of the world, the demand by the populace for more options in television programming also grew. Unlike wireless systems, CATV utilizes cable, and since there are no restrictions on frequency band usage with cable, CATV became one of the leaders in the development of multichannel television. Recognizing the possibilities offered by CATV, the number of service providers quickly increased to fill this new demand, particularly after 1985 when multichannel cable television was introduced.

In the major countries of Europe and North America, the drive for more diversity in broadcast television began about 1985. Those moves have recently led to the convergence of telecommunication and broadcasting. The example of such convergent telecommunications-broadcasting services is using communication satellites for multichannel digital broadcasting, utilizing the existing telephone subscriber network (xDSL[*1], or FTTH, or FTTH + B-ISDN[*2]) to distribute the video services. (Details given in Ch. 7.)

These trends are reducing the barriers that once delineated the fields of communications and broadcasting, ushering in a period where various service providers are in competition to provide the subscriber with more diversity of services. From the standpoint of the end-user, today's such competition simply could abriviate as enriching information pipeline providing many services through which traditional TV broadcasting service, internet services and would-be-data broadcasting services which seems to be rather reluctant to increase in volume. In the future, the subscriber will no doubt have more opportunity to access and select of media that are easier to use, and in more user-friendly formats.

With regards to trends in the technology, since CATV has been a hard-wired system, it has been a long-cherished desire to support interactive media services. Interactive media over CATV started in the 1980's with the development of a viewer-participating narrow-castling[*3] [Ishiguro, 1987], which resulted in Videotex (VTX). With VTX, textual information and still-pictures are accessed from a media-center computer data base for two-way transmission over a communications network. At one point research in this area was very active, but because of ingress noise problems in the cable network and difficulties in operating the two way system, research efforts in interactive media applications have declined along with the declining popularity of VTX.

The Multimedia Period

Entering the 1990's, research and development again started on advanced CATV services, this time taking the form of Video on Demand (VoD)[*4]. The VoD opened up new opportunities for interactive CATV and up several experimental programs, including VCTV[*5], BQ Cable[*6], and FSN[*7], and efforts to perfect these systems are still active. (Such programs also include the Next-Generation Integrated Telecommunications Pilot Project in the Kyoto-Osaka-Nara area[*8], the Experimental Multimedia System (ATMS) in the Tokyo Bay area[*9], and the ACT Center[*10].)

As digital data processing speeds have increased and the prices of software and hardware have come down, personal computers have become commonplace items in most household. As a consequence, the recent growth of the Internet has been quite phenomenal. In fact, we can say that the growth of the Internet has proven the marketability of the interactive services. CATV industry has been ardent about writing standards for CATV cable modems. Cable modem standards for the US were implemented in 1997.

At the same time, the work on international standards for the digital compression and decoding of the video (MPEG-2) which began in 1994 has been progressing on schedule [Yasuda, 1995], as have the developments in digital CATV systems using the MPEG-2 standard. (Refer to Chapter. 2 for details on MPEG-2.)

Digital CATV systems are becoming very popular, which brings about the need for standards in areas such as MPEG-2 source coding, channel coding, services control information, and interfaces. These standards will allow the user to configure a personal system, coordinating equipment such as personal computers, TV sets, video tape recorders, and various audio components. Standards for digital CATV are also being written, both in Japan and abroad. In Japan, the Telecommunications Technology Council (TTC) and civilian standards-setting organizations are together working on developing digital television broadcasting standards that will allow all information contents to be encoded so as to be cross-compatible with all forms of television media (including terrestrial and satellite TV, and CATV).

*1 The system to carry video information utilizing the existing metal cable from a telephone office to subscriber. There are a few systems such as ADSL, DSL and SDSL, and X-DSL is a general term for these ADSL assigns bandwidth up to 3.4 KHz for voice, that 3.4 KHz to 1.1 MHz for video. The X-DSL is susceptible to cross-talk interfence.

*2 This system consists of an optical fiber network extending into the subscriber home. The (pai) system is an interim system replacing the current metal cable until the FTTH system is completed. The completed FTTH system will enable the transmission of wideband videos, and the FTTH intra-city network will enable broadcasting services. B-ISDN stands for broadband-integrated services digital network, and the network supports a 64 kbit/s × p (integer) transmission hierarchy. The transmission medium for high-speed B-ISDN will likely be FTTH.

*3 In this program, viewers are regularly polled regarding television programming preferences, and CATV operators reflect those preferences in their programing. The results of polling are collected and stored by a center computer. This type of application is typical of early programs using interactive CATV, and the Qube System in Columbus, Ohio (USA) is a well-known example.

*4 Video information stored in the media center is available for the user to search and call up from an video data base. Storage media choices are video tape, CD, disk, etc. video information is packetized and storaged in redundant array of disks, and is regenerated upon being called up. 30 minutes to two hours of program contents can be instantly accessed, ready for immediate playback. Advantage point is that the system can accommodate many users. (Refer also to Sec. 7.3.)

*5 Tele Communications Inc. (TCI), in cooperation with the local telephone company (BOC) conducted a trial survey to determine the demand for VoD in the suburbs of Denver, Colorado (USA). Programming offered consisted of theatrical movies and video packages. No system server was used, but VCR cassettes were used for playback on the customer's demand. Non-technical market research.

*6 Trial survey conducted by Time Warner Cable of New York regarding the market for NVoD in the boroughs of Brooklyn and Queens. NVoD programs occupied four channels, and one program was broadcast in different time slots so that the subscriber had only a short wait for the program to start anew.

*7 Experimental program by Time Warner Corp. in Orlando, Florida (USA). Using part of the Time Warner Cable system, multimedia consisting of VoD, cable phone, and cable PHS were offered. (FSN: Full-Service Network) Program designed as a technical test. Time Warner reports the program failed in the area of system management because there introduced and integrated too many cutting-edge services. Test was ended (Spring, 1997) after the initial multimedia fever subsided. Big test – widely publicized.

*8 Experimental program started in 1994 by Next-Generation Integrated Telecommunications Pilot Project. Multimedia services including HDTV, video phone, VoD, and interactive TV were offered in about 500 households in the Kansai Academic City of Seika-machi. VoD initially was analog, and was later switched to digital. Specific services offered include Virtual Mall, News on Demand, Karaoke on Demand, and Game on Demand. This aimed both technical and marketing survey.

*9 Multimedia test conducted in the Tokyo Bay area as the Experimental Multimedia System (ATMS) starting 3/1996. Video phone, VoD, interactive TV were offered to about 1200 households. Tests included distance learning and remote medical consultation. Test ended 9/1997.

*10 CATV-based multimedia experimental facility set up within the Ministry of Posts and Telecommunications' CRL grounds. Purpose is to conduct proof-testing of digital CATV systems that relate to MPT Inquiry No. 74.

6.1.2 CATV Systems

Basic System Structure

As illustrated in Fig. 6.1, the CATV system consists of the Center facility (sending terminal), the transmission path, and the user-terminal (receiving end). The system structure is applicable to both analog and digital systems, the only difference being that the Center facility and user-terminals are digital in the digital system.

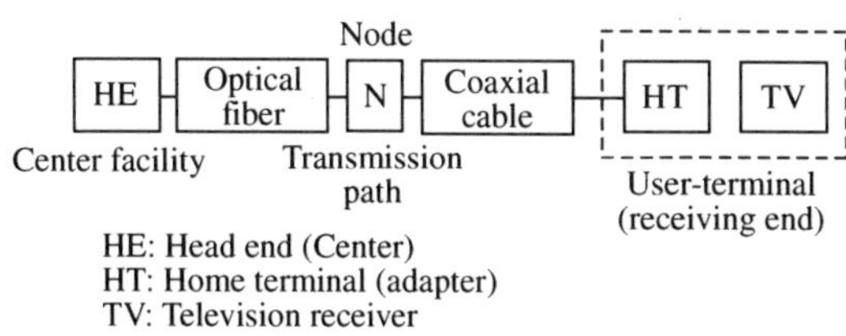

Fig. 6.1   Basic Structure of CATV System

The Center facility houses the receiving antennas and the equipments interfacing the external lines. It also consists of the demodulating and remodulating equipment, the multiplexer, and the terminal control equipment (scramblers, etc.). The transmission facility consists of the transmission network, amplifiers, and distribution apparatuses (tap-off, TO).

The user-terminal configuration can be rather complicated, as it consists of household appliances such as a television set, personal computer, and other devices which serve as the interfaces between the system and the user. The interface between the household appliance and the CATV network is a home terminal (HT, for analog systems), or a set top box (STB, for digital systems), and other various adapters. The form that the user-terminal takes thus depends on how the user wishes to utilize the CATV system.

The Center facility, transmission facility and user-terminal configuration also varies depending on whether the network uses coaxial cables or a hybrid fiber coaxial (HFC), and whether the system is unidirectional or bidirectional. Figure 6.2(a) shows an example of a coaxial-unidirectional system, and 6.2(b) shows the configuration of the HFC-bidirectional system.

Transmission System

As illustrated in Fig. 6.3, the total CATV transmission bandwidth is divided into a low band and a high band, with the low band used as the reverse stream, and the high band as the forward stream. Normally the reverse stream uses the range of 10 ~ 50 MHz, and the forward stream 70 ~ 770 MHz. In bidirectional communications, as again shown in Fig. 6.3, a pair of carrier is assigned in the low band and high band, which form the path for one communications channel. For broadcasting, the multichannel TV signals are multiplexed every 6 MHz slot in the transmission frequency band. This is referred to as a frequency-division multiplex (FDM) .

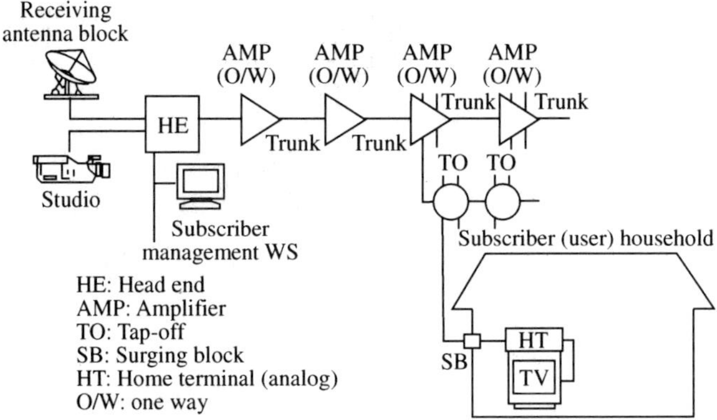

(a)   Coaxial-Unidirectional CATV System

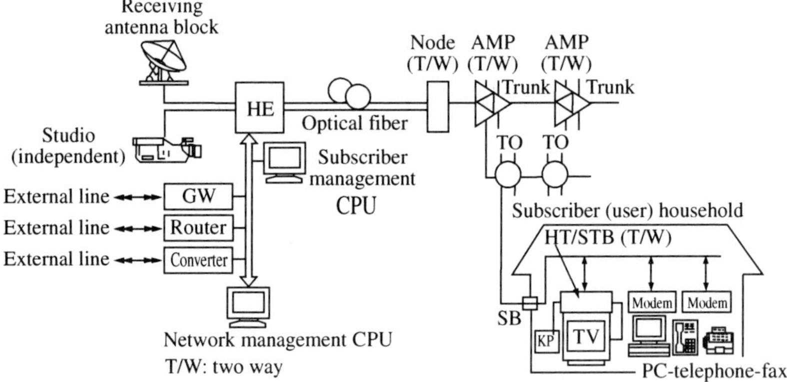

(b)   HFC-Bidirectional CATV System

Fig. 6.2   CATV Center, Transmission, and Terminal Facilities (example)

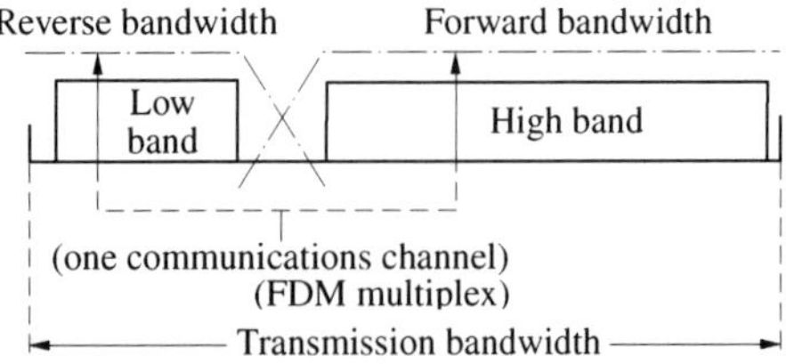

Fig. 6.3   Basic Principle of Bidirectional CATV (A communications channel established by a pair of carrier in low and high bands.)

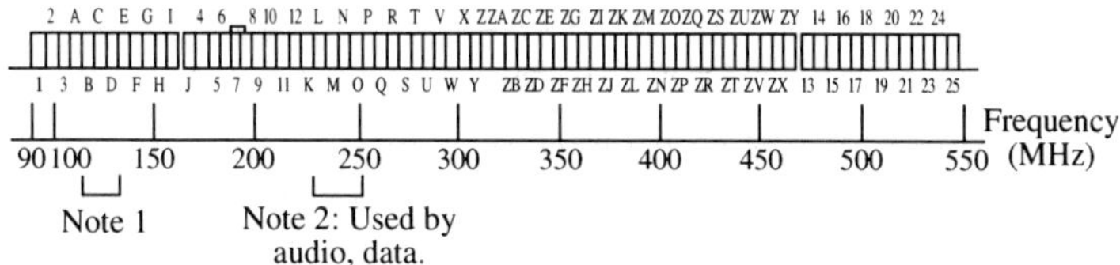

Note 1: Be aware of potential oscillator interference in CHs 4, 5, 6 from CHs B, C, D.
Note 2: CHs L, M, N, (O) can also be moved up 2 MHz, but potential oscillator
interference exists in CHs V, W, X, (Y) in that case.

Fig. 6.4   CATV Channel Arrangement (example) (FDM transmission typical)

Unidirectional and bidirectional amplifiers are shown in Fig. 6.5, (a) and (b), respectively. In the bidirectional amplifier there is dividing filters (DF) at the input and output (including trunk and bridging outputs), and the DFs separate low band (reverse) and high band (forward) streams.

As was mentioned previously, the structure of the transmission path is basically the same for both systems (analog, digital), but with digital CATV all television signals are packet-multiplexed, and every 6 MHz bandwidth can carry 2 ~ 6 channels (normally 4). The number of digital channels that can be multiplexed at a 6-MHz bandwidth depends on the maximum amount of bits generated by the source video encoder, and this is statistically determined by the picture detail, the number of scene changes, and the quickness of motion in the picture.

Transmission Capacities

The nominal system transmission capacity is determined by dividing the total transmission bandwidth by 6 MHz, and subtracting from this value the number of forbidden channels. The forbidden channels are the channels whose use is prohibited because of the interference they may produce, or conversely, be subjected to. Whether a channel gives interference, or receives interference is related to the signal level itself, or to the modulation system of the signal. These various factors must be carefully considered when deciding which channel is the forbidden channels.

Network Structures

The network used by CATV can be the bus type, tree type, or the star type structure (Fig. 6.6). Because of economical considerations, the most common network is a tree type structure using coaxial cables. More recently, use of an optical fiber star network between the head end (HE) and the node is becoming popular, with a coaxial-tree network between the node and the subscriber (refer to Fig. 6.2(b)). This archtecture is known as a Hybrud Fiber Coaxial (HFC). In an HFC network, the coaxial-tree network section is called a cell. The purpose of using an HFC is as follows:

(1) To enhance system performance at the CATV subscriber terminals[*11] while
permitting a larger-scale system to be constructed,

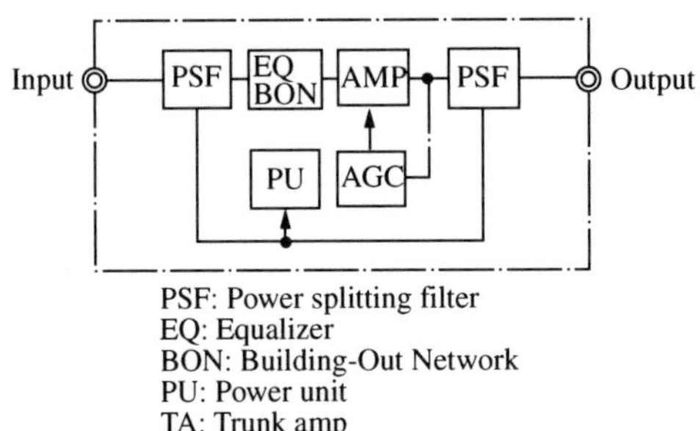

PSF: Power splitting filter
EQ: Equalizer
BON: Building-Out Network
PU: Power unit
TA: Trunk amp

(a)   Unidirectional amplifier

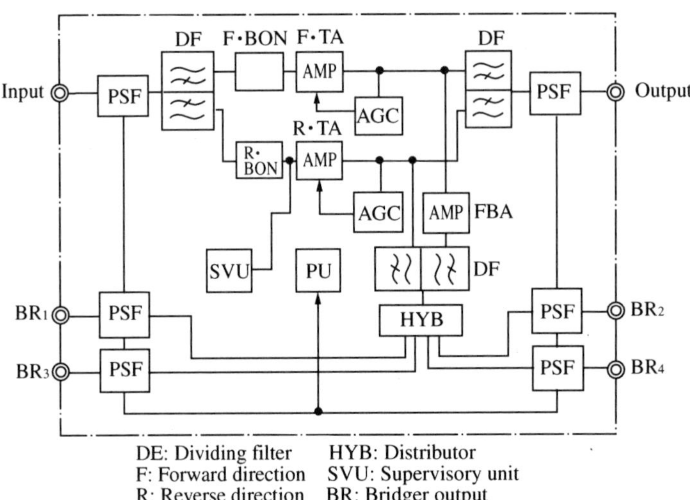

DE: Dividing filter   HYB: Distributor
F: Forward direction   SVU: Supervisory unit
R: Reverse direction   BR: Bridger output

(b)   Bidirectional amplifier

Fig. 6.5   Unidrectional and Bidirectional Amplifiers for Coaxial Systems (example)

(2) to lighten traffic congestion for bidirectional services[*12], and
(3) to reduce ingress noise in the system. (Refer to Sec. 6.3.4.)
In order to accomplish these objectives, the size of the cell must be restricted to from 500 to 2000 households (including potential subscribers).

---

*11 Refers to the performance specified by the electrical parameters of the CATV system. In single unit houses, specified at the surging block (SB) output terminal; in multidwelling (apartments, condominiums, etc.), specified at the wall outlet terminal.

*12 Where there are more total subscribers than communications lines in the CATV system, subscribers can use the system on a time-sharing basis. In a large-scale system enough lines must be available to prevent congestion. In an HFC system, the number of subscribers in a cell can be summarily reduced, which allows the number of lines between the node and the Center to be reduced. This permits a heavy of traffic to be accommodated without congestion. CATV telephone service quality indices for the HFC system (500 telephone subscribers per cell): Average call volume = 0.03 ~ 0.08 erl, distribution rate = 100%, dropped call rate = 0.01.

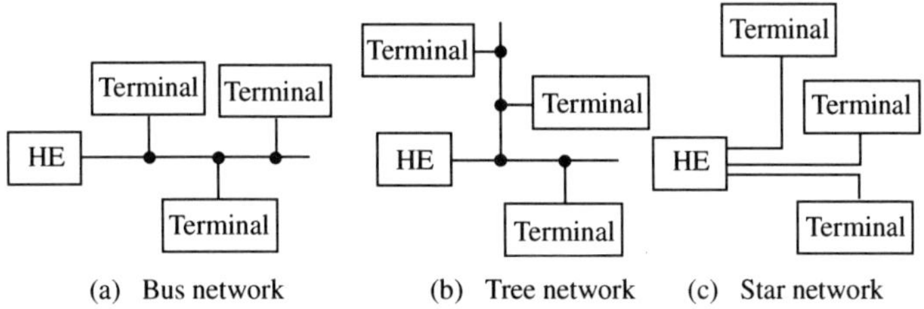

Fig. 6.6   Types of Network Configurations

CATV Terminals

As briefly noted above, the configuration and the functions of the CATV terminals depend primarily on the type of services the end-user desires. Since most households have available the television set, VCR, or personal computer needed to provide the man-machine interface, we only need to determine which functionalities these appliances require to receive CATV services.

Except in special cases, the functions normally required by a CATV terminal includes a tuner function capable of tuning or selecting the desired channel from the multiple channels available, a conditional access function (very important), and a communications controller for bidirectional communications. The communications controller is used to process the information contained in the transmitted/received signals, and takes several different forms. Additionally, digital CATV systems require the copy protection, of which control signals are transferred from STB to the TV receiver by utilizing the vertical blanking interval (VBI), and an electronic program guide (EPG) function which displays programming information on the TV screen.

The configuration of bidirectional terminals varies greatly depending on the services the system is set up for: for example, whether the telephone line or CATV line will be used for the reverse stream, what type of communications control is used for the reverse stream. Whether or not security measures (encryption function) are required will also affect configuration and functions, as well as hardware and software requirements.

There are also some significant differences in conventional analog terminals and the digital terminals now under development. For example, with analog terminals the multichannel selector requires only the capability of selecting the 6-MHz FDM channel (the physical channel). On the other hand, with the digital terminal the user must be able to select specific information within the physical channel. This specific information is packet-multiplexed within the physical channel, and is called the logical channel. Figure 6.7(a) shows the structure of an analog terminal (bidirectional HT), and 6.7(b) shows a digital terminal (bidirectional, multichannel, STB).

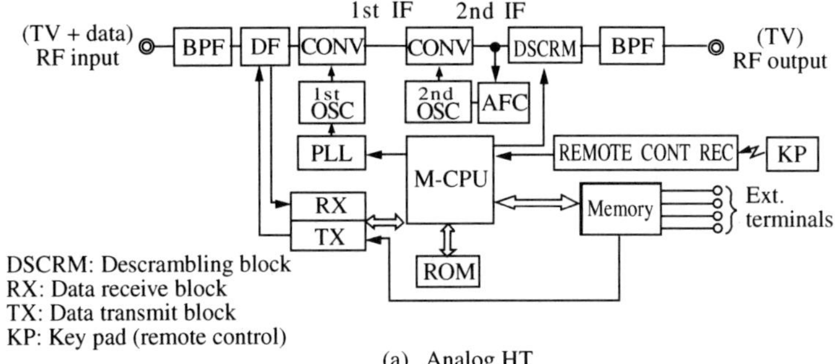

(a)   Analog HT

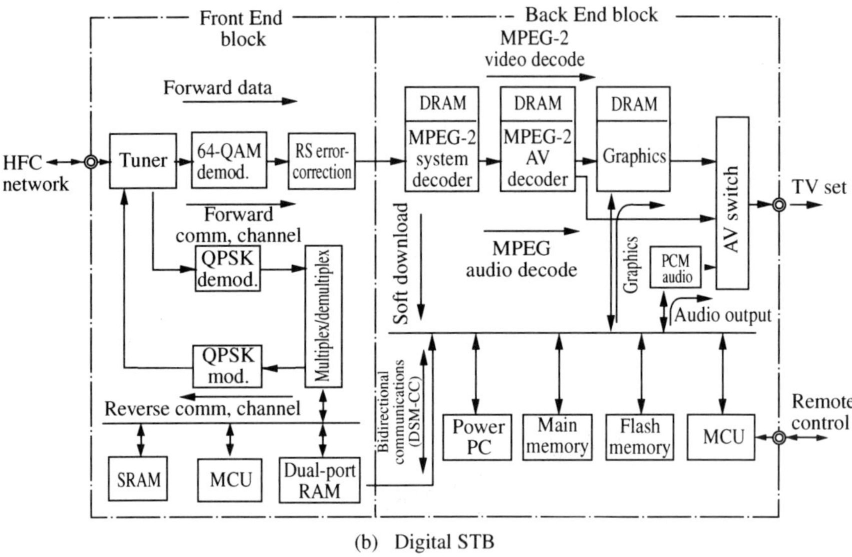

(b)   Digital STB

Fig. 6.7   Example of Terminal Configurations for Analog and Digital Terminals

6.1.3 CATV Transmission Network

In this section we will discuss the transmission path in a bit more detail. We shall start with a description of the system structure of the transmission path, and then discuss the features and characteristics of the devices comprising the transmission path.

Transmission Network

The description given here is based on the hybrid fiber coaxial (HFC) system [MPT, 1994], the configuration of which is illustrated in Fig. 6.2(b). An example of the cell structure down after from the node is shown in Fig. 6.8. (Figure 6.8(a) shows an optical fiber node.)

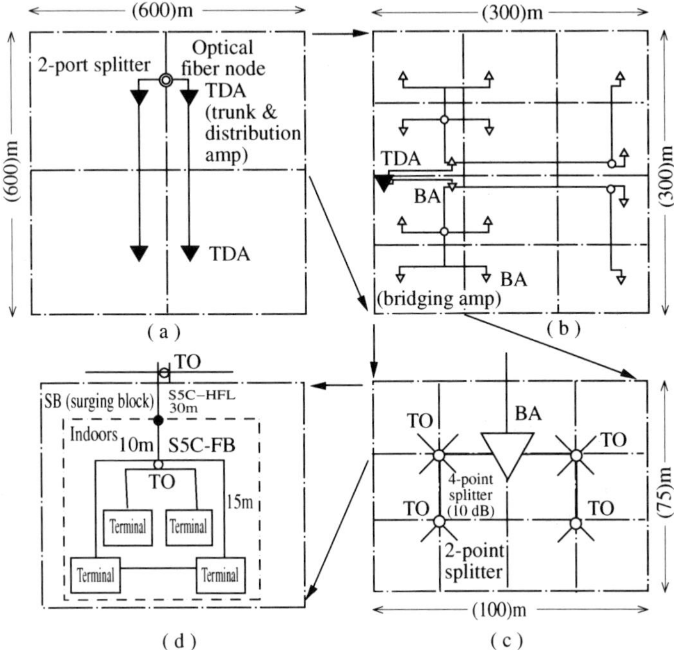

Fig. 6.8   CATV Transmission Network Model

The node output signal is applied to the trunk and distribution amplifier (TDA, 2-stages in Fig. 6.8 (a)) and bridging amplifier (BA, 2-stages in Fig.6.8.(b)) where transmission losses in the coaxial cables are compensated. The signal is then passed to the splitters and tap-offs (TO) for distribution to the subscribers (Fig. 6.8(c)).

A general diagram showing the routing between the outdoor TO to the subscriber's terminals is given in Fig. 6.8(d). The signal is distributed to each terminal from the indoor TO.

The reverse and forward signals can be transmitted over one coaxial cable by FDM. Each amplifier has a forward and reverse amplifier block, with DF used to separate and route the signal to the appropriate amplifier and control block. The signals pass again DF for recombining. (Refer to Fig. 6.5.) In an optical fiber network, some systems use separate lines, different wavelengths of light or multiplexing the signals through a signal fiber for the forward and reverse streams.

Coaxial Cables

The structure of the coaxial cable is shown in Fig. 6.9. The coaxial cable consists of a center conductor surrounded by an outer conductor, with an insulating material between the two conductors. The exterior of the outer conductor is usually jacketed. Low-density polyethylene foam is typically used as the insulator because of its low

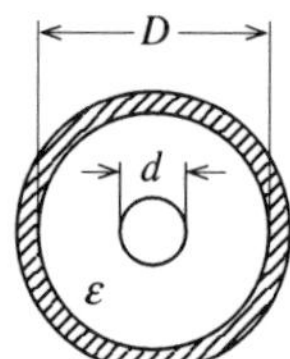

Cable	5C	8C	12C
d	1.02	2.1	2.9
D	4.57	8.3	11.3
Loss (dB/km) 770 MHz	167	84	63

D : Inner diameter of outer conductor (mm)
d : Outer diameter of center conductor (mm)
ε : Equivalent relative dielectric constant of insulator (air = 1)

Fig. 6.9   Coaxial Cable Cross-Section (and specifications)

dielectric constant (equivalent relative dielectric constant, approx. 1.2).

The losses in a coaxial cable are calculated by

$$\alpha = 360 \cdot (K_1 / d + K_2 / D) \cdot \sqrt{f} / Z_0 + 90.9 \cdot f \cdot \sqrt{\varepsilon} \cdot \tan \delta \quad \text{(dB/km)},$$

$$(6.1)$$

where,

Z_0 : characteristic impedance
d : outer diameter of center conductor (Fig. 6.9)
D : inner diameter of outer conductor (Fig. 6.9)
ε : equivalent relative dielectric constant of insulator
f : frequency (MHz)
$\tan \delta$: dielectric tangent loss of insulator (polyethylene foam: $2{\sim}1.5 \times 10^{-4}$)
K_1 : relative conductivity, center conductor (copper: 1.00, aluminum: 1.28)
K_2 : relative conductivity, outer conductor (copper: 1.00, aluminum: 1.28)

The first term in the above equation represents conductor losses. Conductor loss due to conductor resistance increases in direct proportion to $\sqrt{f}$. The second term is dielectric loss (Joule heat loss), which is small in value compared to the first term, but increases in high frequency. In the frequency domain bands used by CATV, conductor losses are dominant.

The amount of attenuation in a coaxial cable varies according to conductor temperature, and temperature characteristics are also related to the dc resistance of the conductor. In Tokyo, an example of the internal temperature of a cable might range from $70\,^{\circ}\text{C}$ in the sun on a summer day, to $-3\,^{\circ}\text{C}$ before sunrise on a winter day. This $73\,^{\circ}\text{C}$ difference in temperature will account for a change in attenuation level of approximately 14%.

Distributors and Splitters

The distributors and splitters used in the system are normally ferrite-core broadband hybrid transformers with flat frequency response characteristics. The theoretical distribution loss D for an n-ports distributor is given as follows:

$$D = 10 \log(1/n) \quad \text{(dB)}. \tag{6.2}$$

Splitters are located at appropriate points in the transmission line, and function to shunt a portion of the signal power to another part of the circuit. By the conservation of energy laws, ideally, insertion loss should consist of only the amount of power tapped off by the splitter. Coupling loss C (dB) is relative to the number of splitting ports, n. The theoretical insertion loss I (dB) between the input and output terminals is expressed in the following:

$$I = 10 \log(1 - n \cdot 10^{-(C/10)}) \quad \text{(dB)}. \tag{6.3}$$

Because of the difference in load resistance between input and output, and distributed capacitance, actual distribution and insertion losses can be 1 ~ 1.5 dB larger than theoretical values.

When a signal is applied to the output terminal of a distributor or a splitter, due to the reciprocal rule of the passive components in the circuit, the signal level appearing at input will be lowered only by distribution and coupling loss. This means that distributors and splitters can also be used as signal combiners, and thus, a TO in the forward link can function as a distributor or splitter, and in the reverse link as a mixer.

Amplifiers

The types of amplifiers used in the CATV system are the trunk and distribution amplifier (TDA), which is designed for long distance transmission with minimum degradation in quality, and the extension amplifier (EA), which is designed to economically serve as many subscribers as possible. The function of the amplifier is to compensate $\sqrt{f}$ cable transmission losses (fis. frequency), to compensate cable losses and cable loss changes due to temperature changes, and to compensate flat losses due to the splitters and distributors inserted in the trunk.

The level relationships internal to the TDA are shown in Fig. 6.10(a). The TDA block diagram consists of a building out network (BON) with the same $\sqrt{f}$ characteristics as the coaxial cable, an equalizer (EQ) with inverse $\sqrt{f}$ characteristics, an attenuator (ATT) with flat losses. The level diagram in Fig.

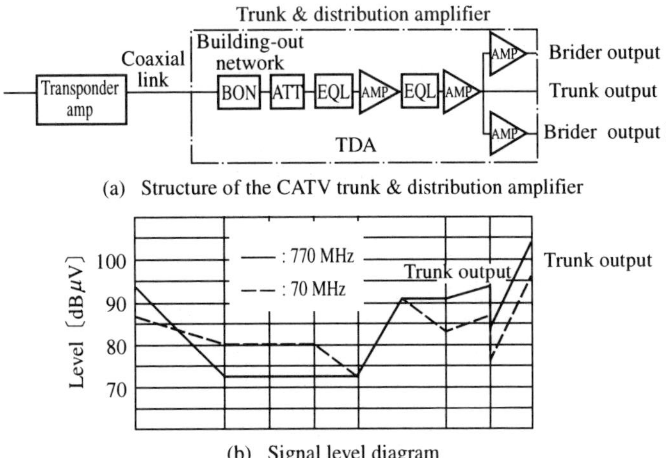

(a)   Structure of the CATV trunk & distribution amplifier

(b)   Signal level diagram

Fig. 6.10   CATV Trunk & Distribution Amplifier Structure and Signal Level Diagram

6.10(b) illustrates an example where the existing cable loss of 21 dB (at 770-MHz) is compensated. In cases where the cable length of the preceding stage is shorter or losses are smaller than 21 dB, the BON can be used to compensate the shoeter cable lenghth at the first-stage, and at the second-stage the equalizer (EQ) and the attenuator (ATT) are applied to compensate the flat loss. The diagram in Fig. 6.10 is an example of a half-tilt system in which the input level of the high and low band edge is adjusted to a half of the full tilt.

The usual method of compensating temperature changes in transmission losses is by using a pilot signal automatic gain control (pilot AGC). This system controls all transmission bands in unison, rather than individually controlling each signal. The AGC function is used in the TDA and TA (trunk amplifier). A tilt compensation type AGC is normally utilized.

Using this configuration, the input levels to all amplifier modules in each stage could be held constant. Then if this could be realized, in case of a n-stage cascaded amplifiers with the same nonlinear and noise characteristics, the total nonlinear distortion and thermal noise values simply becomes a value that is an n-multiple of a stage (refer to 6.4.3).

Fiber Transmission Path (Optical Link)

As of 1997, the most widely used modulation scheme in HFC systems is the AM-FDM system. In this system, all frequency-division multiplexed (FDM) electrical signals in the coaxial network are modulated by a laser diode (LD) modulates to a corresponding light signal (Fig. 6.11). The AM-FDM signal modulated by LD is reconverted by photo diode (PD) to an FDM electrical signal at the node terminating the optical fiber cable. This system features good compatibility with coaxial cable networks, and is very cost-effective.

Figure 6.11 illustrates the concept of the LD modulation system. Increasing the level of modulation improves *C/N*, but degrades distortion characteristics. Conversely, a lower level of modulation degrades *C/N*, but results in better distortion characteristics. Thus, modulation levels must be adjusted appropriately. When modulation level for each carrier is *m*, and *n* is the number of TV signals to

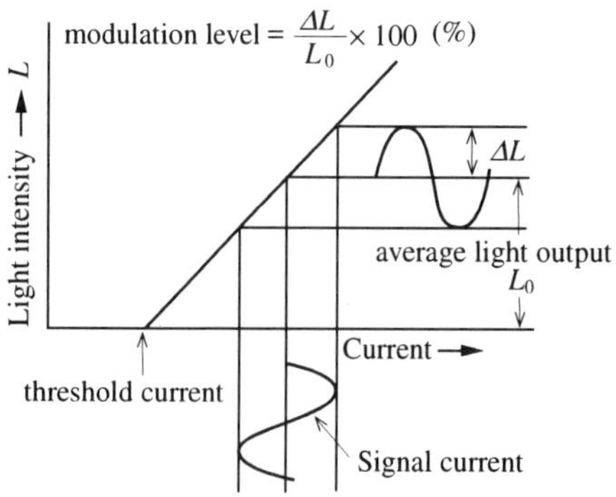

Fig. 6.11   LD Modulation Concept

be transmitted, given that each signal has a random phase, total modulation is defined by $M = m \cdot \sqrt{n}$. Generally M takes a value of about 30%.

The LD, PD, and the output-stage amplifiers are known sources of noise. Noise from the LD stems from modal noise and feedback-induced noise, and the ratio of DC to the noise component in the output signal is expressed as relative intensity noise (RIN). Noise from the PD is predominantly the shot noise.

The optical fibers used in the system are normally silica fiber with a zero-dispersion at the wavelength of 1.3 or 1.5 μm. Transmission losses are very low; at a wavelength of 1.3 μm, loss is approximately 0.5 dB/km, and at the 1.55 μm wavelength, loss is approximately 0.3 dB/km. Losses of this order enable signal transmission over long distances.

6.2 Digital CATV

In CATV, the information signal is accessed from a number of different media sources and then transformed at the head end (HE) to a suitable form for CATV transmission. The system between the HE and the terminals is generally designed specifically for CATV, and its fundamental architecture will remain the same for digital CATV. In the future, as work progresses on the digitalization of the various media sources, things will gradually evolve to the point where the information presented to the HE will be fully digital. At the same time, it is difficult to imagine that suddenly one day all subscriber's appliances will become digital. Thus, CATV must continue to transmit both analog and digital information over a given period of time referred to as the simulcast period.

6.2.1 Information Source Coding

Digital CATV will be based on the same MPEG-2 standards (for video,audio: see Ch. 2 for details) as used for digital satellite and digital terrestrial broadcast television [MPT, 1995], [MPT, 1996]. MPEG-2 compression standards are specified for encoding in four areas: source coding, video coding, audio coding, and signal multiplex. The MPEG-2 source coding standard is essentially based on MPEG-1 (H.261), but has been enhanced with features for field/frame, and dual prime motion compensation. ITU-T recommendations H.262 and IEC 13818-2 Video standards [ISO/IEC, 1994] cover processing of the bitstream from the point of source coding block output until the PES (packetized elementary stream) is created. Audio processing from the source coding block to conversion to PES is as specified by ISO/IEC 13818-3 Audio [ISO/IEC, 1994]. Generation, interpretation, and decoding of the transport stream (TS) and program stream (PS) created from the video-audio PES is specified by ISO/IEC 13818-1 Systems standard [ISO/IEC, 1994].

With regards to the TS between transmit and receive, since satellite, terrestrial, and CATV broadcasts are carried by different modes of transmission, the multiplex and the modulation systems for specific media are not prescribed by the standard. For CATV, the standards are based on ITU-T SG9 recommendations J.83 and J.84 [ITU-T, 1987] (Table 6.1), and the Japanese system conforms to Annex C in the table. A brief description of each parameter will be given in the next subsection.

Note that service information identifiers and some of the other details are left to service providers. Of these items, where uniformity in application within a specific medium is desired, practices are specified by Ministerial ordinance (compulsory standard). Other items are either discretionary, or set by the Industry Standards Organization.

Table 6.1   ITU-T Recommendation J.83, Annex A, B, C, and D Systems

ITU-T	ANNEX C	ANNEX A	ANNEX B	ANNEX D
Country	Japan	Europe	USA	
Transmission	MPEG-2 TS (transport stream) (187 + 1 byte: 1 byte used for sync)			
TS bit-rate	not specified (29.162 Mbit/s)	not specified	26.970 Mbit/s	38.78 Mbit/s
Error-correction	RS(204, 188) code		RS(128, 122) code	RS(207, 187) code
Energy dispersion	$1 + X^{14} + X^{15}$		$X^3 + X + \alpha^3$	$1 + X + X^3 + X^6 + X^7 + X^{11} + X^{12} + X^{13} + X^{16}$
Interleaving	convolutional, 12-byte depth		128-symbol convolutional	26-segment convolutional
Bit arrangement	quadrant rotation, Gray coding		quadrant rotation + trellis-coding	absolute value
Differential encoding	2 MSBs		I, Q, 1-bit each	none
Trellis-coding	none		14/15 convolutional	none
Transmit bit-rate	not specified (31.644 Mbit/s)	not specified	30.342 Mbit/s	43.05 Mbit/s
Modulation	64 QAM	16/32/64 QAM	64 QAM	16 VSB
Symbol-rate	not specified (5.274 Mbaud)	not specified	5.057 Mbaud	10.76 Mbaud
Roll-off rate	13%	15%	18%	11.5%
Bandwidth	6 MHz	8 MHz	6 MHz	

6.2.2 Multiplexing System

Multiplexing

The purpose of multiplexing is to convert the transport stream (TS) packets from multiple programs to a single TS. Figure 6.12 shows a typical HE system primarily covering the multiplexer, the scramble encoder and 64 QAM modulator. The process of multiplexing will be explained, referring to that diagram.

We start with the input of multiple (normally 3 ~6) MPEG-2 transport streams to the multiplexer. Since the timing of each TS packet is different, the first task is to synchronize the signals. Since each TS packet contains the original service information provided by the program supplier, or for a particular medium, the next step is to convert this information to a form appropriate for CATV transmission. Specifically, this means attaching a PAT (program association table), a PMT (program map table), or a NIT (network information table) descriptors. Also, since each program supplier or media source uses a different method for conditional access (discussed below), and perhaps also a different subscriber management system, a CAT (conditional access table) descriptors is also attached at this time. Since the process of changing the descriptors takes time, resulting in a time lag, a new time is written into the time stamp.

The next step is to scramble the signal in the scramble encoder (Fig. 6.12), add the scramble key (K_s), and output the signal to the channel coding block (64-QAM modulator).

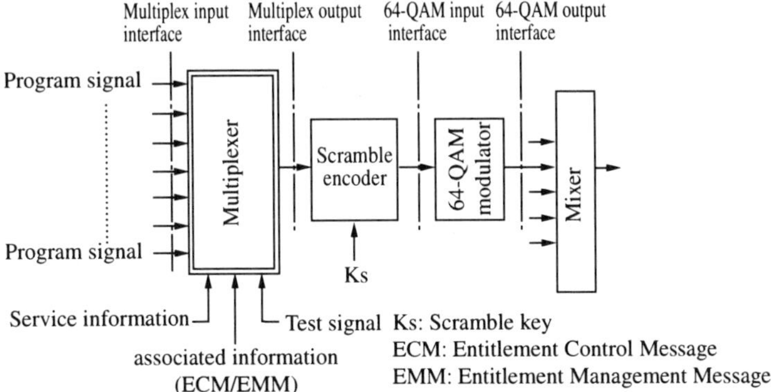

Fig. 6.12   Digital CATV Head End Structure

In statistical multiplexing, the bit rate of a program can be varied according to what is required to reproduce a picture of acceptable viewing quality. Generally the bit rate of a program is variable at the output of the source coding block; in video programs where there is little motion in the picture, and there are few scene changes, the bit stream can be set to a low transmission bit rate. On the other hand, programs with such a sports and other action scenes require a high bit rate. Utilizing the above, the instantaneous bit rate of each program can be variable (VBR:variable bit rate) to place as many programs as possible into the CBR (constant bit rate) of 31.644 Mbit/s which is carried on a 6 MHz bandwidth. Thus realized the statistical multiplexing. Statistical multiplexing allows the optimum bit-rate to be set according to the nature of the program, which permits a maximum number of programs to be transmitted in the 6-MHz bandwidth. One cautionary note: statistical multiplexing is possible only when there is feedback from the multiplexer block to the source coding block can be realized, which becomes impossible when too much geographical distance separates these two blocks. In such cases, the bit rate of for each program should be set for maximum rate.

There could occur such case, depending on the transmitted media, as that the physical channels themselves is set for constant bit-rate (CBR), while in the individual logical channels, bit-rate of a program is varied to and without notification. When this is the case, another cautionary note: in the event that program streams are selected from different physical channels to make one physical channel at the HE, we must be able to recognize the bit rate of the program received at HE site.

Operational Management of Service Information and Identifiers
The digital signals entering the HE of the CATV are tagged with service information provided specifically by the program supplier, or media. If a digital signal is received from a different program supplier (or media platform), and is sent through to the set top box (STB) without re-writing the specific information, the STB cannot make the proper program selection. To avoid such a situation, the service information or the service information identifier furnished by each program provider or platform must be re-written. This can be done easily for a particular program provider. However, unless this modification process is elaborately planned and excuted, the signal provided by a particular program provider will be receivable by some STBs of a particular vender, but will not be received by other vender. Consequently, the service information and identifiers should be divided into areas having a common code, and areas specific to a service provider. The common code area is understood throughout the system. It is also desireable ultimately to have a unique system of identifiers nation-wide (Japan) for managing the digital signals from CS, BS, terrestrial broadcasts, and all media sources used by CATV.

With regards to the operations-management of identifiers, these matters are as specified by Ministerial (MPT) ordinance, by standardization organizations, and by service provider specifications. In order to avoid any confusion in the operations of the system, these procedures should be recorded and be publicly accessible.

The procedures specified by Ministerial ordinance were promulgated in 1996 in the Ministry of Posts and Telecommunications Bulletin No. 78 [MPT, 1996], and the parts specified by the standardization organization were published by the Japan CATV Engineering Association [JCTEA, 1997]. The management information specific to CATV industory is scheduled to be under the control of a provisional information management organization within the Japan Cable Television Association (JCTA).

Conditional Access
All information transmitted in broadcast form should be available at all subscriber terminals. However, this does not mean that all information is accessible by all terminals. Some information is available only to designated terminals that have conditional access (CA) to that information. Conditional access is closely related to the functionalities provided in an STB, and to the subscriber management system of each CATV operator. For operators both conforming to international standards and realizing, of STBs that are interconnectable and interoperable are important

considerations.

In this regards, the following items are taken into consideration in Japan:

(1) The use of block encryption (ISO 9979/0009: Multi 2) for the scramble subsystem.

(2) The two subsets of CA associated information (ECM and EMM) are used; ECM is a common information combined the program associated information and the control information to turn on/off descrambling. EMM is a management information including the contract information for individual cable subscriber and Work Key (K_w) to de-encript.

These systems should be standardized uniquely within the industry. ECM and EMM are transmitted from the multiplexer in the section format, extracted by the STB decoder for transfer to the IC card. When the subscriber selected any program and if he is authorized to receve it, the K_s is sent back to the STB. The IC card (CA module) function and program selection flow is illustrated in Fig. 6.13.

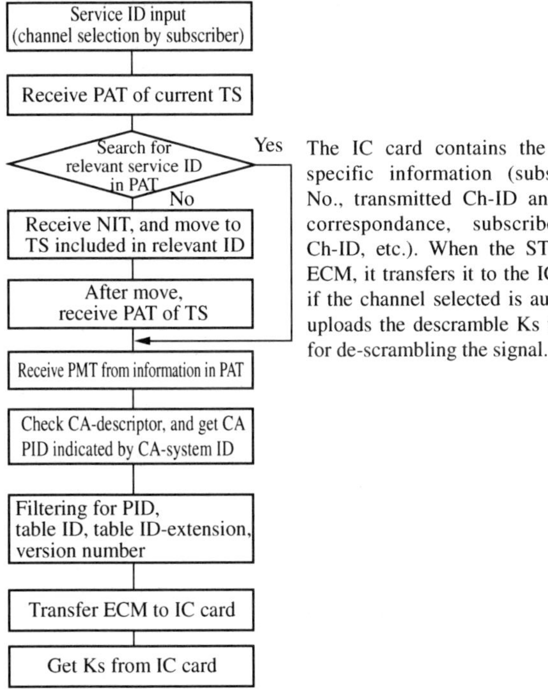

The IC card contains the operator's specific information (subscriber ID No., transmitted Ch-ID and program correspondance, subscriber-contract Ch-ID, etc.). When the STB receives ECM, it transfers it to the IC card, and if the channel selected is authorized, it uploads the descramble Ks to the STB for de-scrambling the signal.

Fig. 6.13   Program Selection Flow at the STB

By utilizing the IC card as the IC module and loading the operator's specific information into the IC card, the operator can be provided with STB by any vender in interoperable mode and subscriber of some operator can move with his STB to subscribe to other operator's CATV, where he receive only a new IC card. If so prescribed circumstance was established, the STB itself except the IC card could be common in oparation, which means that the STB can be interoperable between both venders and operators. Thus realizing the interoperable STB is inevitable for realizing the "open cable" or open marketing of the STB.

6.2.3 Channel Coding

The configuration of the channel coding and decoding blocks for CATV is shown in Fig. 6.14. The transmission channels for the various media each have different channel characteristics, and for each channel, there exists an optimum interleaving, FEC, and modulation scheme (refer to Chapters 4 and 5). This means that the characteristics of the digital signals input to the HE will be quite diversified.

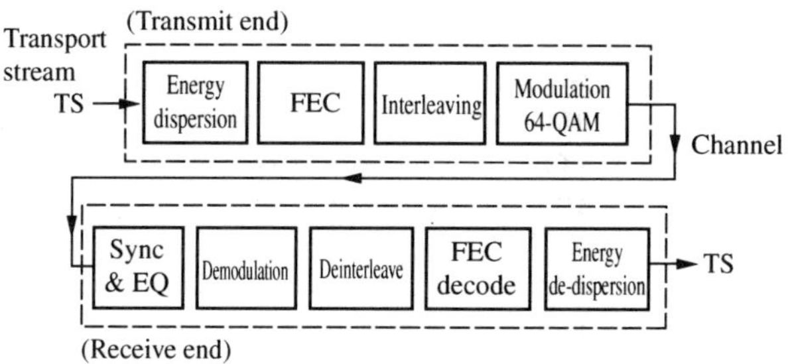

Fig. 6.14   CATV Channel Coding – Decoding Block Structure

In CATV (ITU standard, recommendation J.83 Annex C), as shown in Fig. 6.15(a), energy dispersion starts with an initial value of [100101010000000], and uses the generator polynomial $x^{15} + x^{14} + 1$ to add a pseudo-random signal. Shown in Fig. 6.15(b), interleaving is in byte units, and is accomplished as a 12-block skip convolutional interleave, followed by adding the RS(204, 188) FEC code. Also, as shown in diagram (c), modulation of the carrier is by a differentially encoded quadrant rotation, Gray coded 64-QAM system, with the Q-channel leading the I-channel by 90-degrees. Roll-off factor is 0.13, transfer rate (reference) is 31.644 Mbit/s, and symbol rate is 5.274 Mbaud.

Also, the packet length of the digital signal sent by the ATM (or STM) to the CATV head end is different (54 byte for ATM), so the ATM packet length must be converted to a TS packet, and after adding the RS(204, 188) outer code, the signal is processed through energy dispersion and interleaving.

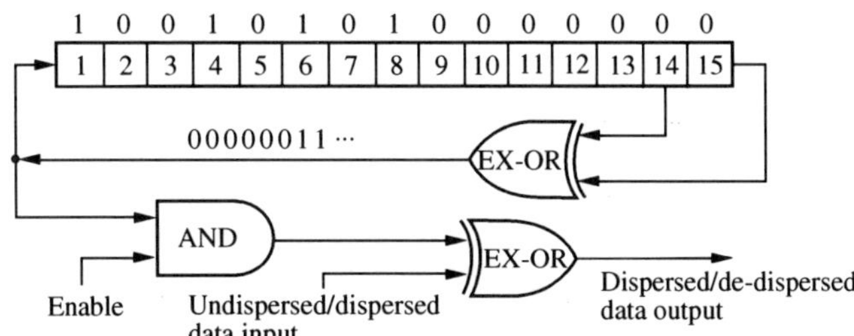

(a) Randomization (Energy dispersion)

Interleave depth	N $(=12)$
Sync	Routing SW (period N)
Required memory capacity	$1/2 \times N \times L$ N $L\,(=17\times11)$

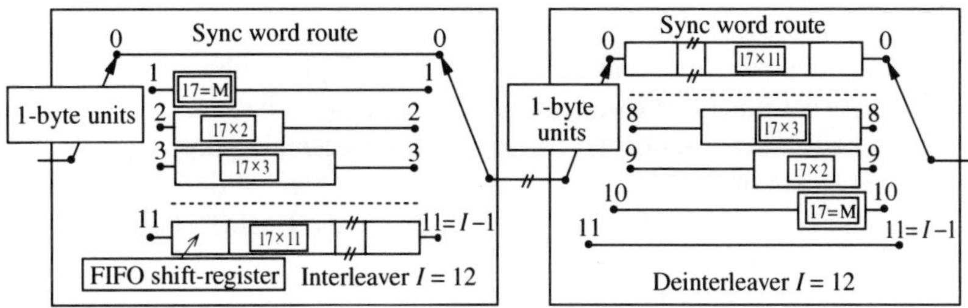

(b) $n = 12$ convolutional interleaver

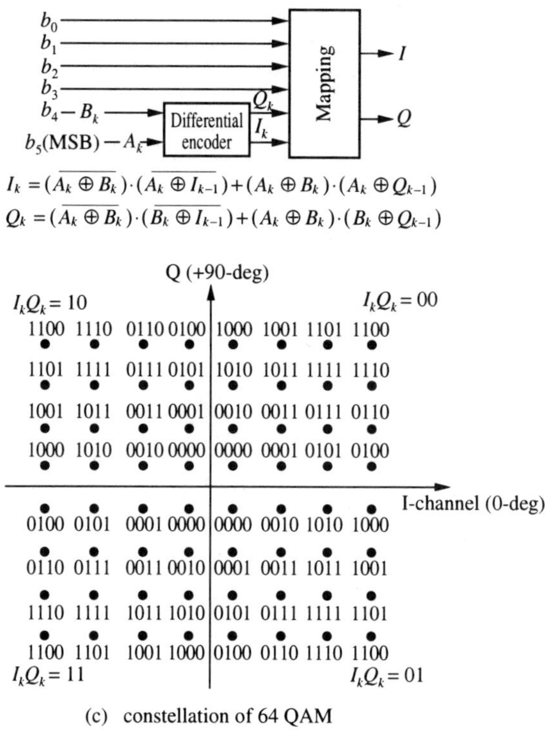

$$I_k = (\overline{A_k \oplus B_k}) \cdot (\overline{A_k \oplus I_{k-1}}) + (A_k \oplus B_k) \cdot (A_k \oplus Q_{k-1})$$

$$Q_k = (\overline{A_k \oplus B_k}) \cdot (\overline{B_k \oplus I_{k-1}}) + (A_k \oplus B_k) \cdot (B_k \oplus Q_{k-1})$$

(c)   constellation of 64 QAM

Fig. 6.15   64-QAM (J.83 Annex C) Channel Coding and Decoding Blocks

6.2.4 *The Set Top Box (STB)*

The set top box (STB) functions as the video and associated information decoder for the CATV network, while at the same time serving as the interface between the CATV system and the TV set, video recorder and audio equipment, the computer, and any other subscriber equipment. The STB can take a number of different forms, depending on what functions with regards to service are expected.

Standards were established for an STB whose main puropose was to meet the immediate demands of the time (which was to receive digital, multichannel broadcasts) in May, 1997 by the Japan Cable Television Engineering Association in JCTEA STD-004-1.0 [JCTEA, 1997].

In order to receive multichannel digital broadcasts, subscriber equipment must be capable of receiving FDM physical channels spaced every 6-MHz in the band ranging from 90 MHz to 770 MHz, in addition to receiving specified selectable logical channels within the physical channels. When the subscriber activates the STB, it reads the service information and determines which physical channels and logical channels are selectable. At the same time, it extracts the ECM, and if receive is authorized, the K_s is uploaded from the IC card (CA module) to the STB, and the signal is de-scrambled.

In digital CATV, since the STB must be capable of selecting both physical and logical channels, at least two decoders are required for recording into a video recorder. Recording does present some problems with digital systems. Record enable/disable controls, and controls for the number of copies (generation control) are required on the video recorder equipment, and the control signal must be sent from the STB to the receiver as a VBI[*13] transmission. Also, services such as text for the visually handicapped, and alpha-mosiac pictures likewise must be sent from the STB to the receiver multiplexed on the VBI signal. With regards to bidirectional STBs, as the technical requirements and conditions needed by the operators are now more clear, further investigations have been scheduled by the Standard and Specifications-Committee and the Japan Cable Laboratories (JCL) (as of Aug. 2000).

6.3 CATV Transmission Quality

CATV transmits an AM-NTSC or AM/FM-MUSE (Multiple Sub-Nyquist Sampling Encoding: transmitted from satellite as Hi Vision) analogue television signal and digital signals modulated by 64-QAM, FSK, PSK, TDMA, or CDMA, using FDM multiplexing system. Since the major part of the signal energy is concentrated in the carrier (which is typical of AM-NTSC), in multichannel FDM the generation of composite beats are unavoidable (see subsection 6.3.2). When using different types of modulation schemes, the need to carefully investigate how these beat noises affect every signal, and inter-signal interference relationships is also evident. Also, when the coaxial tree network serves as the basic structure of the CATV network, the system is susceptible to ingress noise in the reverse stream.

How a channel is affected by the various types of noise differs depending on the modulation scheme, and whether the signal is analog or digital. These factors also determine how interference acts to degrade signal quality.

Noise appears in analog television signal transmissions as non-periodic noise (thermal noise, or "snow"), and periodic noise as a beat. In the case of digital television, thermal noise and periodic noise are additive, resulting in an additional increase in bit error rate (BER). Also, when analog television signals are affected by noise, picture quality gradually becomes worse, until at some point the noise masks the signal completely. On the other hand, with digital television, picture quality stays at a constant level as long as BER remains below a given threshold value; once that value is exceeded, however, reception is completely lost. (This is known as the cliff effect.)

6.3.1 The Effects of Thermal Noise

Thermal noise, as measured in degrees Kelvin (ToK), is an intrinsic property of all substances. Thermal noise is variously known as *noise*, *white noise*, or *Gaussian*

---

*13 VBI: Vertical Blanking Interval. Switching signals for data transmission, character multiplex, emergency broadcasts, and mono/stereo/bilingual broadcasts are transmitted on the VBI signal.

noise. Thermal noise affects an analog video signal by producing "snow" in the picture, and in digital signals modulated by 64-QAM is demonstrated as degradation of BER characteristics.

Figure 6.16 illustrates the theoretical values and actual measured values for BER vs. *C/N* in a 64-QAM system. (This example is from a proof-testing in the Telecommunications Technology Council, Inquiry No. 74 report [MPT, 1996].) The data shown in the Figure is applicable for direct coupling of the IF stages of the transmitter and receiver (bypassing the CATV transmission path), but illustrates how in the high *C/N* region, actual BER values depart from theoretical values (the floor effect). The floor effect is thought to be primarily due to internal nonlinear distortions generated within the receiver.

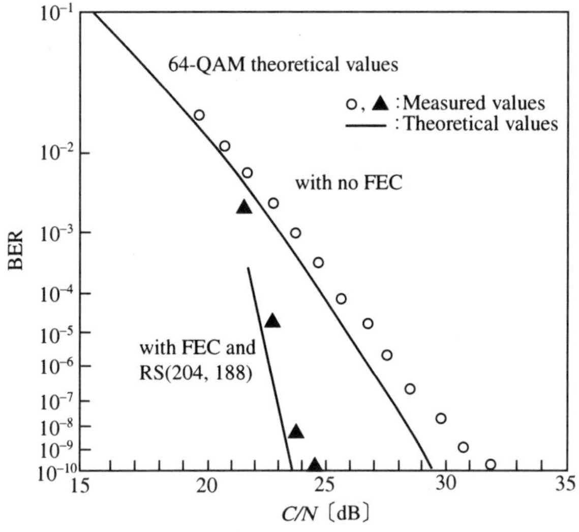

Fig. 6.16   BER vs. *C/N* Characteristics for 64-QAM (An example showing theoretical values and measured values.)

6.3.2 The Effects of Nonlinear Distortion
Second-order Distortion

When continuous waves (CWs) with frequencies of $f_1 \sim f_n$ are transmitted through a circuit having nonlinear characteristics, coupled waves with frequencies of $f_i \pm f_j$ are generated as second-order intermodulation products, which results in distortion. As Fig. 6.17 illustrates, the coupled waves that drop to the same frequency combine and grow to a resultant vector length. When these coupled waves fall in the vicinity of another carrier (within the same band), a beat interference is generated in the case of an analog signal, and BER is degraded in a digital signal.

In a 64-QAM system, where second-order distortion is adequately small, it can be treated as a part of the thermal noise equivalent (which is a part of the total noise power within the band), but when second-order distortion is a large value,

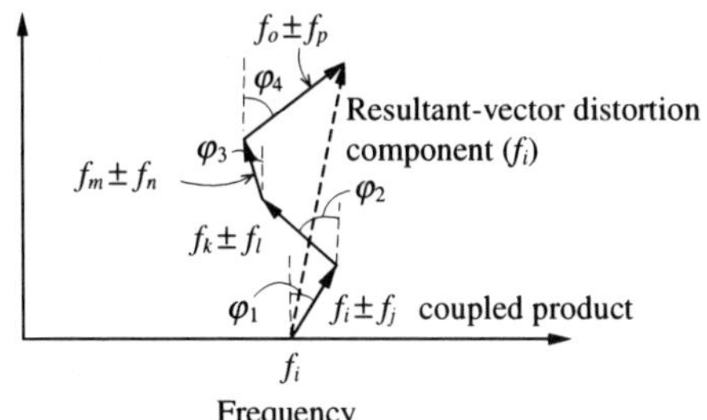

Fig. 6.17   Second- and Third-Order Coupled Waves Combine as Vector Sums

experiments have shown that in the high *C/N* regions, the BER vs. *C/N* curve (*N* = thermal noise) departs sharply from theoretical values, and it then becomes difficult to convert the distortion noise power to a thermal noise equivalent (refer to Fig. 6.30 in 6.4.1).

Second-order distortions are defined as the amplitude ratio of the coupled wave to the carrier. Theoretically, these are, expressed in decibels, of the amplitude ratio of the resultant-vector coupled wave component and the carrier of the channel under test. Practically, the measurement is performed using a multicarrier signal generator (spurious level: less than -100 dB), Which generates m-1 carriers, then the carriers are imposed on the CATV system under test, where m is the channel capacity. Under this set up, a distortion component fallen on the channel under test is measured in dB at the subscriber terminal. The measured distortion by the above is known as composite second-order (CSO) *beat*.

The basic measuring procedure is illustrated in Fig. 6.18(a). Here, if a signal with a frequency f_j, and an amplitude a_j is applied to the circuit it will produce harmonic distortion at frequency $2f_j$ (the second harmonic) and amplitude a_i. By measuring the difference in the level (t_2, in dB) between a_i and a_j, this measurement can be applied to a formula for calculating the second-order intermodulation (IM_2) (dB) (the double-frequency method). Frequency characteristics will depend upon the selection of f_j, and measurement results tend to vary.

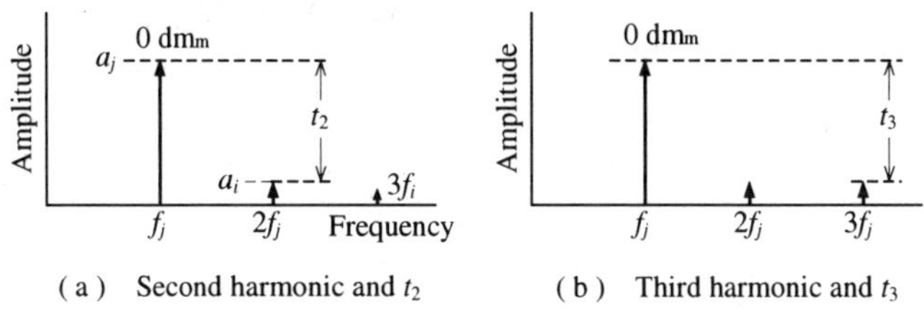

(a)   Second harmonic and t_2 (b)   Third harmonic and t_3

Fig. 6.18   Nonlinear Distortion

Third-Order Distortion

Like second-order distortion, when CWs with frequency of $f_1 \sim f_n$ is transmitted through a nonlinear circuit, coupled waves are generated having frequencies of $f_i + f_j - f_k$ as third-order intermodulation products, which results in distortion. In analog video signals, third-order distortion produces a beat interference, and BER is degraded in the digital signal. In a 64-QAM system, the effects of third-order distortion are the same as second-order distortion. Third-order distortion is also defined as an amplitude ratio of the coupled wave to the carrier, and similarly, the measurement consists of specifying (in dB) the amplitude ratio of the resultant-vector coupled wave component and the carrier of the test channel. Here too, distortion is measured by using a multicarrier signal generator to inject (m) -1 carriers on the CATV under test. The third-order distortion is then read from the test channel at the receiver. This distortion is known as the composite triple (CTB) *beat*.

The basic measuring procedure is illustrated in Fig. 6.18(b). Here, if a test signal with a frequency f_j is applied to the circuit, it will produce harmonic distortion at frequency $3f_j$ (the third harmonic). By measuring the difference in the level (t_3, in dB) between f_j and $3f_j$, this measurement can be used in a formula for calculating IM_3 (dB)[*14].

Compared with CSO, the number of CTB coupled waves which drop to the same frequency is greater (particularly the $f_i + f_j - f_k$ type), and the effects of noise are more severe.

6.3.3 The Effects of Phase Noise
Phase Noise

In general, the frequency region in which amplitude vs. frequency characteristics are changing is also the region in which phase is changing. In the discrete channel amplifier of the head end (HE) and in the band-pass filter (BPF) of the receiver terminal, phase change is significant at both edges of the 6-MHz band. However, in the CATV transmission line with wide-band amplifiers, phase change is gradual and small across the the total transmission bandwidth (70 ~ 450/770 MHz). Phase change in the selected channel, as observed from the decoder block of the receiver terminal, is the sum of the phase change in the BPF and the transmission line.

Prior to the late 1960's when each channel used its own amplifier, the input in FDM was isolated into each channel, amplified by a discrete channel amplifier, then recombined into FDM signals. In a discrete channel amplifier, rotation of phase in the BPF cannot be avoided. Where discrete channel amplifiers are used in a multiple-stage repeater configuration, the BPFs must necessarily be placed in a multistage cascade. This results in a phase difference developing between the luminance and color signals, which appears as color smearing in the picture.

The above problems are not present in the wideband systems in use today. Turning our attention specifically to the CATV transmission channel, as shown in Fig. 6.19(a), a reflected component ripple is riding on the otherwise gently sloping phase characteristics of the transmission equipment. If a 6-MHz slice (one channel

---

*14 When a carrier (f) is applied to a nonlinear circuit, harmonics that are integer multiples of the fundamental frequency (i.e., 2f, 3f) are produced as distortion. The theory and methods for determining the resulting distortion of these second- and third-order harmonics can be found in the reference [MPT, 1975].

from the wideband transmission channel) is taken from the graph in (a), the graph in (b) shows relatively flat characteristics. If this is superimposed on the characteristics of the HE and the receiver BPF, the resulting waveform appears as shown in diagram (c), and this becomes the causal for group delay.

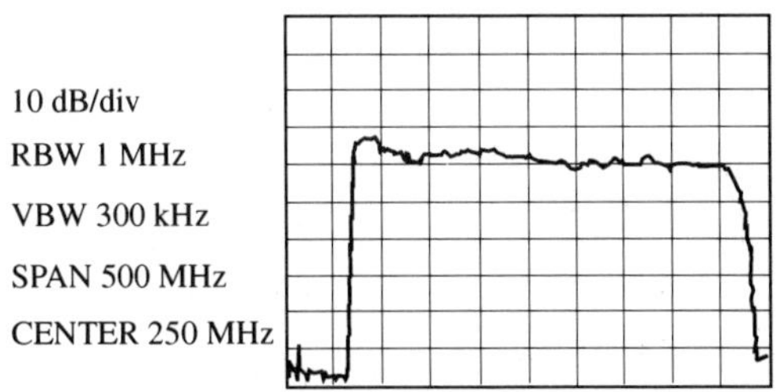

(a)   Amplitude characteristics across total transmission band (example).

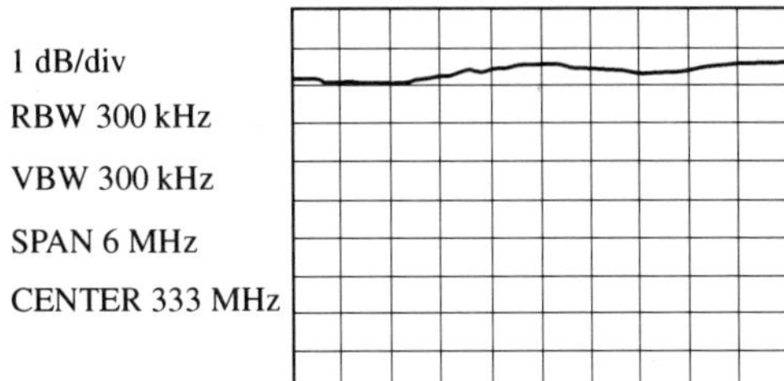

(b)   Amplitude characteristics in a single channel.

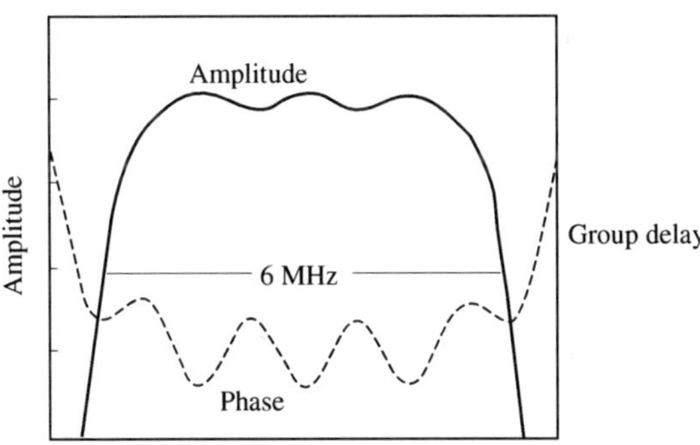

(c)   Effects on head end and band-pass filter.

Fig. 6.19   Amplitude vs. Frequency Characteristics measured at Receiver Terminal

As a specific example of group delay, when the Hi Vision signal is transmitted as an AM signal, the subsampling frequency and the characteristic peak of group delay periodically overlap, resulting in vertical lines in the received picture. The detection limit is between 10 and 20 ns. With an M-ary signaling system such as 64-QAM, since the carrier has an amplitude and phase component, if there is a shift in phase, the post-decoded symbols will be rotated away from their prescribed constellation point arrangement, resulting in a degradation of BER performance.

An equalizer (EQ) is utilized in the STB to mitigate phase noise. There are two types of equalizers: the fixed type equalizer is a circuit having the inverse characteristics of the received signal amplitude-frequency characteristics, while the adaptive equalizer samples the amplitude-frequency characteristics, and based on sampling results, generates the inverse characteristics needed to smooth the received signal. The results comparing use and non-use of the adaptive equalizer for 64-QAM are illustrated by BER vs. *C/N* in Fig. 6.20. This diagram makes clear the effectiveness of the adaptive equalizer.

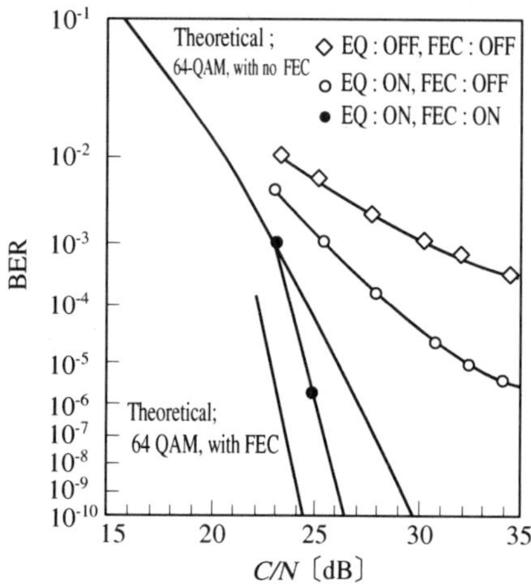

Fig. 6.20   BER vs. *C/N* Characteristics for Equalizer Use (64-QAM)

The Effects of Delayed Current Noise (Due to Reflected Components)
The CATV transmission channel consists of many different pieces of equipment connected in tandem. A small amount of the signal is reflected at each input and output terminal. Although relatively small, this "doubling back" of the reflected component finally reaches the intended terminal device affected by path length differences, and carrying a phase shift. Since amplifiers have one-way directional characteristics, when the reflected component reaches an amplifier output it is prevented from going further upstream, and is reflected back toward the next-stage the amplifier input. At passive devices such as tap-offs (TO), however, the reflection is passed on through to the next upstream section (attenuated only by insertion loss), and then returns as another reflection component. When these reflection components start overlapping, a small, short period ripple builds up across the channel bandwidth (with amplitude vs. frequency characteristics), which rides on the ripple seen in Fig. 6.19(c). As a result, the group delay frequency characteristics also contain a ripple component. In an analog CATV system, this

delayed current noise (appearing as the ripple in the amplitude vs. frequency characteristics) rarely becomes a source of interference, but in digital CATV systems, it can possibly contribute to degrading BER performance.

6.3.4 Interference Caused by Undesired Signal(s)

Ghosts

Ghosts are a type of interference that cause echo in analog sound and video broadcasts, cause overlapping (doubling) of pictures (ghost images). As illustrated in Fig. 6.21 (a), when the broadcast signal(s) reflected off buildings or other obstructions are received at the same time, the reflected signal has a delay corresponding to the path length difference (d), and the image that the reflected signal carries is to the right of the primary image carried by the direct signal. When there are multiple reflected waves, there are also multiple delayed pictures (ghosts), and if phase is shifted, there will be smearing of the color in the picture. Ghosts are a particular problem in terrestrial broadcasts, but much less so in CATV. CATV has been a popular solution to ghost interference in areas where obstructions are numerous.

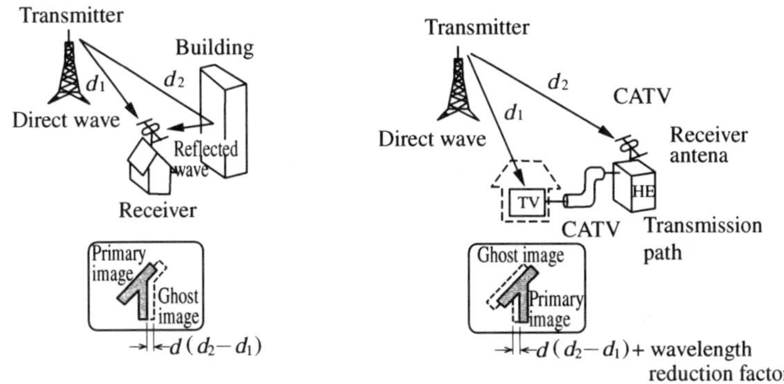

(a) Ghost interference in terrestrial broadcast.   (b) Direct wave interference in CATV.

Fig. 6.21   Ghost Interference in Terrestrial Broadcasts and CATV

The ghost problem illustrated above also has an effect on CATV systems. As shown in Fig. 6.21(b), due to the dielectric in the coaxial cable network, the signal propagated through the network is delayed relative to the direct wave propagated through space (air). Consequently, when the air-propagated wave (as the ghost component) enters the system, it forms a secondary image offset to the left of the primary image. Factors which allow entering air broadcasts into the cable signals include cracks in the cable, loose connectors at wall outlet terminals, TV input terminals, or at the TV tuner input.

The Effects of Monotonic Interference

Monotonic interference is generated by adding one or more continuous waves (CWs), either within the transmission band, or out-of-band, and is one of the primary parameters used for evaluating the effects of noise. With analog video signals, this results in a beat interference, and with digital video signals, is one of the factors which degrades BER performance. Figure 6.22 illustrates the effect of monotonic interference on an analog NTSC television signal. Generally this effects can be different by the energy distribution within the band of the video signal, or by the human perception against the beat interference. Note that the effects an out-of-band CW has on the NTSC signal largely depends on the receiver input filter characteristics. With regards to the effects of CW on the 64-QAM system: since the energy dispersion function spreads the carrier wave energy widely across the bandwidth, the profile becomes flat (Fig. 6.23).

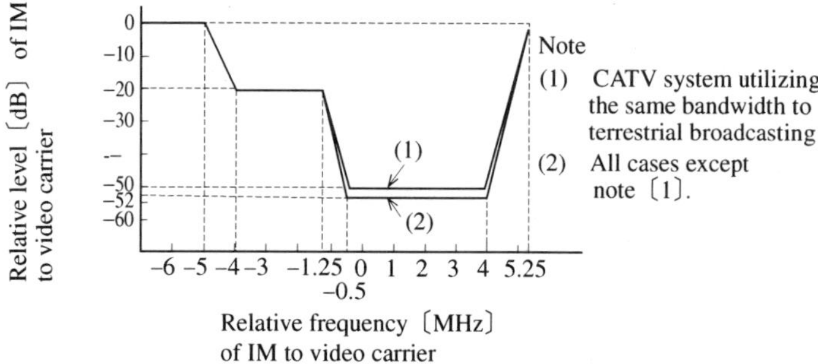

Fig. 6.22   The Effects of Monotonic Interference on the NTSC Television Signal

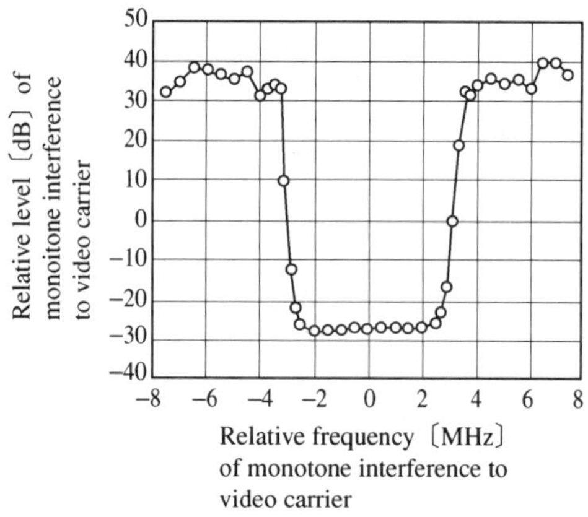

Fig. 6.23   The Effects of Monotone Interference on the 64-QAM [Noda, 1996]

The Effects of Adjacent Channel Interference

When a different signal is transmitted in an adjacent channel, the (receiving) channel under test will be affected by the adjacent channel (as the interference recipient), and at the same time, the test channel will affect the adjacent channel (as the interferer). How a channel behaves as an interferer, or an interference recipient, depends on the modulation systems employed in both channels. When channels are allocaded with adjacent frequency slots, it is therefore necessary to check on how their modulation systems perform in relation to one another.

Additionally, the interferer-interference recipient relationship will vary depending on the selectivity of the receiver input circuit. Consequently, all terminals found in a typical home (including home terminals (HT), STBs, TV receivers, audio receivers, etc.) that will likely be connected to the CATV network must be inventoried according to class, model (old or new), and significant statistical sample data should be compiled for reference regarding interferer-interference recipient relationships. In the future, as CATV terminals and household appliances (TV sets, VTRs, etc.) transition to digital systems, and personal computers take on the role of receivers, we can expect significant changes in the average home with regards to receiving equipment. As for interferer-interference recipient relationships, one should always strive to keep abreast of the changes occurring in this area.

In the following, tests results are discussed regarding interferer-interference recipient relationships in 64-QAM systems.

As indicated in Table 6.2(a), the receiver terminals tested consists of various TV receivers (coded A through H), and measurements were made to determine the CATV-ready models[*15], and non-ready models. Selectivity of the receivers vary depending on whether the reception bands (channels) are low or high, but results reflect testing of channels 9 and 34.

The 64-QAM signal was set for the "high, low, and both sides" of the NTSC signal adjacent channel, and the effect that the 64-QAM signal (as an interferer) had on the NTSC signal is tabulated in Table 6.2(b) [Maeda, 1996]. We also measured the interference effect that the NTSC signal had on the 64-QAM signal. Based on these test results, the interferer-interference recipient relationships of the 64-QAM and NTSC signals, operating under adjacent channel conditions were determined, and allowable level ranges were established for the various signals. This is illustrated in Fig. 6.24. We performed the same tests, and constructed the same diagram for the 64-QAM and MUSE-AM adjacent channel relationships, and show those results in Fig. 6.25. Likewise, the 64-QAM, MUSE-FM relationships are illustrated in Fig. 6.26.

---

*15 These are television receivers designed and manufactured for multichannel CATV compatibility. The receivers feature wide tuning ranges, and are designed for good immunity to adjacent channel interference.

Table 6.2   An example of Interference between 64-QAM and NTSC signals

(a) Receiver used for subjective test

		TV receivers			HT is used or not prior to receiver
	species	CATV ready or not	screen size (inch)	Date of Manufacture	
A	NTSC	yes	29	late '92	yes
B	NTSC	yes	25	early '93	yes
C	NTSC	yes	25	late '95	yes
D	NTSC	yes	28	late '95	yes
E	HD-ready	yes	36	early '95	
F	HD-ready	yes	32	late '95	
G	NTSC	no	25	late '91	
H	NTSC	no	29	early '88	

(b) Test results as an interferer

(1) DU ratio measured using CATV ready TV, on VHF Ch. 9   (units: dB)

Adjacent	A	B	C	D	E	F	Ave.
Upper	2.8	−0.5	−2.8	2.8	1.5	−2.3	0.3
Lower	4.8	1.3	−1.0	2.3	1.0	−3.0	0.9
Both	6.5	3.5	3.5	3.7	0.8	1.5	3.3

(2) DU ratio measured using CATV not-ready TV, on VHF Ch. 9   (units: dB)

Adjacent	G	H	Ave.
Upper	10.8	4.3	7.5
Lower	14.0	9.5	11.8

(3) DU ratio measured using CATV ready TV, on C-Ch. 34 (fv : 289.25MHz)   (units: dB)

Adjacent	A	B	C	D	E	F	Ave.
Upper	5.0	2.3	−4.3	3.5	4.5	−0.5	1.8
Lower	9.0	3.8	4.8	4.3	4.5	2.3	4.8
Both	9.8	6.5	7.5	5.7	4.0	3.3	6.1

(4) DU ratio measured using HT & CATV ready TV, on C-Ch. 34 (fv : 289.25MHz) (units: dB)

Adjacent	A	B	C	D	Ave.
Upper	2.7	4.3	3.0	4.8	3.7
Lower	8.7	6.3	5.0	3.8	6.0
Both	8.8	5.3	3.0	4.3	6.5

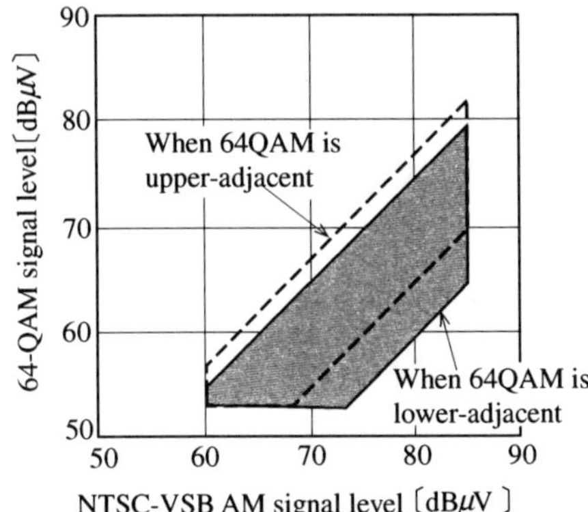

Fig. 6.24   Allowable Signal Level Range for 64-QAM and NTSC considering adjacent signal interfernce. From [Maeda, 1996].

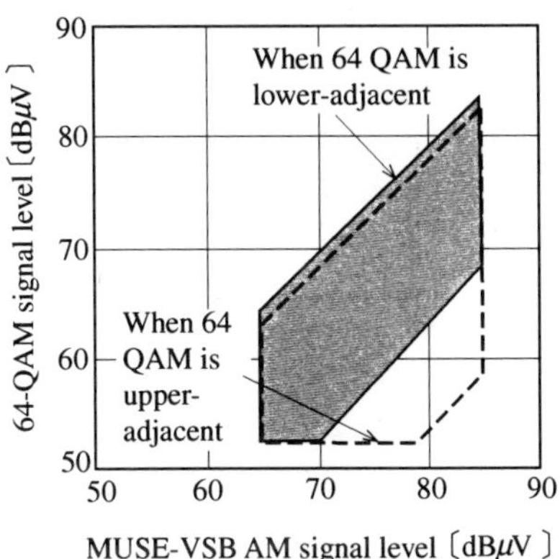

Fig. 6.25   Allowable Signal Level Range for 64-QAM and MUSE-AM considering adjacent signal interfernce. From [Maeda, 1996].

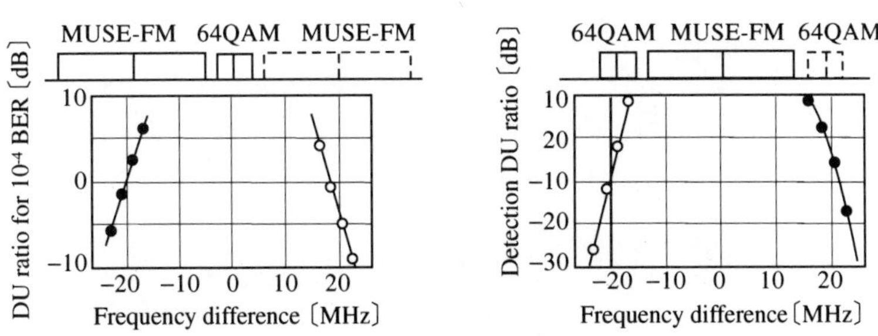

(a)   Interfrence to 64 QAM by MUSE-FM   (b)   Interfrence to MUSE-FM by 64 QAM

Fig. 6.26   Allowable Signal Level Range for 64-QAM and MUSE-FM considering adjacent signal interfernce.

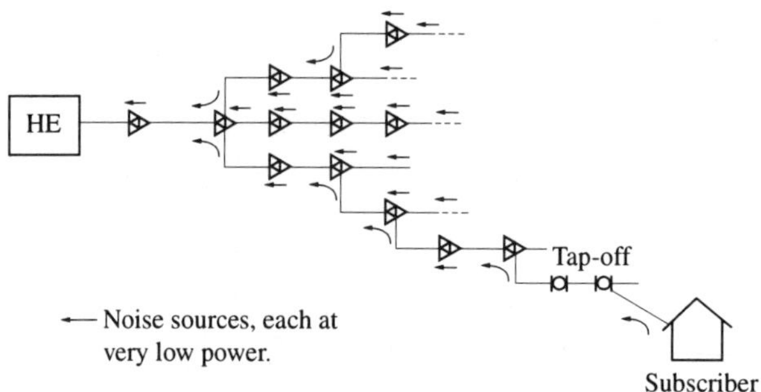

Fig. 6.27   Cumulative Noise Entering System at Each Block (ingress noise)

The Effects of Ingress Noise

CATV systems with a tree (bus) type network typically serve numerous subscribers and terminals. As shown in Fig. 6.27, each user terminal or piece of equipment in the transmission path is susceptible to the entry of very low power noise, but this noise accumulates in the reverse stream toward the center terminal (HE). As a consequence, C/N and $C/N + S/I$ is degraded at that input of the center terminal. This is referred to as ingress noise, and an example of the observed pattern of this noise is shown in Fig. 6.28. As you will notice in the pattern, ingress noise varies with time, and also occurs in bursts.

Figure 6.29 shows all the different points at which ingress noise can enter the system. Noise not only enters the system in the reverse stream, but also affects the forward stream. However, since the forward stream noise does not have a cumulative effect, it is not a great a problem as in the reverse stream, but the presence of the burst noises has still been observed.

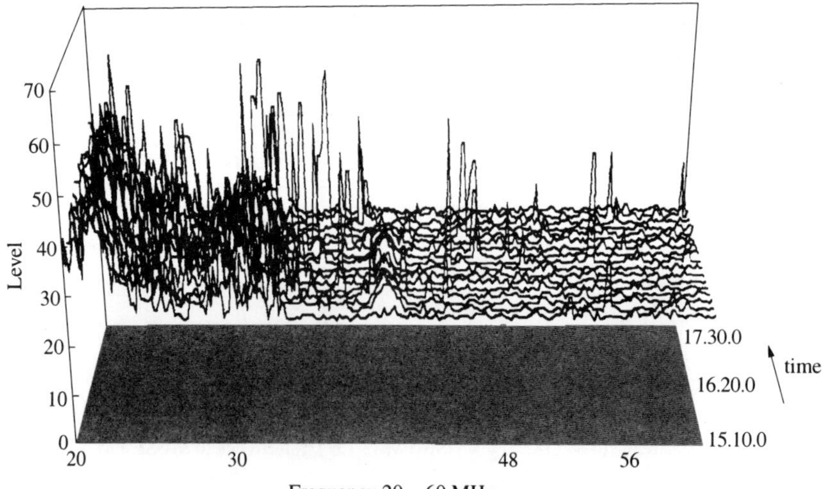

Fig. 6.28   Ingress Noise Pattern Actually Measured (example)

One of the reason that ingress noise is so troublesome in CATV system is that it does not occur in a regular pattern, so is difficult to quantify. Thus, given its nature, rather than focusing energy on its complete elimination, it will probably be better to turn attention on how to successfully transmit high-quality data in an environment that includes ingress noise. One method that has been used is to employ multiple carriers within the transmission bandwidth, and when interference occurs in some carrier(s), temporarily halt transmission and switch to other interference-free carrier(s). (This is called frequency hopping.) The use of code-division multiplexing (CDMA), which is relatively immune to noise, has also been proposed.

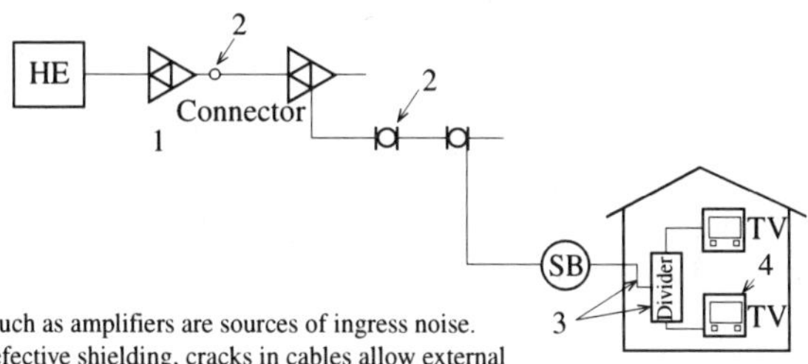

1. Active components such as amplifiers are sources of ingress noise.
2. Loose connectors, defective shielding, cracks in cables allow external
   noise to enter the system.
3. Cracks in indoor cables, indoor dividers and distributors, impedance
   mismatching of tandem outlet unit, and the noise generated by various
   electric appliances are sources for ingress noise.
4. The noise from the input terminal of TV receiver which includes noise
   generated inside of TV and ones picked up external noises are sources
   for ingress noise. Also digital appliances which generate harmonics of
   the clock for CPU are soueces for ingress noise.

Fig. 6.29　Ingress Noise Sources, Entrance Points

6.4 CATV Link Budget

In addition to the noise in the CATV transmission channel, as was discussed in subsection 6.3.2, periodically occurring noise in the form of *CSO* (IM 2) and *CTB* (IM 3) may also be present. In an analog signal (NTSC), *CSO* and *CTB* appear as a beat signal, and in a digital signal, they result in a degradation of BER performance. Hum modulation (HM) and group delay are also factors in degrading BER. Of these two factors, with respect to the fact that group delay can be dealt with by using an adaptive equalizer in the input section of the receiver, group delay affects BER differently than other forms of noise.

With regards to influence on the digital signal, *C/N* also has a different effect on BER than does *CSO* or *CTB*. These effects also differ depending on whether there is just one *CSO-CTB* component, or there are multiple components within the 6-MHz band.

Given this background, we shall now discuss the CATV link budget, and contrast it to the link budget for an NTSC system.

6.4.1 64-QAM Transmission Characteristics

The Effects of Noise

The constellation of 64-QAM is shown in Fig. 6.15(c). In a noisy channel, the signals form a scatter pattern around the constellation points, rather than converging as they should. This scatter pattern makes the conditional decision process less accurate, increasing error probability, and degrading BER.

If an error is detected in the data sequence at the receiver end, the receiver sends a request to the transmitter to retransmit the data segment or packet containing the error. However, in a broadcast type information transmission, retransmit requests are not possible. In such cases, error detection and error correction bits are added to transmitted bit sequence for use by the receiver in detecting and correcting any errors. In digital CATV, 16 bytes out of the 188 byte transport stream (TS) are used for error detection and correction in the Reed-Solomon RS(204, 188) code.

The error probability equation (theoretical) for random noise in 64-QAM is expressed as follows:

$$P_{e,64\text{QAM}} = \frac{7}{24}\,\text{erfc}\sqrt{\frac{CNR}{42}} \tag{6.4}$$

$$\text{erfc}(x) = \frac{2}{\pi}\int_x^\infty \exp^{-u^2} du \;. \tag{6.5}$$

The calculated, measured, and theoretical bit error rate for random noise when using the RS(204, 188) code for FEC are illustrated in Fig. 6.16.

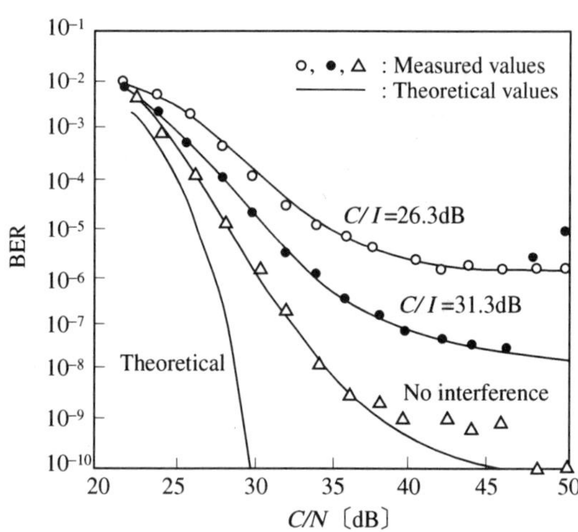

Fig. 6.30   BER vs. C/N for an Extremely Large Monotonic Interference Signal (Theoretical, Example of Measured)

The Effects of CSO and CTB

A certain level of CSO and CTB exists in the CATV system between the transmission channel and the decoder input of the receiving terminals. The receivers also have a considerable amount of nonlinear distortion, and tests indicate somewhat high level of nonlinear distortion in the STB. In an actual CATV system, it is also difficult to isolate the thermal noise and nonlinear distortion when measuring effects on BER performance.

In channels having a *C/N* component, adding a high level monotonic noise signal (~ *C/I* = -30 dB) causes measured BER to deviate sharply from theoretical values in the low *C/N* region (Fig. 6.30). Normally in CATV, *C/I* for a single coupled-wave component is -55 dB or less. Tests have also shown that this departure from theoretical values becomes even worse when multiple monotonic noise signals are added. This is believed to be the cause of the difficulty in converting CSO and CTB interference to a *C/N* equivalent value.

6.4.2 The Objectives of Transmission Quality

Using the subjective test method for evaluating the quality of reproduced video and sound in conventional analog CATV, there are prescribed performance parameters defining interference factors such as *S/N*, *CSO*, *CTB*, HM, and ghost noise. In testing detectable limits, allowable limits, and endurable limits, the values stipulated for the subjective test are 50% value of viewers who attend the test. Table 6.3 shows the relationship between *S/N* and the subjective test. For link budget, generally detectable limits are used. In column (a) of Table 6.4, allowable performance is indicated (MPT, approx. allowable limits), in column (b) desired performance is indicated, and in column (c) the industry standards (EIAJ) are indicated.

In digital CATV, since the effects of *C/N*, *CSO*, *CTB*, and Hum modulation (HM) all combine to determine BER, the BER value at the input of the STB's demodulation-decoder circuit (with EQ and FEC in an ON status) determine system performance. Error detection BER ranges from $10^{-9} \sim 10^{-10}$, and with 64-QAM (EQ: ON, FEC: ON), the measured value for equivalent *C/N* (= *C/(N + I)*) is approximately 25 dB[16]. (This is illustrated in Fig. 6.16.)

This equivalent *C/N* value is applicable to 64-QAM and converting it to the NTSC system, is approximately 30 dB. Since 64-QAM transmits at -10 dB relative to NTSC, the equivalent *C/N* with respect to the NTSC signal level is 40 dB. Adding a margin of 3 dB to account for the nonlinear distortion-induced floor

---

[16] The 25 dB value in Fig. 6.16 includes *CSO*, *CTB*, and other nonlinear distortion and noise components, measured without using an equalizer or FEC, and converted from a *BER* of 10-4 to a *C/N* value. Using this as a standard *C/N* value and adding an equalizer and FEC (RS(204, 188)) results in a measured BER of better than 10-9. Thus, if "raw" *C/N* is about 25 dB, experimental results confirm that there will be no problems with a digital broadcast signal.

Table 6.3   *S/N*, *C/N*, and Subjective Testing

Line-of-sight distance	Subjective test grade	Unweighted *S/N*[*2] [dB] flat	*C/N* conversion[*3] [dB]	
6H[*1]	5	49.8	56.8	(detection limit)
	4	43.5	50.5	(allowable limit)
	3	36.8	43.8	(endurance limit)
	2	30.3	37.3	
	1	23.7	30.7	

*1 Line-of-sight distance generally 6H.
*2 Based on DOC. CMTT/14.
*3 adjusting in dB using $C/N = S/N - 20 \log (m\sqrt{2} \cdot \sqrt{B_0}/B) = S/N + 7.04$,
   where, $m = 0.625$, $B = 3.75$ MHz, $B_0 = 4$ MHz.

Table 6.4   CATV System Performance

Broadcasting system / Organization / Item	Standard television broadcasting system						
	Domestic standards (1987 TTC, etc.)		(c) EIAJ-JCTEA			IEC	System B performance target (proposed)
	(a) Required performance (allowable limit)	(b) Desired performance (detective limit)	Current performance standard	Immediate performance target	Future performance target		
C/N	>38 dB	>42 dB	40 dB	43 dB	46 dB	42 dB (3.33 MHz)	>46 dB
(reference) Weighted *S/N*	>37 dB	>41 dB					>45 dB
IM distortion	< −50 dB (Note 1)	< −55 dB (Note 1)	< −52 dB	< −55 dB	< −61 dB	< −57 dB	*CSO, CTB* specified separately
Composite 2nd-order IM distortion (*CSO*)	–	–	–	–	–	–	<−55 dB
Composite 3rd-order IM distortion (*CTB*)	–	–	–	< −53 dB	< −56 dB	–	< −55 dB
Cross-modulation (*XM*)	< −40 dB	< −46 dB	< −42 dB	< −46 dB	< −55 dB	< −46 dB	unspecified
Subscriber terminal level	60 ~ 85 dBμ	70 ~ 85 dBμ	62 ~ 85 dBμ	58 ~ 85 dBμ	68 ~ 85 dBμ	57 ~ 85 dBμ	65 ~ 85 dBμ

Note 1: Bandwidth frequency characteristics prescribed by regulations. Figures are minimum values.

effect, assuring an equivalent *C/N* of greater than 43 dB is the desired target. With regards to *C/N* in an actual transmission channel, obtaining a thermal noise *C/N* equal to or better than 48 dB, and a *CSO/CTB* of equal to or less than −55 dB should not be difficult; thus, realizing an equivalent *C/N* of equal to or better than 43 dB should be equally possible. Experiments have confirmed that this value is more than sufficient.

6.4.3 Formulating the Link Budget

The *C/N* Budget

The primary sections making up the CATV system consist of the head end block, transmission system (link), and the receiver system (indoors section + receivers). The *C/N* parameters for those sections use the notation C/N_H, C/N_L, and C/N_R, respectively, and the total system C/N_T is calculated as

$$C/N_T = -10 \log (10 E (-C/N_L/10) + 10 E (-C/N_R/10)).$$

We now have

$$C/N_H = Ei_H - F_H - 0.9 \quad (\text{dB}), \tag{6.6}$$

where Ei_H is the HE input level, and F_H is the HE noise figure. Now

$$C/N_L = Ei_L - F_L - 0.9 - 10 \log N \quad (\text{dB}), \tag{6.7}$$

where Ei_L is the amplifier input level, F_L is the amplifier noise figure, and N is the number of amplifiers cascated. Next,

$$C/N_R = Ei_R - F_R - 0.9 \quad (\text{dB}), \tag{6.8}$$

where Ei_R is the receiver system input level, and F_R is the noise figure of the receiver system.

Column (a) of Table 6.5 illustrates an example budget for a system using

Table 6.5   *C/N* Budget (example)

	(a) All coaxial	(b) HFC	
E_{iH}	80.0	80.0	dBμV
F_H	7.0	7.0	dB
C/N_H	72.1	72.1	dB
E_{iL}	70.0	70.0	dBμV
F_L	9.0	9.0	dB
N	20	5	–
C/N_L	47.1	53.1	dB
C/N_{opt}	–	53	dB
E_{iR}	70.0	70.0	dBμV
F_R	12.0	12.0	dB
C/N_R	57.1	57.1	dB
C/N_T	46.7	49.2	dB

Note: The standard value for C/N_{opt} is 53 dB.

only coaxial cables.

We shall now construct a link budget for the fiber-coaxial hybrid system (HFC) discussed in Subsection 6.1.3. We will use the same performance parameters as above, and assume that there are five cascated amplifiers in the coaxial cable system ($N = 5$). Total system C/N is

$$C/N_T = -10 \log (10\, E\, (-C/N_L/10) + 10\, E\, (-C/N_R/10) + 10\, E\, (-C/N_{opt}/10)).$$

An example of this budget is listed in column (b) of Table 6.5. Compared with an all-coaxial system, we see a total improvement of about 2.5 dB in the HFC system.

The *CSO* Budget

CSO for the transmission system and the receiver system (indoor transmission system + receivers) carries the notation CSO_L and CSO_R, respectively. Total system *CSO* is calculated by

$$CSO_T = -10 \log (10\, E\, (-CSO_L/10) + 10\, E\, (-CSO_R/10)).$$

We write

$$CSO_L = CSO_0 - 10 \log n - (\alpha_1 + \beta_1) \quad \text{(dB)}, \tag{6.9}$$

where CSO_0 is the *CSO* per amplifier, n is the number of amplifiers cascated, α_1 is the total amount of level fluctuation in the repeaters in dB, and β_1 is the total equalization drift in dB.

$$CSO_R = CSO_1 - 10 \log m - (\alpha_2 + \beta_2) \quad \text{(dB)}, \tag{6.10}$$

where CSO_1 is the *CSO* per indoor amplifier, m is the number of indoor amplifiers connected in tandem, α_2 is the amount of level fluctuation in the indoor system, and β_2 is the equalization drift in the indoor system. Since the head end *CSO* (CSO_H) is applicable to a single channel amplifier, CSO_H can be considered as infinity. In the case of the HFC system, where the *CSO* of the optical fiber portion of the system is considered CSO_{opt}, we have

$$CSO_T = -10 \log (10\, E\, (-CSO_L/10) + 10\, E\, (-CSO_R/10) + 10\, E\, (-CSO_{opt}/10)).$$

Example *CSO* budgets are shown in Table 6.6. *CSO* for the optical fiber sections of the system are applicable for transmitting in 50 channels, and other than different values for n, all other conditions are the same as for the coaxial cable sections.

The *CTB* Budget

CTB for the transmission system and the receiver system (indoor transmission system + receivers) carries the notation CTB_L and CTB_R, respectively. Total system *CTB* is calculated by

$$CTB_T = -20 \log (10\, E\, (-CTB_L/20) + 10\, E\, (-CTB_R/20)).$$

We write

$$CTB_L = CTB_0 - 10 \log n - (\alpha_1 + \beta_1) \quad \text{(dB)}, \tag{6.11}$$

where CTB_0 is the *CTB* per amplifier, n is the number of amplifiers cascated, α_1 is the amount of level fluctuation in dB, and β_1 is the equalization drift in dB.

$$CTB_R = CTB_1 - 10 \log m - (\alpha_2 + \beta_2) \quad \text{(dB)}, \tag{6.12}$$

where CTB_1 is the *CTB* per indoor amplifier, m is the number of indoor amplifiers cascated, α_2 is the amount of level fluctuation in the indoor system, and β_2 is the equalization drift in the indoor system. Since the head end *CTB* (CTB_H) is applicable

Table 6.6   *CSO* Budget (example)

	(a) All coaxial	(b) HFC	
CSO_0	85.0	85.0	dB
n	20	5	–
α_1	0.5	0.5	dB
β_1	1.5	1.5	dB
CSO_L	70.0	76.0	dB
CSO_{opt}	–	62.0	dB
CSO_1	65.0	65.0	dB
m	2	2	–
α_2	1.0	1.0	dB
β_2	2.0	2.0	dB
CSO_R	59.0	59.0	dB
CSO_T	58.7	57.2	dB

to a single channel amplifier, CTB_H can be considered as infinity.

For an HFC system, where the *CTB* of the optical fiber portion of the system is considered CTB_{opt}, we have

$$CTB_T = -20 \log (10\,E\,(-CTB_L/20) + 10\,E\,(-CTB_R/20) + 10\,E\,(-CTB_{opt}/20)).$$

Example *CTB* budgets for both all-coaxial and HFC systems are shown in Table 6.7. *CTB* in the optical fiber sections of the system can be normally assured as $CTB_{opt} =$

Table 6.7   *CTB* Budget (example)

	(a) All coaxial	(b) HFC	
CTB_0	90.0	90.0	dB
n	20	5	–
α_1	0.5	0.5	dB
β_1	1.5	1.5	dB
CTB_L	60.0	72.0	dB
CTB_{opt}	–	65.0	dB
CTB_1	75.0	75.0	dB
m	2	2	–
α_2	1.0	1.0	dB
β_2	2.0	2.0	dB
CTB_R	63.0	63.0	dB
CTB_T	55.3	56.3	dB

65 dB. Other than different values for n in the coaxial part of the system, all other conditions are the same.

$C/(N + I)$
In an actual CATV system, in addition to the thermal noise (N) of C/N, CTB and CSO are present in the 6-MHz band in quantities which we designate as p and q, respectively. Taking $I = CSO \times p + CTB \times q$, we convert CSO and CTB to a thermal noise equivalent, and calculate $C(N + I)$. In the total transmission band, CTB_T and CSO_T carry the largest CSO and CTB values, so the CSO and CTB in any 6-MHz band will be smaller than this maximum value. Also, the values for p and q will differ according to the channel plan.

We can assume that all video signal carriers are harmonically related (HRC) as an integer multiple of 6 MHz, and that if there is one coupled wave component each for CTB_T and CSO_T, each of their magnitudes will be added to the power of the signal. If the above calculation results give $C/N_T = 48.4$ dB, and we substitute $CSO_T = -58.7$ dB, and $CTB_T = -55.3$ dB in the above equation, we arrive at $C(N + I) = -47.6$ dB, which means that $C(N + I)$ degrades the C/N value by a bit over 1 dB.

6.4.4 Cautionary Notes Regarding Budgets
As was noted previously, CSO and CTB are measured values. Also CSO_0 and CTB_0 are measured valures. Therefore CSO_0 and CTB_0 include the improvement by the tilt in the amplifier in the system performance and the degradation by the deviation from the ideal amplitude vs. frequency characteristics. Based on these CSO_0 and CTB_0 values, calculations of total system performance in an n-stage cascaded system will often give values that do not correspond to measured results. The calculated values shown above tend to be worse than measured results. The reasons behind this lie in the way that the second- and third-order coupled wave vectors add, and that as system stages are increased, frequency characteristics are cumulative in nature. Consequently, when measuring CSO and CTB for the total n-stage cascaded system, compiling measurement data for future reference is recommended.

Also, for a wired television broadcasting system (CATV), system performance is specified at the output terminal of the surging block (or the output of the wall outlet). Specifying performance up to the surging block is a simple matter, as is assuring performance at the actual site. However, where there is a large system beyond the surging block, depending on system performance within the building, performance at the HT or STB may vary widely. It is therefore best to have a margin of about 3 dB. With digital CATV, it is particularly difficult to assure an acceptable level of performance at the STB input in cases where attenuation within the indoor system is greater than anticipated, large reflection components are present, or external (ingress) noise is entering the system. One of the factors adding to this difficulty is when people other than CATV service provider personnel design and maintain the system, which is a good reason for establishing standards for in-house wiring systems.

6.5 Developing Digital CATV and Establishing Standards

6.5.1 Transmitting Digitally over CATV

Efforts to transmit digitally over CATV systems started early, and broadcasts of digital data and digital audio have already been successfully implemented. The development of practical digital CATV telephones has also been realized, but expected demand for these systems has been tepid. The advent of digital satellite broadcast television, however, has been a primary factor in the movement to digitalize CATV. Also, developments in the multimedia system which the computer industry has been promoting presents an opportunity to bring CATV services to the personal computer terminal, and in the future, we can expect good progress in the task of developing cable telephones and cable modems (and establishing applicable standards).

6.5.2 International Standards for Digital CATV

ITU-T SG 9 made quick work of the international standards for digital CATV, completing the standards recommendations for J. 83 (digital CATV) and J. 84 (digital MATV) in 1996 [ITU-T, 1996]. J. 83 establishes coding systems and transmitter-receiver protocols based on the MPEG-2 system (which should assure commonality across the media spectrum), and standardizes channel coding procedures (refer to Table 6.1). Annex A was recommended for Europe, B and D for North America, and Annex C for Japan. Annex A, B, and C specify M-ary modulation (i.e., 64-QAM), while Annex D covers 16-VSB systems.

With regards to general multimedia terminals, rapid progress is being made by DAVIC (Digital Audio-Visual Council, the industry body establishing technology standards for applications such as video-on-demand, etc.) in establishing network-to-network and network-to-terminal interface standards.

In the IEC, TC 86 has made good progress on the work of setting standards for optical fiber cables, and a technology committee (TC 100) for multimedia was set up in 1995. That committee was further divided into SC 100 A (receivers), SC 100 B (storage media), SC 100 C (AV systems), and SC 100 D (cable systems). SC 100 D consists of WG 3 (measuring methods and performance standards), WG 4 (electromagnetic radiation and countermeasures), and WG 5 (distribution of satellite broadcasting signal).

6.5.3 Domestic (Japanese) Standards for Digital CATV

Being ushured by the ITU-T SG 9 groups working on digital CATV standards, the digital transmission committee members of Cable Television Council (set up by JCTEA) commenced research in digital CATV systems. In the meantime, CRL has set up experimental facilities called as the ACT Center [Iguchi, 1997], and JCTEA have used both them and field CATV system facilities to conduct proof-testing of digital CATV transmission systems. The results obtained in those experiments were also made part of the response to Inquiry No. 74 issued by the TTC. Japan's response to ITU-T SG 9 was to adopt the J. 83 Annex C part of ITU-T recommended standards, and in accepting this, revisions were made to the appropriate ministerial ordinances (12/1996). The Specifications and Standards Committee, which is Japan's standards setting organization set up in JCTEA by the public sectors, then set forth to write standards for detailed items, and system standards were made official in May of 1997 [JCTEA, 1997].

The recommendations covering MPEG-2 generic coding [ISO/IEC, 1994] specify both general and specific details, but the Specifications and Standards Committee settled the undecided detailed items, and in May, 1997, released the Japan Cable Television Engineering Association Standards for the domestic television industry [JCTEA, 1997].

With regards to the cable internet (internet services provided over CATV), the work in writing standards for cable modems is still in progress (as of 12/1997).

6.6 Conclusion

CATV has had a long history in the development of analog transmission technologies, and the technology in that area has almost reached perfection. However, attention now is quickly shifting to the area of digitally transmitting TS packets based on the MPEG-2 compression standards. In this chapter, we have primarily covered only the areas of the immediate technologies which have resulted in the realization of the current digital CATV system, and shall save future technologies for another time. However, many of the technologies we will see in future digital CATV will have common threads to today's technologies. For those readers desiring a more comprehensive understanding of CATV, you are invited to consult the reference list.

Chapter Questions

Q1: CATV systems typically use the tree type network structure. Explain why.

Q2: CATV systems are being developed to accommodate multichannel media. List the reasons for this, and explain the problems confronting these developmental programs.

Q3: CATV is also viewed as one of the potential communications service mediums. Explain why. Also describe the problems encountered in primarily operating as a broadcasting service, but at the same time providing a communications service.

Q4: In an HFC system, given the conditions listed below, determine the maximum number of coaxial line amplifiers that can be cascaded in the system, while assuring a $C/N \geq 48$ dB, $CSO \leq -55$dB, and $CTB \leq -55$dB :

In the fiber section: $C/N = 55$ dB, $CSO = -60$ dB, $CTB = -65$ dB (at the node output).

In the coaxial section trunk amplifier system (1 amp): NF = 10 dB, $CSO = 80$ dB, $CTB = 85$ dB.

In the trunk distribution amplifier (bridger output): $NF = 12$ dB, $CSO = 74$ dB, $CTB = 78$ dB.

For both the trunk and bridger, $\alpha = 0.5$ dB, $\beta = 0.5$ dB.

Q5: Solve Q4 under the same conditions, but adding a margin of $C/N = CSO = CTB = 3$ dB.

Chapter References

[ATMS, 1998] Tokyo Maruchimedeia Shisutemu Kyougikaihen: "Tokyo-to Rinkai Fukutoshin Maruchimedeia Jikken Hyouka Houkokusho" (1998.1) (in Japanese)

[Iguchi, 1996] M. Iguchi, et al.: "Jisedai CATV Shisutemu no tame no Koutaiiki Intarakuteibu CATV Jikkenshisetsu no Gaiyou", Eizou Jouhou Medeia Gakkaishi, Vol. 51, No. 3, pp. 405-409 (1997) (in Japanese)

[Ishiguro, 1984] I. Ishiguro: "CATV", Denshi Jouhou Tsuushin Gakkaishi, Vol. 67, No. 7 (1984.7) (in Japanese)

[Ishiguro, 1986] I. Ishiguro: "CATV no Genjou to Shourai", Terebijyon Gakkaishi, Vol. 40, No. 9 (1986.9) (in Japanese)

[Ishiguro, 1987] I. Ishiguro: "Souhoukou CATV", Denshi Jouhou Tsuushin Gakkaishi, Vol. 70, No. 6 (1987.7) (in Japanese)

[ISO/IEC, 1994] "Information Technology-Generic Coding of Moving Pictures and Associated Audio : Systems," ISO/IEC 13818-1, "Information Technology-Generic Coding of Moving Pictures and Associated Audio Information : Video Recommendation ITU-T H. 262," ISO/IEC 13818-2. "Information Technology-Generic Coding of Moving Pictures and Associated Audio : Audio," ISO/IEC 13818-3 (Nov. 1994)

[ITU-T, 1987] ITU-T SG 9 recommendation J. 83 ANNEX C: "Digital multi-programme System C," J. 84 ANNEX C: "Digital multi-programme SMATV System C" (1987)

[JCTEA, 1997] (sha) Nippon CATV Gijutsu Kyoukai: "JCTEA STD-001-1.0 Deijitaru Yuusen Terebijyon Housou Gentei Jushin Houshiki", "JCTEA STD-002-1.0 Deijitaru Yuusen Terebijyon Housou Tajuuka Souchi", "JCTEA STD-003-1.0 Deijitaru Yuusen Terebijyon Housou Bangumi Hairetsu Jouhou no Kihon Kousei oyobi Shikibetsushi no Unyou Kijun", "JCTEA STD-004-1.0 Deijitaru Yuusen Terebijyon Housou Jushin Souchi" (1997.6) (in Japanese)

[Maeda, 1996] M. Maeda, et al.: "Dentsuu Gishin Junkyo 64 QAM Shingou to Anarogu Henchou no Rinsetsu Channeru Densou Jikken Keeburu Kyougikai Jisshou Jikken Houkoku – sono 2", Terebijyon Gakkai Gihou, Vol. 20, No. 35 (1996.6)

[Miyagawa, 1986] H. Miyagawa Kanshuu: "CATV", Denshi Tsuushin Gakkaihen, Ohmsha (1986.8) (in Japanese)

[Yasuda, 1995] H. Yasuda: "1. MPEG no Keii to Tenbou", Terebijyon Gakkaishi, Vol. 49, No. 4, pp.409-415 (1995) (in Japanese)

[MPT, 1975] Yuuseishou Denpa Kanrikyoku CATV Gijutsu Kenkyuukai: "CATV Gijutsu Kenkyuukai Houkokusho" (1975.3) (in Japanese)

[MPT, 1994] Hikari Keeburu Terebi Shisutemu ni Kansuru Chousa Kenkyuukai: " Heisei 5-nendo Hikari Keeburu Terebi Shisutemu ni Kansuru Chousa Kenkyuukai Houkokusho Dai 1 hen Hikari/Doujiku Haiburiddo Keeburu Terebi Shisutemu" (1994.8) (in Japanese)

[MPT, 1995] "Heisei 7-nendo Denki Tsuushin Gijutsu Shingikai Toshin: "Shimon Dai 74-gou [Deijitaru Housou Houshiki ni Kakaru Gijutsuteki Jouken] no uchi 12.2 ~ 12.75 GHz wo Shiyou suru Eisei Deijitaru Housouhoushiki (27 MHz Taiikihaba wo Shiyou suru mono) no Gijutsuteki Jouken" (1995.7) (in Japanese)

[MPT, 1996] "Heisei 8-nendo Denki Tsuushin Gijutsu Shingikai Toshin: "Shimon Dai 74-gou [Deijitaru Housou Houshiki ni Kakaru Gijutsuteki Jouken] no uchi Yuusen Terebijyon

Housou ni okeru Deijitaru Housouhoushiki no Gijutsuteki Jouken" (1996.7) (in Japanese)

[MPT, 1996] Heisei 8-nen Yuuseishou Kokuji Dai 78-gou: "Eizou no Asshuku Tejun oyobi Soushutsu Tejun, Onsei no Asshuku Tejun, PES Paketto, Sekushyon Keishiki oyobi TS Paketto no Soushutsu Tejun, Densou Seigyou oyobi Shikibetsushi no Kousei narabi ni Sukuranburu no Tejun narabi ni Kanren Jouhou no Kousei oyobi Soushutsu Tejun" (1996) (in Japanese)

[Noda, 1996] T. Noda, et al.: "Dentsuu Gishin Junkyo 64 QAM Deijitaru Yuusen Terebijyon Housou Jisshou Jikken keeburu Terebi Kyougikai Jisshou Kikken Houkoku – sono 1", Terebijyon Gakkai Gihou, Vol. 20, No. 35 (1996.6) (in Japanese)

[Noda, 1996] T. Noda, et al.: "64 QAM ni yoru MPEG 2 Gazou no HFC Densou Jikken", Denshi Jouhou Tsuushin Gakkai Tsuushin Sosaietei Taikai, B-873 (1996.10) (in Japanese)

[Terebijyon Gakkai, 1975] Terebijyon Gakkaihen: "Gazou to Zatsuon", Nippon Housou Shuppan Kyoukai (1975.1) (in Japanese)

[ITU-T, 1987] ITU-T SG9 Recommendaction J. 83 ANNEX C : "Digital multi-programme System C," J. 84 ANNEX C : "Digital multi-programme SMATV System C" (1987)

Chapter 7
Future Trends in Broadcasting Services

The appearance of new services in the broadcasting industry are closely correlated with new developments in digital broadcasting technologies. These developments have swept away most of the boundaries that once delineated communications and broadcasting as separate fields, and integrated services such as interactive television and Video on Demand are now quickly becoming a reality. Advances in computer technologies have certainly played a major role in supporting the changes in communications and broadcasting, and must be considered an integral part of any new services. In this chapter, we shall discuss the current new services, as well as provide a view of the broadcasting services we can expect to see in the 21st century.

7.1 The Integration of Communications and Broadcasting

For many years, communications and broadcast television were developed under different rules as separate branches in the field of electro-communications. Electro-communications has a history going back to the 19th century, and while the technological characteristics of the different systems have been based on required transmission capacity and functionalities, the field of electro-communications has had a large role in supporting the communications needs of a modern society. Television and radio can be considered as most representative of broadcasting systems, which under the law are generally defined as methods which provide the one-way transmission of information to the public. All other areas of electro-communications fall under the category of communications. The telephone is a typical communications instrument with which normally two people have a two-way exchange of information, but communications also encompasses one-way radio systems and point-to-multipoint communcation systems.

Details surrounding the concept of integrated communications and broadcasts have been debated since about 1990. There are two aspects to this concept, the first being technological in nature. In providing integrated services, the events that have made it possible for both communications and broadcasting to share a common information transmission channel were digitalization of the equipment used by the systems involved, and the development of wideband technologies used in information transmission channels. Specific examples of these types of services are CATV-telephone networks, pagers multiplexed on the FM broadcast signal, and television signals broadcast from communications satellites. The Telecommunications sector of the International Telecommunications Union (ITU-T) is currently working on standardizing the Broadband Integrated Services Digital Network (B-ISDN), under which broadcasting services will be provided over an integrated communications network.

The second aspect of integrated services has to do with the way systems are operated. Some new services occupy a middle-ground between communications and broadcasting; some communications must be open to the public, yet secure from third-party eavesdropping. Typical examples of this are the electronic bulletin boards and home pages posted on the internet. On the other hand, there are broadcasting services with mass-media characteristics, open to direct access by the public, but at the same time restricting who can access the service. Pay TV (pay per view) and special information channels over multichannel CATV and satellites are two such examples.

Figure 7.1 illustrates some examples of services that we should see in the near future. Note that these services are arranged according required transmission bit-rates, and communications modes [MPT, 1996]. Conventional broadcasting services are listed on the left, communications services on the right, and new services in the center.

In the section that follows, we shall briefly describe the technological aspects of integrated communications and broadcasting.

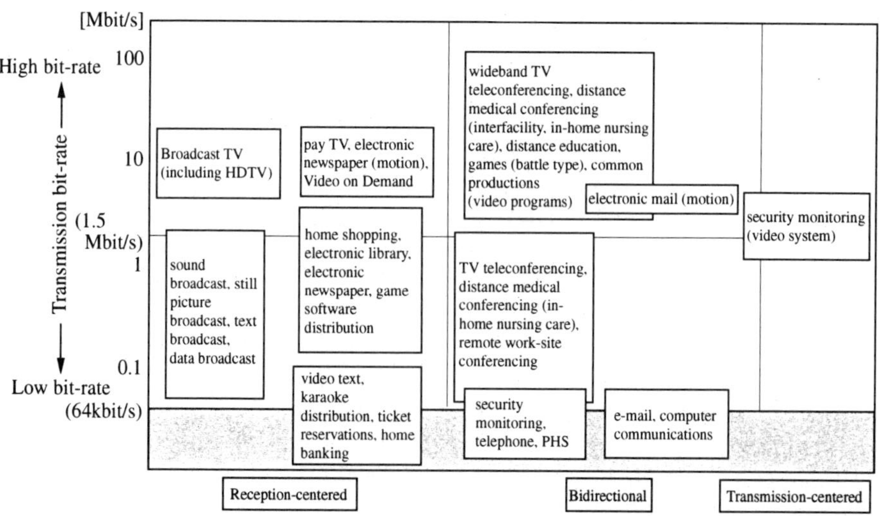

Fig. 7.1   Bit-Rates and Transmission Modes of Typical Information Communication Services

7.1.1 Converting the Communications Network to Digital

The current telephone network is a circuit-switched system, with each channel occupying a 3 kHz audio band coded for PCM transmission at a rate of 64 kbit/s (8 bit, 8 kHz sampling). In Japan, the telephone repeater system is now fully digital. The potential of digital services offered to consumers has had a positive effect on digitalization of the subscriber system, and converting the subscriber system to digital is making steady progress as ISDN (integrated services digital network) services are introduced. Figure 7.2 illustrates the increase in the number of ISDN subscribers in Japan [MPT, 1997].

Additionally, the growing increase in the number of people accessing the internet has produced considerable interest in high-speed access lines, and the use of these types of lines has increased rapidly since 1996. The ISDN system can use existing telephone lines (paired cables), or subscriber lines consisting of coaxial or optical fiber cables can be installed for a system that conforms to the 2 Mbit/s group service international standard. Basic rate ISDN consists of two 64 kbit/s line switching channels (the so-called B-channels) and a packet switching channel (the D channel, 16 kbit/s).

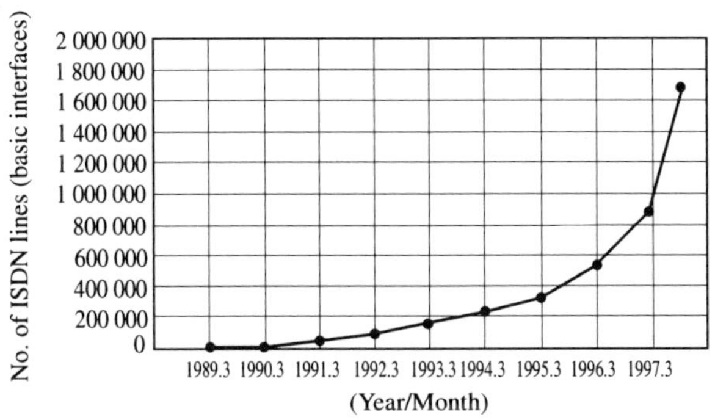

Fig. 7.2   The Increase in ISDN Lines

The TV-telephone services operated over the ISDN system were developed under international standards [ITU-T, 1993-1]. However, at bit-rates of the basic rate system, image quality is less than satisfactory. As was discussed in Chapter 2, in order to transmit pictures of "standard television quality," information transfer rates on the order of 4 ~ 10 Mbit/s are required using the MPEG-2 compression standard. (Note that network information transfer rates are commonly expressed in bandwidth.) In other words, it is quite impossible to achieve any kind of acceptable image quality using an unmodified telephone network; network bit-rates must exceed at least a few Mbit/s.

Starting around 1996 many countries commenced commercial services over the B-ISDN network, whose basic technology is centered on a type of packet transmission system called ATM (asynchronous transfer mode). With ATM, all information is included in short, fixed-length packets (cells) of 53 bytes which are transmitted asynchronously. The network does not use any type of retransmit controls, but accommodates real-time transmissions and multimedia functions simultaneously. Information transmission is handled by logical channels called *virtual channels* (VC) on both ends of the network, and within limits, VC transfer rates can be varied as needed. This means that the bit-rate can be set to match the amount of information being transferred, and a single network can be used for video, sound, and data, and has the built-in flexibility to meet future needs. With regards to user-network interfaces at the ends of the system, speeds of 155.52 Mbit/s and 622.08 Mbit/s are standard [ITU-T, 1991].

B-ISDN is also capable of transmitting high-quality images. However, switching over from the telephone network is not a simple matter, as the subscriber system must consist of lines (i.e., optical fiber cables) that can support wideband transmissions. One such system being studied is termed fiber to the home (FTTH). With this system, optical fiber cables are strung from the switching node into the subscriber's home. With fiber to the curb (FTTC), optical fiber is brought to the neighborhood of a group of subscribers, and each subscriber can then connect into the system using coaxial cables. These are just two systems being studied. In May of 1994, the Policy Council of the Ministry of Posts and Telecommunications (MPT) announced an Information Communications Infrastructure Development Program [MPT, 1994], setting a target date of 2010 for completion of a nation-wide optical fiber system. Work is now in progress based on that policy decision. Sometime near the start of the 21st century, anyone in Japan desiring to become a subscriber should have access to an optical fiber subscriber line. Of course, whether or not people are going to want to become subscribers will depend on whether they are attracted to the services offered, and on how much it costs. We are still waiting to see whether the integration of communications and broadcasting services will provide that incentive.

Some interest has been shown in the Asymmetric Digital Subscriber Line (ADSL), a wideband system proposed and developed by Bellcore Technologies, and which uses the existing paired cable telephone network [ANSI, 1995]. A modification of ADSL, designated xDSL is very similar [Humphrey, 1997]. In any case, switching subscriber lines to optical fiber will require a transition period of 10 years. The above proposals are simply a means of providing high-speed services prior to the new technologies being fully implemented. In these systems, the telephone center transmits over wideband channels to the subscriber (forward direction) at bit rates ranging from several hundred kbit/s to 8 Mbit/s, and the subscriber transmits back to the telephone center over narrowband at rates ranging from 100 kbit/s to 1.5 Mbit/s. The asymmetric line system is suited for the distribution of services such as Video on Demand, and sending information (primarily video) from the telephone center. It is also has the potential of providing high-speed internet access, making the process of accessing web sites much faster. The technologies for ADSL are rapidly being developed, but since this type of system uses uses existing telephone lines, problems such as cross-talk and line lengths must still be overcome. Also, since we assume that the future communications protocol will be full-duplex (two-way channels with equal bit-rates), the technologies for ADSL should be developed while keeping in mind that the system will likely be replaced in the future.

7.1.2 The Influence of Computer Technologies

The digital coding of communications and broadcasting systems has not been the only factor involved in the reform of information transmission technology. New computer technologies have had an enormous effect in this area, and technological advances in computers are sure to continue playing a big role in the development of future broadcasting systems.

The first electronic computers were all-purpose, stand-alone mainframes, with all data processed in a central location. Mainframes maintained a predominant position in the computer market until the 1980s. The origins of the computer network can be traced back to ARPAnet, an experimental system developed in the US in 1969. By the 1990s, computer networks had become very popular, and decentralized data processing using small, high-performance computers (personal computers (PCs)) and work-stations had largely replaced data processing with mainframes. With the globalization of the internet, growth in the use of computer networks has continued at a phenomenal rate.

Growth in the development of computer technologies is strongly related to advances made in the semiconductor industry. Improvements in large-scale integration techniques have resulted in dramatic increases in processing speeds, and in the capacity of memory devices. These factors are very important in the real-time transmission of motion pictures because of the high speeds involved in coding the images. As information storage devices continue to grow smaller, and at the same time increase in memory capacity, it also becomes easier to store the massive amount of data required to process video images. Semiconductor storage mediums are now used in centralized data bases which can service decentralized processing stations. Thus, because they support the decentralized processing concept, computers have become an essential element in the digitalization of the communications network.

High-performance computers also play an important role in the supervision and maintenance of the communications network structure, and provide the "intelligent" functions used by the different media services. With regards to the computer's role in broadcasting systems; computers were introduced early to broadcasting stations and studios, where they are used to edit programming materials and perform many other vital maintenance functions. It is thus evident that broadcasting and communications rely heavily on computers in the everyday tasks that must be performed at the station.

The new services offered by the integration of communications and broadcasting have only been made possible because of computer technologies. With the assistance of computer technologies, broadcasting has now broken out of its traditional confines, and the opportunity now exists to truly expand the number of new broadcasting services available.

7.2 Interactive TV Services

Providing a higher level of broadcasting services takes many different forms. In the area of improved quality for example, we have the upgrading of TV from standard television (NTSC system) to HDTV. There is also UDTV (ultra-high definition TV), in which the number of scanning lines and frames is several times greater than in standard TV. The development of three-dimensional television is also now a possibility. Another area is in multichannelization, a feature which expands the convenience and attractiveness of television through a wider choice in programming. There is also the area of bidirectional communications, some of the features of which we shall describe below.

Standard broadcasting is a service in which many viewers select and receive a program blanket-transmitted over a unidirectional feed by the television station. The concept of providing a means for viewers to communicate back to the television station dates back a long time. The idea here is to have a viewer participation type programming format modeled after the disk jockey radio programs in which listeners phone in to join the program in live conversation. Until recently, this type of direct participation was limited to only one, or at most, a few people. Other viewers are relegated to the status of spectators. To attempt any more than this would be to invite tying up the telephone lines into the television station and disrupting the programming format. The concept of a bidirectional broadcasting service can be described as having the following components:
(1) The mass-calling format, where the opinions of many viewers or information provided by many viewers can be gathered by the station and added to the program in real-time live broadcasts.
(2) The Video on Demand format, where individual viewers can express their choices regarding programming or information services.

Bidirectional broadcast service can also be classified according to link structure. By using a telephone network a link can be established between the television program and viewers, or alternatively, a CATV network with bidirectional capabilities can be used. In the following subsection, a system using a telephone communications network to provide bidirectional capability through a reverse link is described. More details on Video on Demand services in a bidirectional system are provided in Section 7.3.

7.2.1 Mass-Calling Service

In November of 1993, mass-calling services using the telephone network were initiated in Japan. Viewers utilize this service by calling in to the television station on the reverse link, and the plan was to have viewer-participation type programs in which viewers can vote by telephone, or viewers expressing a certain opinion can be selected to participate directly in the program. The mass-calling service mechanism functions similar to toll-free calling in its use of a high-level technology referred to as the Intelligent Network (IN) [ITU-T, 1993-2]. In conventional systems, each telephone exchange had its own network service control program, but with intelligent network this has been modified so that the computers performing the service control function (switching of line connections between the information feed networks) are located at the service control nodes. With the added flexibility realized in this system, additions and changes to the service can be made by modifying the program at the service control node, which avoids having to change programs in the several thousand telephone exchanges throughout the country. In toll-free calling the centralized management system is used to manage control information at the service control node, and this permits the system to perform such dynamic functions as identification of originating calls by time or area, and then automatically switching calls to the destination number as appropriate.

In mass-calling, this structural function is used in reverse. In this case, the service control information is downloaded to every subscriber switch in the country, a feature designed to prevent congestion in the network. (See Fig. 7.3.)

In a polling-call conducted with mass-calling, subscribers from all over the country can dial in using a special contract subscriber number that starts with a header (e.g., 0180). The subscriber switch in which the user is registered uses this number to identify the call as participating in mass-calling services. The normal process of making the connection to the destination number is foregone, and instead, a connection is made to a voice announcement. At that time, the subscriber switch reads the polling (voting) code number appended to the subscriber-dialed number, and compiles the results. This ends the polling-call process.

The service control nodes then gather the data from the subscriber switches from all areas, and under the direction of the television station (under contract), releases the compiled results nation-wide. The polling method is used to query all

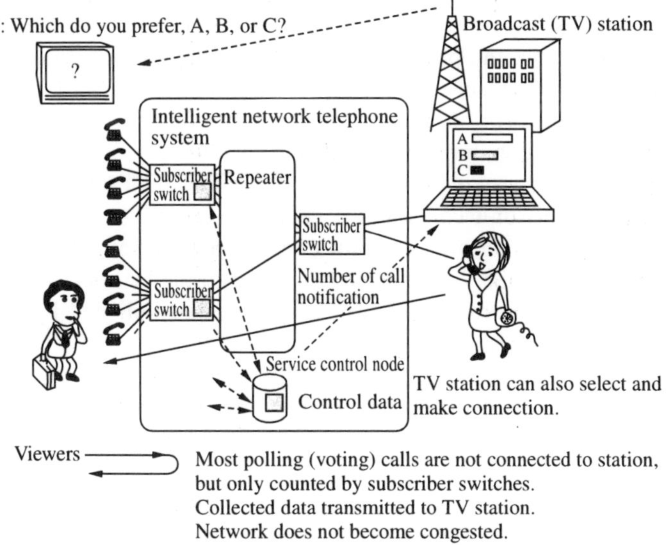

Fig. 7.3   The Mass-Calling System

switches in a predetermined sequence to determine whether or not the switch contains data. Thus, in mass-calling, the large volume of calls that would normally be directed to the television station are processed at the originating switchboard, and since only polling data is collected, the planned congestion problem can be avoided. Mass-calling moreover can select from among the viewer calls any individual caller, and connect that person through a designated telephone line for a direct conversation with the television station.

With the introduction of mass-calling technologies, many viewers can use the bidirectional network to express opinions regarding programming, which helps the program planning directors do a better job. It also resolves viewer complaints regarding busy phone lines, plus it avoids the congestion that occurs when too many people are trying to make calls simultaneously, which satisfies the network manager's needs.

7.2.2 Intertext Broadcasting

In the November, 1995 report released by the Ministry of Posts and Telecommunication's Telecommunication Technology Council, a technological standard was announced for using the vertical blanking interval (VBI) slot for the transmission of data in terrestrial broadcasts. Of the VBI scanning lines not normally appearing on the screen, the use of four intervals (10H, 11H, 12H, 13H) (equivalent to 5,280 byte/s) are permitted for data transmission purposes. Terrestrial data broadcasts (TDB) differ from the multiplexed broadcasting of textual data (Teletext) which started in 1985 (using the 14H ~ 16H and the 21H VBI) in that with TDB, the selected method of coding offers some flexibility. Terrestrial data broadcasts permit hand-written information to be transmitted, as well as textual data. *Script* can also be transmitted to the viewer's receiver (by Intertext) which describes on the screen the procedure for utilizing the various services available. This system was initiated in October, 1996, and multiplexed character broadcast provides the potential for an easy-to-use user interface, as well as providing a telephone calling function through the telephone network.

Intertext broadcasts require a special unit either built into the TV set, or as an external unit. This unit contains the character broadcast receiver circuit and modem circuit, along with a program execution function. It also stores and processes the data that the viewer intends to send to the broadcasting station over the telephone lines, the transmission of which is accomplished by the response server and the center server (see Fig. 7.4).

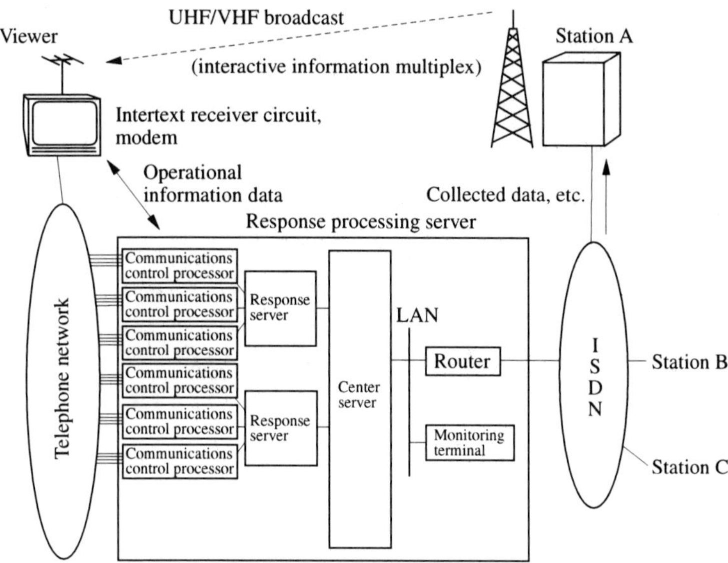

Fig. 7.4   Intertext Broadcasting Structure

7.3 Video on Demand Service

Video on Demand (VoD) is a service which transmits moving or still picture video information to viewers at their demand. The mode of transmission is normally viewed as a one-on-one type communication, customized for individual clients. However, in the sense that the same information can be distributed to many viewers, it also fits the definition of a broadcasting service. Note though that this is not a blanket broadcast of the same information; rather its main advantage is that the same information can be provided to different viewers in different time slots. VoD has been around since the 1970s, and was developed and tested in specialty fields of application such as medicine, fashion, and education. The earlier systems used analog recording media such as video cassettes and laser disks, but the system we shall discuss here is digital VoD, which is a recent development. (See Fig. 7.5.)

A system similar to VoD is Near Video on Demand (NVoD), which we will describe very briefly. NVoD service consists of one program running in several different time slots, from which the viewer can access as desired and when desired. The same program is transmitted over multiple channels in time slots spaced five to 10 minutes apart, so while it is not a purely on-demand system, it does provide a service which seems to satisfy viewers, and is considered to be one of the more promising services for multichannel TV. Since there are no technological restrictions as to the number of simultaneous access lines connected to the server, NVoD has the potential of being more popular than full VoD when it comes to newly released movies and other popular programs that will attract many viewers. The NVoD service provider will need to plan program scheduling, and distribute the

Fig. 7.5   Example Video on Demand Menu Screen [Courtesy of Foundation for MultiMedia Communications]

service selection menu as appropriate.

As a service, the user network environment for VoD can be generally divided into three different types. The first is as one of the new services of CATV, which uses a bidirectional, wideband network. The second is as a communication network (Broadband ISDN, B-ISDN) service, and the third is as a LAN application, or an internet service. In the description that follows, we shall assume that the system occupies either the first or second user environment.

7.3.1 System Structure

The basic structure of the Video on Demand (VoD) system is illustrated in Fig. 7.6. The system essentially consists of, ① the video server which stores the program, and transmits the video data as the viewer demands it, ② the editing system which edits and digitally codes the video information, then feeds it to the server, 3) the Set Top Unit (STU) located in the subscriber's home, over which the service request is made, and which functions to regenerate the video signal, and 4) the network connecting the system.

Access Network

Figure 7.7 illustrates the network topology of a typical Video on Demand (VoD) system, and Table 7.1 lists the types of transmission systems used for VoD. The hybrid fiber coaxial (HFC) system uses optical fiber cables for the trunk lines between the center and the hub, while the branch lines from the hub to the system subscribers are coaxial cables. This system offers the advantage of low replacement cost when switching over from the existing CATV network to a newer system, but bidirectional communication capabilities are limited. With the passive double star (PDS) system a single line serves the coupler. An optical coupler (a passive divider, 8 or 16 taps) at the hub is used to split off the signals which feed the subscriber lines. The hub performs the channel selection function in the active double star

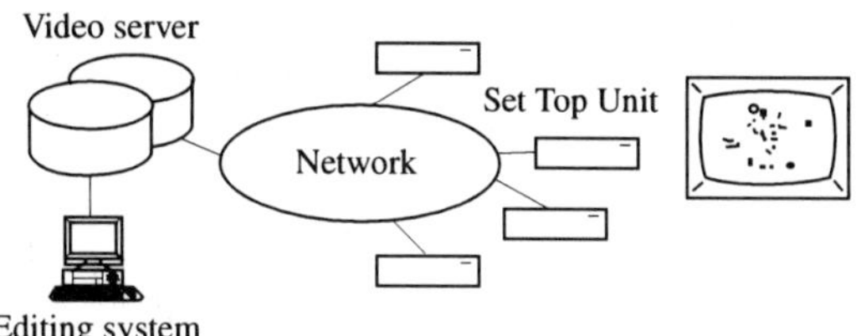

Fig. 7.6   Video on Demand System Structure

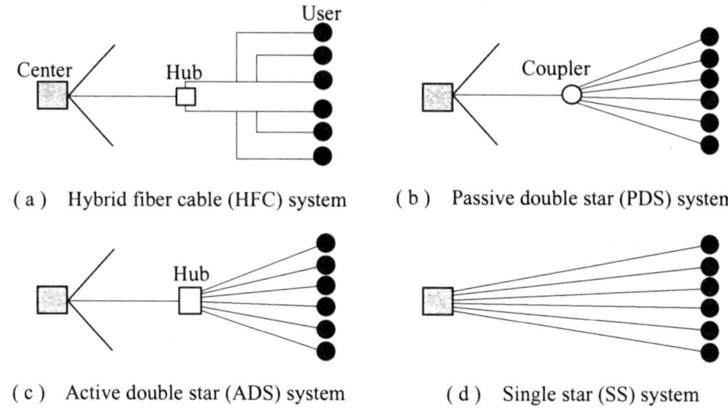

(a)   Hybrid fiber cable (HFC) system   (b)   Passive double star (PDS) system

(c)   Active double star (ADS) system   (d)   Single star (SS) system

Fig. 7.7   Typical Access Network Topologies

Table 7.1   Transmission Systems (examples)

Transmission		Hybrid Fiber Coaxial (HFC)	Optical fiber (PDS-FDM/ATM)	Optical fiber (SS-ATM)
Signaling systems	Bidirectional, multiplexed	FDM only in coaxial section	Time-compressed muliplex system*	Wavelength-division multiplex (WDM) system

	Multiplexed signal	FDM system
		  Analog modulation (AM, FM)
		  Digital modulation (QAM)
			WDM system
			  Video distribution: FDM
			  Bidirectional: TDM/ATM
				TDM system
				ATM

No. of receivable video channels		Analog modulation: approx. 50 Ch. Digital modulation: approx. 200 Ch.	Analog modulation: approx. 50 Ch. Digital modulation: approx. 350 Ch.	Unlimited (but number of simultaneously receivable channels depends on transfer rate)
Maximum info. transfer rate, bidirectional service		about 64 kbit/s	64 kbit/s ~ 50 Mbit/s	28 over 50 Mbit/s

Note: * Signal compressed in time to less than one-half original length. Bidirectional communications alternates between transmit and receive (also called ping-pong transmission).

(ADS) system, which results in fewer operations required at the subscriber end. The single star (SS) system has the same topology as the telephone subscriber system, and potentially offers high-speed communications services.

Video Server

Within the computer system, the server exercises unified management control over information and the system functions. Its name also implies that it is the hardware and software which serves simultaneously all the clients (user terminals) in the network desiring some type of service. The video server in the VoD system stores and manages multimedia information (primarily video), and any of the programs can be accessed at any time by the numerous system clients. Its required functionalities are, 1) a large information (program) storage capacity, 2) multiple input and output ports, and 3) it must be able to output simultaneously different programs from each I/O port. Some typical server structures are shown in Fig. 7.8. Note that for a high-capacity recording medium, the server uses either a hard disk or a semiconductor memory. The hard disk is typically an array of disks, which provides an economical high storage capacity, and superior durability. RAID is the acronym for Redundant Array of Inexpensive D*isks*, and in this system, information is stored on the disks by striping redundant data in blocks across multiple disks (data striping). The RAID system provides a highly reliable means of transferring information at high speeds, and the same information can be read out of multiple ports virtually simultaneously.

One of the ways which the video server can be structured to make the best use of memory capacity is to store popular programs that will be subject to frequent and simultaneous accessing on hard disk or in semiconductor memory, and store programs that will be accessed less frequently in high-capacity library type storages such as magnetooptic disks. In the future, interconnected servers through the Wide Area Network (WAN) will bring down the cost of storing materials that are not frequently used, and should greatly increase the number of programs that the system can store.

The video information stored in the video server is normally compression coded. When materials are edited, they can be in uncompressed form, or they can be coded according to the Motion-JPEG standard, which is an extension of the still-picture JPEG compression standard. In either case, video is normally transmitted in the MPEG-2 compression standard format.

Editing

Editing is the applications development subsystem of Video on Demand service, and consists of the video editing, coding, and authoring equipment. The video editing unit is used for cropping, editing, and computer graphics (CG) composition on the video materials held in the various storage media. The coding unit is used to apply the MPEG-2 data compression code. The authoring unit is used to make up the program menu screen used by the STU, and also supports the management system in registering programs.

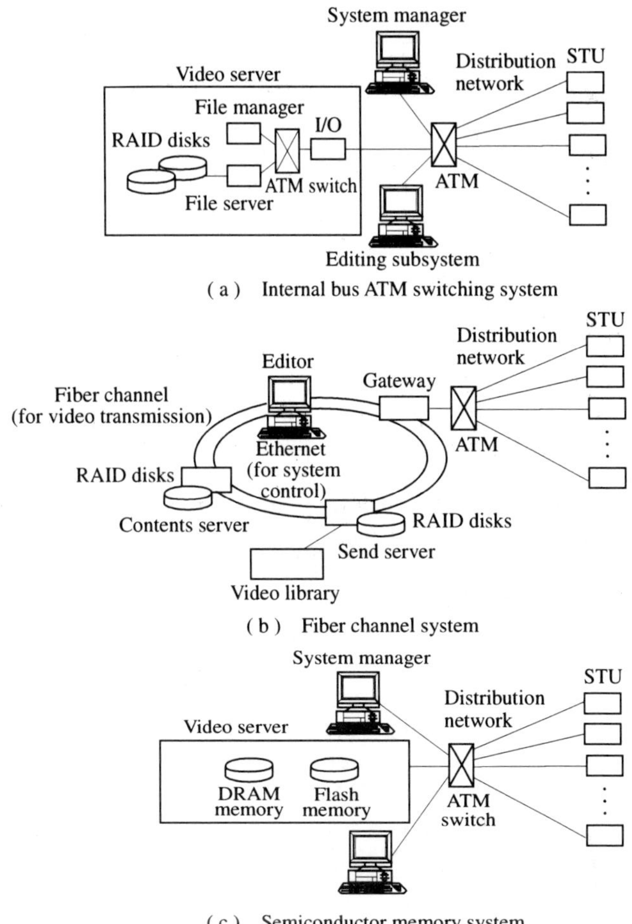

(a)   Internal bus ATM switching system

(b)   Fiber channel system

(c)   Semiconductor memory system

Fig. 7.8   Video Server Systems (example)

Set Top Unit (STU)

The set top unit (STU) is used in the VoD system to select and perform control functions on the various programs. As its name implies, it is part of the TV set in the subscriber's home. The STU is in the form of a personal computer or a workstation, and its functions depend on the network transmission system being used. Its general tasks are to manage control signal protocol and to decode the received video data. The STU also stores service data such as the program menu in an internal hard disk. The technology needed to automatically update the service information and programs distributed by the service provider center is currently under development.

7.3.2 Network Technology Developments and Related Problems

At the same time that Video on Demand technologies are being developed, the future form that the system should take is also being investigated. While we can readily recognize that factors such as picture quality, speed, and storage capacities are priority items, the transmission network is also very important. Rather than just limiting certain services to specific service areas, tying networks together will greatly increase the variety of applications services that can be offered as a whole. Utilizing a high-capacity storage type server network spread over a wide area also results in the creation of a system which can provide both communications and media services.

In North America, the debate over whether to put the VoD system on a network has been going on for a long time, and now an "open" type system structure is being studied. On the international scene, the industry is currently looking toward DAVIC (Digital Audio Visual Council), a private standards organization launched in 1994 with MPEG ties. DAVIC has as participants 140 major companies and organizations world-wide, and is moving to develop interface standards similar to those used by the reference model shown in Fig. 7.9. By 1997, DAVIC standards and specifications for MPEG-2 motion picture services were just about complete. At this point, proving out the specifications, which will be accomplished by verification of the interoperability of the various participating organization's designs, should result in the DAVIC standard being adopted as the international guideline in this field. Also, another group within DAVIC, working in conjunction with IETF (Internet Engineering Task Forces) has embarked on research in multimedia systems in the internet environment, and have commenced investigations into IP-based systems [DAVIC, 1997].

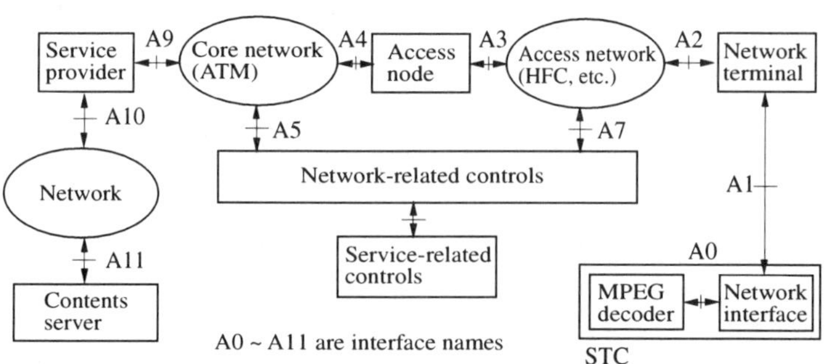

Fig. 7.9   DAVIC System Reference Model

Efforts are also being made to develop new service media that utilize the video server network. In a joint effort, the MPT's Communications Research Laboratory and NTT's Multimedia Network Laboratory have developed an electronic mail service in which high-quality motion pictures can be attached to an e-mail message sent over the video server network. This service is called VA-mail [Iwama, 1997]. The video server provides the storage capacity needed for the large amounts of data involved with motion pictures, and this project has demonstrated the potential for a new public communications medium. VA-mail additionally requires some action on the part of the participant, which some people find more attractive than merely being a passive recipient of information (Fig. 7.10).

The original intent of the VoD system network was as a "leisure-time" service featuring video entertainment, shopping and such. However, the new VA-mail system is more community service-oriented, and is expected to feature programs dealing with a wide range of useful subjects, including education, medical news, traffic, the environment, and government administration matters.

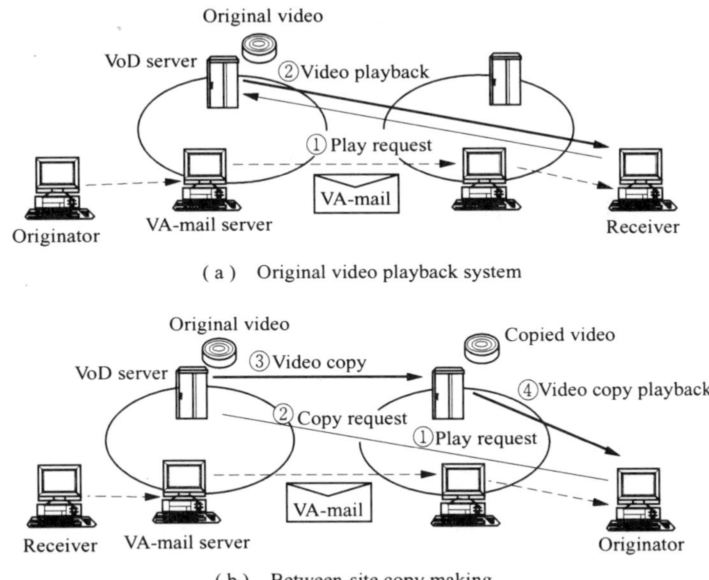

(a)   Original video playback system

(b)   Between-site copy making

Fig. 7.10   VA-Mail System Concept

7.4 Internet Broadcasting

The internet (a computer network based on TCP/IP protocol) has grown so much around the world, that with regards to infrastructure, it is now second only to the standard telephone network in size. Spurred by the advancements made in developing audio/video compression coding techniques, research and development in real-time sound and video which will go out over the internet is now in full swing. The internet telephone is already a practical reality. Efforts are now beginning on developing broadcasting services which will be provided over the internet. The original internet was constructed as an information storage and routing type network, and did not have the capability of continuously transmitting sound or video. Its purpose was therefore considerably different than what we are now trying to accomplish, and in order to fully utilize the internet's advantages for real-time broadcasting, technologies suited for the internet are now being developed.

7.5 ISDB

In this section, we would like to briefly return to a discussion of Integrated Services Digital Broadcasting (ISDB), a concept that holds the potential to change the future course of developments for digital broadcasting. ISDB is a high-level system that has been developed using the fundamental technologies of digital broadcasting, and as such, expresses the total concept of the new broadcasting services. With ISDB, station tasks ranging from the editing and production of programming materials, to the transmitting and reception of signals and the provision of viewer services are all based on digital techniques. This new concept builds on the broadcasting services realized so far, and includes interactive TV and many other new integrated services. ISDB's role in future digital satellite broadcasts was discussed in Section 5.5, and you may refer to that section for more information.

Planned interactive viewer functions provided by ISDB include "instant view" programming such as the latest news and weather reports automatically fed to a home server built into the television receiver, the reception of supplementary program information developed as an extension of character multiplexed broadcasting, in addition to utilizing the communications network for services such as participation format programs, shopping, and so on.

While some of these projected services have already individually been realized, the biggest selling point for ISDB is the concept of the viewer being able to freely select from a group of inter-related services. Research and development is also progressing on an integrated service type television set that will serve as an "intelligent terminal" for ISDB.

The goal in developing ISDB is to produce a comprehensive system through which the viewer can obtain from a single system any type of service desired.

7.6 Conclusion

From a technological standpoint, digital coding techniques and computer technologies are breaking down the divisions between communications and broadcasting. The discussion in this chapter has centered around services such as interactive TV, Video on Demand, and other new broadcasting services which will continue to be realized in the 21st century.

Chapter Questions

Q1: What are the differences between broadcasting and communications? Do these differences create any problems?

Q2: What are the problems involved in utilizing the telephone network in a program production?

Q3: List the structural elements in the Video on Demand system.

Q4: Most motion pictures (video) used in Video on Demand are coded using the MPEG-2 standard. How does MPEG-2 work, and what advantages are offered using this standard?

Chapter References

[ANSI, 1995] ANSI : T1. 413-1995 "Telecommunications-Network and Customer Installation Interfaces-Asymmetric Digital Subscriber Line (ADSL) Metallic Interface (1995)

[DAVIC, 1997] DAVIC : URL : http://www.davic.it/

[Humphrey, 1997] M. Humphrey & J. Freeman : "How xDSL Supports Broadband Services to the Home," IEEE Network (Jan./Feb. 1997)

[Isobe, 1997] T. Isobe: "Intelligent Receiver," Journal of the Institute of Image Information and Television Engineers, Vol. 51, No. 9 (9.1997) (in Japanese)

[ITU-T, 1991] ITU-T : Recommendation I. 121 "Broadband aspects of ISDN" (1991)

[ITU-T, 1993-1] ITU-T : Recommendation H. 261 "Video coded for audiovisual services at px64 kbit/s" (1993)

[ITU-T, 1993-2] ITU-T : Recommendation I. 312/Q. 1201 "Principles of intelligent network architecture" (1993)

[Iwama, 1997] M. Iwama, et al.: "Development of VAmail system: video e-mail system with distributed video servers," IEICE Trans. of Communications, Vol. J80-B-I, No. 6, pp. 330-337 (6.1997) (in Japanese)

[MPT, 1994] Telecommunications Council's Report;" Reforms toward the Intellectually Creative Society of 21 st Century-Program for the Establishment of High-Performance Info-Communications Infrastrure" , May 1994. See the following URL; http://www.mpt.go.jo/policyreports/english/telecouncil/Report1993No5/contents.html

[MPT, 1996] Telecommunications Technology Council's Report; "Network Evolution for the Age of Media fusion of Telecommunications and Broadcasting", April 1996.

[MPT, 1997] MPT White Paper;"Communications in Japan 1997".See the following URL; http:// www.mpt.go.jp/policyreports/english/papers/index-e.html

[Yanagimachi, 1997] A. Yanagimachi: "Digital broadcasting of next-generation: services and content production of integrated services digital broadcasting: ISDB," Journal of the Institute of Image Information and Television Engineers, Vol. 51, No. 9 (9. 1997) (in Japanese)

Answers to Chapter Questions

Chapter 1

A1: High-efficiency coding (such as MPEG-2) of the video and audio information reduces the amount of data to a volume 30- to 40-times smaller than the original content. Digital modulation is also very immune to noise and interference, so the system is well-suited for repeat transmissions at the same frequency. Refer to Sec. 1.3.1.

A2: Refer to Secs. 1.3.1 ~ 1.3.3.

A3: Required digital technologies include high-efficiency coding (such as MPEG-2), forward error correction (FEC), and digital modulation/demodulation. Refer to Sec. 1.2.2.

Chapter 2

A1: Interlaced scanning only requires one-half the bandwidth of sequential scanning without any loss in picture quality. Refer to Sec. 2.1.1.

A2: Refer to Fig. 2.5 and Sec. 2.2.1.

A3: Code transforms include the discrete cosine tranform, Hadamard transform, Fourier transform, wavelet transform, among others. Refer to Sec. 2.2.1.

A4: The pictures which MPEG-2 compression works on are the I-, P-, and B-pictures. Refer to Secs. 2.2.1 and 2.2.3.

A5: Because of the cliff effect. See Sec. 2.3.1.

A6: Circuit (or channel) quality and other transmission conditions vary depending on the transmission medium (terrestrial, satellite, or cable), and the most suitable modulation varies accordingly.

A7: The cliff effect describes the graph of the received signal level. Once received signal level drops below a certain level, picture quality drops sharply (cliff-like). Refer to Sec. 1.3.2.

Chapter 3

A1: Refer to Sec. 3.1.5.

A2: Refer to Sec. 3.1.

A3: First, the arrangement of four bits is decided. Use the diagram in Fig. 3.18 as the 1st quadrant, reverse the Q-channel arrangement for the second quadrant, and reverse the I-channels arrangements for the 3rd and 4th quadrants. Assign

Gray coding to the two high-order digits the remainder in each quadrant.

A4: Corresponding to eqs. (3.30) and (3.31) are $C_1 = \alpha^{12}A_1 + \alpha^{10}A_2 + \alpha^4 A_3$ and , $C_2 = \alpha^{11}A_1 + \alpha^5 A_2 + \alpha A_3$ repectively. This yields $C_1 = \alpha = (0010)$ and $C_2 = \alpha^5 = (0110)$.

A5: Initially starting at state S_0, the transmitted codeword is $x = (11, 10, 00, 01, 01, 11)$. Refer to Fig. 3.26 to construct the trellis-code state transition diagram, and refer to Sec. 3.3.3 to perform the Viterbi decoding.

A6: As one example, consider using a coding rate-2/3 punctured code and a shortened RS code in combination.

A7: Refer to Sec. 3.4.1 and Fig. 3.38.

A8: Refer to Fig. 3.39. Now go back to the signal spacing diagram for 16-QAM in Fig. 3.18, and start by dividing into sets of (0001), (0100), (0010), (0111), (1011), (1110), (1000), (1101) and (0000), (0101), (0011), (0110), (1010), (1111), (1001), (1100).

Chapter 4

A1: Problems include its poor performance when nonlinearities exist in the channel, and the complexity added by the sync circuit and FFT block required in the receiver. Refer to Sec. 4.2.5.

A2: OFDM has good immunity to delayed waves because of its ability to transmit using multicarriers. With multicarriers the symbol length can be made longer, and the longer the symbol length is relative to the delay time of the delayed wave, the lower the chance of intersymbol interference occurring. Refer to Sec. 4.2.2.

A3: SFN problems include difficulty in relaying broadcast signals, and the generation of delayed waves with longer delay times. See Sec. 4.1.2.

A4: Mobile receiver antennas are shorter and have lower gains. Fading is also a problem with mobile applications. Refer to Sec. 4.4.5.

Chapter 5

A1: Refer to Sec. 5.1.1.

A2: Note that the difference between the overall C/N value in the link budget and the required C/N used to assure signal quality is the link margin. Refer to eq. (5.12).

A3: The receiver's figure of merit, the various types of interference, and rain attenuation are factors in this case. Refer to Secs. 5.4.1 and 5.4.2.

A4: Refer to Sec. 5.2.2.

Chapter 6

A1: The tree type network is the most suitable for distributing the energy of the high-frequency CATV signal to the various subscriber receivers (through the signal splitters and distributors). Shorter cable lengths can also be used to serve the subscribers in the network, so it is thus more economical.

A2: The CATV downstream bandwidths range from 70 MHz to 450, 770, or 1000 MHz, so channel capacity is very high. The problem lies in maintaining an acceptable link (circuit) quality. Refer to Sec. 6.3.

A3: Since CATV is a cable system, it can also be used for data transmission and voice (telephone) communications. With regards to problem areas, see Sec. 6.3.

A4: Refer to Sec. 6.4.3 regarding the formulation of a link budget. Consider only the performance of the optical fiber and coaxial sections of the network, but leave the subscriber terminals out of the computations. Calculate the maximum values for C/N, CSO, and CTB which satisfy the conditions of the problem posed. The minimum value obtained using these three values should provide the answer asked for. The maximum number of cascade-connected amplifiers is 18 in this case.

A5: Add 3 dB (absolute value) to the values for C/N, CSO, and CTB posed in the question above, and solve using the same procedure as for Q4. In this case, the number of cascade-connected amplifiers drops to two. Refer to Sec. 6.4.3.

Chapter 7

A1: Technically speaking, there are no differences. Refer to Sec. 7.1.

A2: Too many simultaneous phone calls into the station congest the telephone lines. Refer to Sec. 7.2.

A3: The primary components of the system are the video server, network, and the set top unit (STU). Refer to Sec. 7.3.1.

A4: MPEG-2 is the favored compression coding system because of its ability to highly compress the information content, and because it is the most widely-used standard. See Sec. 7.3.1.

Authors' Profile

Tadashi Shiomi

Graduated from Kyoto University Faculty of Engineering in 1970. Received the Masters Degree from Kyoto University in 1972, and entered employment with the MPT Communications Research Laboratory (CRL) that same year. Appointed as a Research Fellow with the Jet Propulsion Laboratory (JPL, USA) at California Institute of Technology in 1982 ~ 1983. Previously served as the Director of the Kansai Advanced Research Center of CRL, then as Director of the Communication Systems Division. Currently serves as the Director of the Planning Division at CRL. Ph.D. in Engineering.

Mitsutoshi Hatori

Graduated from Tokyo University Faculty of Engineering in 1963. Received the Masters Degree from Tokyo University in 1965, and the Doctorate Degree in Engineering in1968. Appointed as Professor in Tokyo University, Faculty of Engineering (Electrical Engineering Department) in 1986, then Professor of Electronic Information Engineering Science in 1991. Since 1999, he has served as Professor at the National Institute of Informatics, where he currently teaches. Appointed as President of the Institute of Television Engineers of Japan in 1998. Professor Hatori has served on numerous committees, including the MPT Telecommunications Technology Council. Included in his many awards are the Institute of Electronics, Information and Communication Engineers Achievement Award in 1995, and the Institute of Image Information and Television Engineers Achievement Award in 1987. He has been an IEEE Fellow since 2000. He also received the Ministry of Posts and Telecommunications Minister's Award in 1990, and again in 1995, and the NHK Broadcast Culture Award in 2000. Ph.D. in Engineering.

Aiichiro Tsuzuku

Graduated from Nagoya University Faculty of Engineering in 1979, and received the Masters Degree from Nagoya University in 1981. Entered employment with the MPT Communications Research Laboratory in 1981, assigned to research on satellite and digital broadcasting services. Appointed as a Research Fellow with the German Postal Ministry Research and Technology Center in 1990 ~ 1991. Currently serves as the Chief of the Broadcasting Technology Section in the Communication Systems Division at CRL. Ph.D. in Engineering.

Shunichi Iisaku

Received his Doctorate Degree from Hokkaido University in 1982, entering employment with Kokusai Denshin Denwa K.K. (KDD) that same year assigned to research in network architectures in the KDD Research Laboratory. Entered employment with the MPT Communications Research Laboratory in 1995 as Chief of the Visual Communication Section in the Intelligent Communication Division, then served as the Division Director at CRL. Since 2000 he has been with the KDD Research Laboratory. Ph.D. in Engineering.

Yukiyoshi Kamio

Graduated from Nihon University Department of Humanities and Sciences in 1981, entering employment with the MPT Communications Research Laboratory in 1982. Previously served as a researcher at the Yokosuka Radio Communications Research Center of CRL. Currently serves as Head of the Radio Transmission Department of YRP (Yokosuka Research Park), Mobile Telecommunications Key Technology Research Laboratories Co., Ltd.

Hajime Fukuchi

Graduated from Tohoku University Faculty of Engineering in 1975, and received the Masters Degree from Tohoku University in 1977. Entered employment with MPT Communications Research Laboratory in 1977. Appointed as a Research Fellow with Bradford Unversity (UK) in 1986 ~ 1987. Assigned as Chief of the Broadcasting Technologies Section (CRL) in 1992, and as Chief of the Planning Section in 1995. Served as the Director of the Photonic Technology Division (CRL), and since 1998, as the Director of the Kansai Advanced Research Center (KARC). Concurrently teaches as Guest Professor, in the Graduate School of the University of Electro-Communications. Ph.D. in Engineering.

Mitsugu Ohkawa

Graduated from Kanagawa University Faculty of Engineering in 1982, and received the Masters Degree from Kanagawa University in 1984. Entered employment with the MPT Communications Research Laboratory in 1984, and was assigned to the Telecommunications Advancement Organization of Japan from 1988 to 1991. Currently serves as a researcher at the CRL Kashima Space Research Center.

Isao Ishiguro

Graduated from Kyoto University Faculty of Engineering in 1965. From 1965 through 1978 he was employed by Furukawa Electric Co. Ltd. engaged in CATV equipment and systems development. From 1978 through 1991 served as Technology Department Director for Tokyo Cable Television Corporation, and from 1991 to 1997 served concurrently in that post and as a Company Director. He has been the Leading Director of the company since 1997.

Masaaki Iguchi

Graduated from Fukuoka Prefectural Yame Technical High School, Electronics Department in 1966, and entered employment with the MPT Communications Research Laboratory that same year. Appointed as a Tohoku University Faculty of Engineering Researcher in 1982. Served as the Planning and Analysis Section Head at the Kimitsu Satellite Control Center of the Telecommunications Advancement Organization of Japan from 1991 to 1994, then served as a senior researcher in the Communication Systems Division at CRL. He has been with the Space Engineering Department at MITSUBISHI ELECTRIC CORPORATION KAMAKURA WORKS, since 1999.

Fumito Kubota

Graduated from the Keio University Faculty of Engineering in 1974, and entered employment with the MPT Communications Research Laboratory that same year. Assigned to research on mobile communication systems and next-generation networks. Currently serves as Chief of the Communication Network Section within the Communication Systems Division at CRL. Concurrently functions in the capacity of Guest Professor at the Japan Advanced Institute of Science and Technology, Hokuriku, and the National Institute of Multimedia Education, Ministry of Education.

Translator

Wayne Hughes is a free-lance translator specializing in science and engineering subjects. He resides in Carlsbad, New Mexico (USA), and is a graduate of New Mexico Institute of Mining & Technology, Socorro, New Mexico, and the Defense Language Institute at Monterey, California.

Index

A

absorption (by raindrops) ·········167
absorption attenuation ·········167
advanced television (ATV) ········37
amplifier,
    bridging ·········215
    high-power ·········182
    low-noise (LNA) ·········183
    solid-state power ·········178
    trunk ·········218
    trunk & distribution ·········215
    TWT (traveling wave tube) ···9, 178
amplitude and phase-frequency
    characteristics ·········184
amplitude modulation (AM) ········56
amplitude probability density ······132
analog satellite broadcasting ······161
antenna,
    super gain ·········46
    super turnstile ·········46
    parabolic ·········46
    rabbit ears ·········48
    whip ·········48
    Yagi-Uda ·········101
Archimedes satellite ·········199
Asiasat ·········164
aspect ratio ·········18
asynchronous transfer mode (ATM)
    ·········253
automatic gain control (AGC) ······218

B

back-off (amplifier operation level)
    ·········120, 178
bandlimited channel ·········182
baseband signal ·········55
baseband system(BB) ·········180
bit error-rate (BER) ········52, 66, 227
bit stream ·········45
blanking pulse·········20
block ·········39
Boltzmann constant ·········184
broadcast- type network ·········158
broadcasting, multichannel ·········164
broadcasting satellite (Yuri No- 1)
    ·········5
BS digital broadcasting systems
    ·········190
building-out-network(BON)·········217

C

cable modem·········207
carrier hole ·········117
cell (of CATV network) ·········211
central-limit theorem ·········109
channel coding block ·········181
chirp signal ·········120
Clark, Arthur C. ·········157
cliff effect ·········13, 46, 227
coaxial cable ·········215
code,
    block ·········74
    concatenated ·········90
    concatenated error correction
    ·········173, 181
    convolutional ·········83, 181
    cyclic ·········75
    difference set cyclic ·········81
    extended ·········81
    Hamming ·········76
    Huffman ·········33

punctured ·····88
Reed-Solomon (RS) ·····79, 181
shortened ·····81
turbo ·····91
zero-run length ·····41
code word ·····73
coding,
   color component ·····23
   composite ·····23
   predictive ·····29
   subband ·····36
   transform ·····37
coding gain ·····74
coding rate ·····73
coherent detection ·····65,143
coherer ·····1
color difference signal ·····21
Communications and Broadcasting Engineering Test Satellite ·····178
   (COMETS)
community reception ·····159
component (color) signal ·····23
composite signal ·····22,23
concatenated,
   code ·····90
   error correction ·····173, 181
conditional access ·····14, 49, 222
coordination distance ·····168
copy-inhibit flag ·····15
cross-polarization interference ·····186
cumulative distribution ·····169
cyclic code ·····75

D

data striping ·····262
decoding,
   erasure ·····90
   hard decision ·····90
   majority logic ·····83
   maximum likelihood decoding ·····84
   soft decision ·····90
   Viterbi ·····85

deinterleaver ·····183
demultiplexing equipment ·····184
depolarization (of signal) ·····167
descrambling ·····184
detection,
   absolute phase coherent ·····65
   coherent ·····65, 143
   differential coherent ·····65
difference set cyclic code ·····81
differentially coded ·····66
differential decoding ·····122
diffraction loss ·····128
digital audio broadcasting (DAB) ·····105
digital satellite broadcasting(DSB) ·····157
digital signal processor(DSP) ·····113
Drec TV ·····164
discrete cosine transform (DCT) ·····29
discrete Fourier transform (DFT) ·····106,108
distributor ·····216
Doppler frequency shift ·····142
double frequency network (DFN) ·····103
double sideband (DSB) transmission ·····70
downlink ·····184
dual polarization
   communication system ·····167

E

Echo experimental satellite ·····159
electronic program guide ·····213
encoder ·····42
encoder system(ENC) ·····180
equalizer(EQ) ·····186, 217, 232
effective isotropically radiated power
   (EIRP) ·····184
effective radius of the earth ·····127
equivalent temperature ·····168

erasure decoding ·······90
error-free transmission ·······186
error locator polynomial ·······78
European Space Agency (ESA) ·······199
Even field ·······19
eye pattern ·······62, 64
experimental broadcasting satellite ·······169
experimental communication satellite ·······169
Experimental Test Satellite VIII ·······200
extended code ·······81
extension amplifier(EA) ·······217

F

fading ·······14, 50, 141
fading,
   flat·······107, 141
   frequency-selective ·······107, 141
   Rayleigh ·······139
Federal Communications
   Commission (FCC) ·······150
field, picture ·······19
figure-of-merit ·······184
roll-off filter ·······112
roll-off factor ·······110
first astronautical velocity ·······158
follow current noise ·······232
forbidden channel ·······211
forward error correction (FEC) ·······8, 181
Fourier transform, inverse ·······108
free-space path loss ·······184
frequency converter ·······183
frequency-division multiplexing (FDM) ·······110, 209
frequency interleave ·······112
frequency selective fading ·······107, 141
Fresnel integral ·······128

G

Galois field ·······75
Gaussian distribution ·······67
Gaussian noise·······66,184
ghost ·······9, 129, 233
graceful degradation ·······118
Gray coding ·······60
group delay ·······231
guard band ·······112
guard interval ·······103, 115

H

Hamming code ·······76
Hamming distance ·······73
Hamming distance, minimum ·······74
hard decision decoding ·······90
HDTV satellite broadcasting, wideband ·······161
head end ·······211
height pattern ·······126
Hi-Vision ·······166
hierarchical transmission ·······189, 195
high definition television (HDTV) ·······13
high definition TV broadcasting ·······166
high efficiency coding ·······8, 27
highly inclined orbit ·······199
home terminal ·······209
horizontal scanning·······17
housekeeping ·······180
Huffman code ·······33
hybrid fiber-coaxial (HFC) network ·······209, 214

I

Ikonoscope ·······3
impulse response ·······63
individual reception ·······159
ingress noise ·······238
in-phase(cosine) component ·······57
integrated services digital

broadcasting (ISDB)········195, 266
integrated services digital
   network (ISDN) ···············260
interactive media ················207
interference (noise) ···············167
interference,
   cochannel and adjacent channel
   ··186
   cross-polarization ···············186
   interchannel (ICI)···············135
   intersymbol (ISI) ··········62, 106
   monotonic·························233
interference protection ratio ········104
interference recipient ············234
interlaced scanning···············19
interleaver························181
interleaving····················89, 113
intermodulation (IM) noise ········184
in-phase (cosine) component ········57
International Electrotechnical
   Commission ···················27
International Frequency
   Registration Board (IFRB) ·········10
International Standard
   Organization ··················27
International Telecommunications
   Union (ITU) ·················6, 23
Intertext ·······················258
inverse Fourier transform ·········108
irreducible polynomial ·············76

J

Japan Cable Television Engineering
Association (JCTEA) ···········248
Japan Satellite Systems, Inc.(J-SAT)
   ··163

K

key information ··················181
KOPERNIKUS satellite ···········195
Koreasat ························164

L

Levels ···························43
linear quantization ···············32
link availability rate ··············189
link budget ··········136, 186, 239
link margin ······················185
local satellite broadcasting ········194
logical channel ··················213
luminance signal·················21

M

macroblock, pixel ················39
mapping (by set partitioning) ········94
mass-calling ····················256
matched filter ···················63
maximum likelihood (ML) decoding
··84
metric ··························84
minimal polynomial ···············76
minimum distance ················74
minimum Hamming distance ········74
modulation,
   amplitude (AM) ················56
   constant-envelope···············185
   frequency (FM) ················56
   quadrature amplitude modulation (QAM)
   ··9, 69
   quadrature phase················22
   trellis-coded ···················93
monotonic interference··············233
motion compensation ··············30
motion vector ····················30
multiplex block···················180
multiplexing ··················8, 220
multiplexing,
   coded-division (CDM) ···········112
   frequency-division ········112, 209
   orthogonal frequency-division ······9
   statistical ·····················221

N

Nakagami-Ricean distribution ······142

near video on demand (NVoD)
······195, 259
network, building-out ······217
network, bus type ······211
network, tree type ······211
Nipkow disk ······3
noise figure (NF) ······48
noise increase (due to rain absorption)
······167
noise (ripple)······232
noise temperature (system, due to
  atmospheric effects) ······167
nonlinear quantization ······32
null (zero amplitude) symbol ······120
Nyquist criterion ······124

O

Okumura's curves (distance attenuation)
······140
on-air relay ······101
optical link ······218
orthogonal transform ······29

P

packet stream ······45
parity check sum······83
Perfec TV ······164
perfect difference set ······82
phase modulation ······56, 58
phase-shift keying (PSK) ······52
physical channel ······213
picture······39
picture element (pixel, pel)······17
picture frame ······19
Pierce, John ······158
pilot signal ······122
pixel ······39
polling method ······257
polynomial,
  code ······75
  error locator ······78
  generator······76

irreducible ······76
minimal ······76
primitive······76
power-limited channel ······182
power margin ······185
predictive coding ······29
Primestar ······164
primitive element ······76
profile scalable ······43
program management system ······180
program selection information ······180
program stream······219
progressive scanning ······18

Q

quadrature (sine) component ······57
quantization ······32

R

rain absorption ······167
rain attenuation
  characteristics ······167
  countermeasures for ······173
  prediction ······169
rain scatter ······167
Rayleigh distribution ······142
reference signal ······65
required C/N······137
roll-off factor ······110
roll-off filter ······112

S

sampling frequency ······23
satellite broadcasting ······157
satellite digital sound broadcasting
······161
satellite station ······7
satellite thruster ······179
scanning line (horizontal line) ······17
scattering ······167
scramble authorization system(SAS)

....181
scrambling (of signal) 181
Script (text) 258
sequence (of GOP) 39
service availability 169
service information 222
set-top box (STB)..................222, 226
shadowing (by obstacles) 50
signal scrambling 181
signal space diagram 57, 144
simulcast period 219
simultaneous broadcasting period ...11
single-frequency network (SFN)
..................13, 102
single sideband (SSB) transmission
..................70
site dirersity 173
slice (of video data) 39
solid-state power amplifier 178
sound broadcasting 165
source coding block 180
Space Communications
Corporation (SCC) 163
spatial scalability 42
specific attenuation 171, 201
spectrum efficiency 12, 137
Sputnik 157
station keeping 179
sub-band coding 36
subscriber management system(SMS)
..................181
symbol (of data) 59
symbol, null (zero amplitude) 120
synchronizing (sync) pulse..................20
syndrome 78
system scalability 42

T

tap-off (TO) 209, 217
Telecommunications Technology
Council (of MPT)..................151
Telestar satellite 159
through repeater 177

tilt 218
time diversity 175
time-division multiplexing (TDM)
..................110
time interleave 112
transmission and multiplexing
configuration control 192
transmission power control (TPC)
..................168
transmitting and receiving (RF) system
..................180
transponder (satellite) 177
transponder nonlinear characteristics
..................184
transport stream (TS) 219, 220
traveling wave tube amplifier (TWTA)
..................9, 178
tree type network 211
trellis diagram..................83
trellis-coded 8-phase shift keying
(TC 8-PSK)..................9, 97, 182
trunk amplifier 217
trunk & distribution amplifier 215

U

uplink 184

V

vestigial sideband (VSB) transmission
..................70
Video on Demand (VoD) 207, 259
Videotex (VTX) 207

W

wave propagation 126, 139, 166
World Administrative Radio
Conference (WARC) 5, 160
Worldspace program 200
worst-month rain attenuation
cumulative distribution..................171

Index   **283**

Y

Yagi-Uda antenna ·······················101

Z

zero run-length code ···················41

Acronyms and Abbreviations

AC-3 (Audio Coding - 3) ············37

ACATS (Advisory Committee on Advanced Television Service) ···150

ADSL (Asymmetric Digital Subscriber Line) ·····························254

AM/AM characteristics ···········178

AM/FM - MUSE (AM/FM - Multiple Sub-Nyquist Sampling Encoding) ···································227

AM/PM characteristics··············178

ATM (Asynchronous Transfer Mode) ··································253

ATSC (Advanced TV System Committee) ·······················150

ATV (Advanced Television) ···37, 149

B-ISDN (Broadband Integrated Services Digital Network) ······206

BA (Bridging Amplifier) ············215

BB (BaseBand) system··············180

BCH (Bose-Chaudhuri-Hocquenghem) code ·····························74, 80

BER (Bit Error Rate) ······52, 66, 227

BON (Building-Out-Network) ······217

BPSK (Binary Phase-Shift Keying) ··································52, 58

BS (Broadcasting Satellite) ·······························5, 163

BS-2a (Broadcasting Satellite - 2a) ·····························163

BST (Band Segmented Transmission) ·····························153

B-picture (Bidirectionally predictive-coded picture) ···············31, 39

CA (Conditional Access) ············222

CATV (Community Antenna TV) ···3

CATV (Cable Television) ······205, 207

CDM (Coded-Division Multiplexing) ·····························112, 239

CM bank (Commerical Message bank) ································15

COFDM (Coded OFDM) ············106

COMETS (COMmunications and broadcasting Engineering Test Satellite) ·····················178, 189

CP (Continual Pilot) ···················153

CS (Communication Satellite) ·····························6, 169

CSO (Composite Second-Order beat) ·····························229

CTB (Composite Triple Beat) ······230

DAB (Digital Audio Broadcasting) ·····························103

DAVIC (Digital Audio Visual Council) ·····························247, 264

DCT (Discrete Cosine Transform) ·····························29, 39

DFN (Double Frequency Network) ·····························103

DQPSK (Differential Quadrature Phase Shift Keying) ·····················144

DSB (Double SideBand) ············70

DSP (Digital Signal Processor) ···113

DVB (Digital Video Broadcasting) ·····························175

DVB-S (DVB-Satellite) ············175

EA (Extension Amplifier) ·········217

ENC (Encoder system)··············180

EP-DVB (European Project on Digital Video Broadcasting)··············164

EPG (Electronic Program Guide) ·····························181, 213

EQ (Equalizer) ···············217, 232

ESA (European Space Agency) ···199

ETS-VIII (Experimental Test Satellite VIII) ·····························200

FCC (Federal Communications Commission) ·················150

FDM (Frequency-Division Multiplexing)·················110, 209

FEC (Forward Error Correction)
··8,181
FFT (Fast Fourier Transform) ······36
FSK (Frequency-Shift Keying) ······58
FTTC (Fiber-to-the-Curb) ·········254
FTTH (Fiber-to-the-Home)
································206, 254
GE-2 (Primestar communication
satellite) ·····························164
GMD (Generalized minimum Distance
decoding) ···························90
GOP (Group of Pictures) ············39
HD-SAT (HD-SAT project)·········196
HDTV (High Definition Television)
·····························13, 24,166
HE (Head End) ·························219
HFC (Hybrid Fiber Coaxial) ······214
HPA (High-Power Amplifier) ······182
HT (Home Terminal) ···············215
IC card ·································223
IEC (International Electrotechnical
Commission) ···············27, 37
IETF (Internet Engineering Task Force)
···································264
IFRB (International Frequency
Registration Board) ··············10
ISDB (Integrated Services Digital
Broadcasting) ···········199, 266
ISDN (Integrated Services Digital
Network) ························253
ISI (InterSymbol Interference)
····································62,106
ISO (International Standards
Organization) ···············27, 37
ITU (International Telecommunications
Union)·························6, 23
I-Picture (Intra-coded picture) ···31, 39
I-signal (In-phase signal) ···········22
J-SAT (Japan Satellite Systems, Inc.)
···································163
JCSAT-3 (Japan Communications
Satellite - 3) ························176
J SkyB ·································165
LNA (Low-Noise Amplifier) ······183

LNB (Low-Noise Block down
converter) ·························183
MPEG (Motion Picture Experts Group)
·································27, 39
MPEG-2 (Motion Picture Experts
Group - 2)
······················38, 42, 180, 226
MUSE (Multiple SubNyquist Sampling
Encoding) ···························163
MV (Motion Vector) ···············30
NF (Noise Figure) ···················48
NTSC (National Television System
Committee)··················3, 22
NVoD (Near Video on Demand)
························195, 208, 259
OFDM (Orthogonal Frequency-
Division Multiplexing)···9, 102, 199
PAL (Phase Alternation by Line)
···································3, 22
PES (Packetized Elementary Stream)
···································219
Pilot AGC (Pilot Automatic Gain
Control) ···························218
PMS (Program Management System)
···································180
PPV (Pay-Per-View) ···············184
PS (Program Stream) ···············219
PSK (Phase-Shift Keying) ·········52
P-picture (Predictive-coded picture)
···································31, 39
QAM (Quadrature amplitude modulaton)
···································9, 69
QPSK (Quadrature Phase-Shift
Keying)·····················9, 52, 58
Q-signal (quadrature phase signal)
···································22
RACE (Research and development in
Advanced Communications
echnologies for Europe············195
RAID (Redundant Array of
Inexpensive Disks) ···············262
RF (Radio Frequency equipment
system) ···························180
RS code (Reed-Solomon code)

·····79, 181
SAS (Scramble Authorization System)
·····181
SBC (Sub-band Coding)·····36
SCC (Space Communications
    Corporation) ·····163
SECAM (Sequential Couleur a
    Memoire) ·····3, 22
SFN (Single Frequency Network)
    ·····13, 102, 145
SFN 2-path model ·····146
SMS (Subscriber Management System)
    ·····181
SNR (Signal-to-Noise Ratio)·····42, 63
SNR scalability ·····42
SP (Scattered Pilot) ·····153
SSB (Single SideBand) ·····70
SSPA (Solid-State Power Amplifier)
    ·····178
STB (Set Top Box) ·····210, 226
TA (Trunk Amplifier) ·····215
TC 8-PSK (Trellis-Coded 8-Phase Shift
    Keying)·····9
TCM (Trellis-Coded Modulation) ·····93
TDA (Trunk & Distribution Amplifier)
    ·····215, 217
TDAC (Time-Domain Aliasing
    Cancellation) ·····37
TDM (Time-Division Multiplexing)
    ·····110, 175
TMCC (Transmission and Multiplexing
    Configuration Control) ·····153, 192

TO (Tap-Off) ·····215
TS (Transport Stream) ·····219, 220
TV character multiplexed broadcast
    (Teletext) ·····6
TWTA (Traveling Wave Tube
    Amplifier)·····9, 178
VC (Virtual Channel) ·····254
VoD (Video on Demand) ·····207, 259
VSB (Vestigial Side Band)
    ·····9, 70, 150
VTX (VideoTex) ·····207
WARC (World Administrative Radio
    Conference) ·····5, 160

Other Abbreviations

$\pi/4$-shift QPSK ·····60
χ^2 distribution (chi-square distribution)
    ·····109
2-PSK (Binary Phase-Shift Keying)
    ·····58
4-PSK (Quadrature Phase-Shift
    Keying) ·····9, 58
8-PSK (8-Phase Shift Keying) ·····52
16-DAPSK (16-Differential Amplitude
    and Phase Shift Keying) ·····144
21-GHz band satellite broadcasting
    ·····193
64-DAPSK (64-Differential Amplitude
    and Phase Shift Keying) ·····144